Conservation Education and Outreach Techniques

Series Editor: William J. Sutherland

Techniques in Ecology and Conservation Series

Series Editor: William J. Sutherland

Bird Ecology and Conservation: A Handbook of Techniques
William J. Sutherland, Ian Newton, and Rhys E. Green

Conservation Education and Outreach Techniques
Susan K. Jacobson, Mallory D. McDuff, and Martha C. Monroe

Conservation Education and Outreach Techniques

Susan K. Jacobson,
Mallory D. McDuff,
and Martha C. Monroe

OXFORD
UNIVERSITY PRESS

Great Clarendon Street, Oxford OX2 6DP

Oxford University Press is a department of the University of Oxford.
It furthers the University's objective of excellence in research, scholarship,
and education by publishing worldwide in

Oxford New York

Auckland Cape Town Dar es Salaam Hong Kong Karachi
Kuala Lumpur Madrid Melbourne Mexico City Nairobi
New Delhi Shanghai Taipei Toronto

With offices in

Argentina Austria Brazil Chile Czech Republic France Greece
Guatemala Hungary Italy Japan Poland Portugal Singapore
South Korea Switzerland Thailand Turkey Ukraine Vietnam

Oxford is a registered trade mark of Oxford University Press
in the UK and in certain other countries

Published in the United States
by Oxford University Press Inc., New York

British Library Cataloguing in Publication Data

Data available

Library of Congress Cataloging in Publication Data

Jacobson, Susan Kay.
    Conservation education and outreach techniques / Susan K. Jacobson,
Mallory D. McDuff, and Martha C. Monroe.
        p. cm.
    ISBN-13: 978–0–19–856772–1 (alk. paper)
    ISBN-13: 978–0–19–856771–4 (alk. paper)
    1. Nature conservation—Study and teaching.   2. Environmental protection—Study and teaching.
I. Mcduff, Mallory D.   II. Monroe, Martha C.   III. Title.
    QH75.J337 2006
    333.72071—dc22

                                                                                    2005029758

Typeset by Newgen Imaging Systems (P) Ltd., Chennai, India
Printed in Great Britain
on acid-free paper by
Biddles Ltd., King's Lynn

ISBN 0–19–856772–3   978–0–19–856772–1
ISBN 0–19–856771–5 (Pbk.)   978–0–19–856771–4 (Pbk.)

10 9 8 7 6 5 4 3 2 1

This book is dedicated to those who have nourished, guided, and inspired us over the years—our families, colleagues, mentors, and students—and to the natural world that sustains us all.

Acknowledgements

We are grateful to the many conservation educators around the world who have contributed wonderful examples and inspiration for this book. For expert review of various chapters, we are deeply thankful to: Julie Ernst, Alice Cohen Goldstein, Tom Harris, Jeff Hardesty, Trina Hofreiter, Perry Jacobson, Margaret Kinnaird, Lawrence Lowry, Lisa Marks, Susan Marynowski, Lauren McDonell, Douglas McKenzie-Mohr, O. Gene Myers, Molly Nicholie, Michelle Prysby, W.J. "Rocky" Rohwedder, Susan Sachs, Ricky Telg, and Richard Wilke. For secretarial and research assistance, we thank Deidra Brown, Eric Carvallo, Elaine Culpepper, Lisa Pennisi, and Sarah Tarrant. We thank students in the 2005 Methods and Materials in Environmental Education class at Warren Wilson College who reviewed early drafts of chapters: Lev Ben-Ezra, Kylie Kraus, Jessica Mosher, Ryan Nepomuceno, Joy Proctor, Annie Ross, Tucka Saville, and Greg Traymar. We are grateful for the collegiality at the University of Florida and Warren Wilson College. In honor of our parents, Betty and Perry Jacobson, Ann and Larry McDuff, and Helen and John Monroe, the proceeds from the book will be used to support conservation education activities.

Contents

Introduction

Think of a challenging conservation problem you have encountered—protecting a rare species, winning support for legislation, cleaning up a river, or sustainably managing a forest. Inevitably, people are part of the problem and public education and outreach will be part of the solution. Effective education and outreach are essential for promoting conservation policy, creating knowledgeable citizens, changing people's behaviors, garnering funds, and recruiting volunteers. The fate of our ecosystems and the plants, animals, and people that depend on them lies with our ability to educate children and adults, in settings as diverse as schools, communities, farms, and forests. The Millennium Assessment Report (2005) revealed that 60% of the vital ecosystem services that support life on earth—such as fresh water, fisheries, air and water cycling, and the regulation of natural hazards and regional climate—are being degraded or unsustainably used. Conservation education and outreach programs are a critical component in changing course toward a more sustainable future.

The goal of this book is to present the many techniques available for creating effective education and outreach programs for conservation. The first chapter presents a framework for designing programs. Chapters 2 and 3 provide the theoretical and practical background for understanding how to effectively address teaching, learning, and behavior change for adults and youth. Subsequent chapters introduce the reader to an exciting array of education and outreach techniques. These include techniques for classrooms and enhancement of school resources, marketing conservation messages, using mass media, and developing on-site programs for natural areas, parks, and community centers. The planning, implementation, and evaluation processes are described for each technique.

The beginning section of each chapter provides an introduction to the techniques included. The final section recommends further resources to steer the reader to more detailed research and guidelines about the material. The format of the book allows you to quickly find information for a specific technique you want to try immediately. Or you can read the book from start to finish.

What is in the book?

Chapter 1. Designing successful conservation education and outreach programs

Systematic planning, implementation, and evaluation are the foundation of effective education and outreach programs. This chapter begins the planning process with guidelines for identifying program needs, objectives, and target audiences.

You weigh possible strategies based on your resources and constraints of time, money, and staff. Planning helps ensure that you target the right audience and use appropriate messages and delivery systems. Implementation involves pilot testing activities and monitoring the ongoing operations. Continuous assessment allows you to modify materials and strategies based on timely feedback and new information. Evaluation of the outcomes tells if your tactics worked. Eight tools, from quantitative surveys to qualitative observation, are described. These allow you to make decisions about the future of the program — should it be continued or cut.

Chapter 2. Teaching and learning

Theories about learning and teaching form the basis for designing effective programs and marketing them to their audiences and administrators. This chapter provides program planners with a choice of techniques from inquiry learning, experiential learning, and cooperative learning. Guidelines are provided for addressing differences between adults and youth. Concepts such as learning cycles, learning styles, constructivism, and multiple intelligences help educators organize programs to meet the needs of any audience. Critical thinking, creative thinking, and systems thinking skills can be practiced and promoted through well-designed programs focused on conservation issues.

Chapter 3. Changing behaviors

The road to behavior change is paved with many theories. The disciplines of education, sociology, and psychology provide guidelines for strategies that can be used to influence people's conservation-related behaviors. This chapter reviews 10 of the most commonly used theories in conservation program development and research. Some are designed to influence how people gain information and learn skills to become responsible citizens, while others explore ways to orchestrate changes in specific behaviors. Both types of theories have important roles to play in the development of conservation education and outreach programs.

Chapter 4. Partnering with schools

Partnerships with environmental organizations and agencies can help create effective conservation education in schools. This chapter presents strategies for building successful programs, including understanding the realities of school systems, using effective communications, serving as a resource to the schools, connecting programming to academic standards, and integrating conservation education into legislation and standards. Education in the schools should include learning about the local environment, but too often, academic demands preclude the chance to study the natural world outside the classroom door. When conservation education links academic demands to the study of natural and social systems and their interactions, the winners include students, teachers, administrators, and the environment. To implement conservation education within schools, a variety of successful approaches are described, such as environment-based education, education for sustainability, and action projects.

Chapter 5. Making conservation come alive

Making conservation come alive can mean discovering the natural world around us through a neighborhood scavenger hunt, or researching the perspectives of an industry group for a role-play. Many of the techniques in this chapter emphasize the experiential approach to conservation education, such as hands-on activities, field trips, and wilderness skills. The aim of these techniques is to immerse the participants in exploring the outdoors or an environmental concept. Other techniques bring conservation alive through a minds-on approach, such as storytelling, games, case studies, role-playing, and contests. Most of the techniques in this chapter involve an element of fun, from a field trip exploring a wetland to a storytelling session on forests. Planning these techniques involves both research and logistics. This chapter contains helpful hints for implementation, including tips for engaging an audience in a story or developing a role-play. Every technique engages the audience in learning through direct experience.

Chapter 6. Using the arts for conservation

Using the arts for conservation can help attract new audiences, increase understanding, introduce new perspectives, and create a dialogue among diverse people. The arts—painting, photography, literature, theatre, and music—offer an emotional connection to nature. This chapter provides examples of using the arts to inspire people to take action. Planning art activities requires reaching out to artists and the art community, audiences with whom scientists and educators may seldom interact. Conservation problems require creative solutions. It makes sense to access more ways of knowing the world in order to take care of it.

Chapter 7. Connecting classes and communities

Effective conservation aims to integrate, rather than compete, with the needs of the human communities that share the landscape with biological communities. This chapter includes tips for planning, implementing, and evaluating techniques to connect classrooms and communities with conservation. Conservation education techniques, such as service-learning, issue investigation, and project-based learning involve students, teachers, and community members in finding creative approaches to issues such as backyard habitat restoration and solid waste management. Community-based research, citizen science, and mapping are techniques developed to work with students and adult learners. The techniques described in this chapter bring real conservation issues to the forefront of communities and classrooms and ultimately help achieve conservation goals.

Chapter 8. Networking for conservation

Networking involves aligning your interests with other individuals, groups, and communities to increase the success of a conservation effort. Talking with festival attendees at an information booth, forming a long-term partnership, or using negotiation skills to resolve a conflict between groups all require making connections. Networking can create or use existing environmental clubs and groups, as well as promote conservation objectives through public presentations, workshops, and professional posters. The skills to organize conferences and special events are necessary for developing connections among larger audiences to promote conservation messages and ultimately achieve conservation objectives. The techniques in this chapter provide opportunities to network and attain success both within and beyond the conservation community.

Chapter 9. Marketing conservation

A number of marketing techniques can help increase conservation-related behavior. These include using press releases and modeling to promote specific behaviors. Signs, billboards, advertisements, and other techniques provide memory prompts for a target audience. Techniques using feedback and commitment help sustain the behaviors. Working with local leaders to identify behaviors and consider the incentives and motives that support or deter each behavior helps educators choose the most effective combinations of techniques. This chapter provides many examples of how a variety of techniques have been used and evaluated to achieve behavioral and change.

Chapter 10. Getting out your message with the written word

Using the printed word for education and outreach is an essential technique for accomplishing conservation objectives of many organizations and agencies. Harnessing the power of mass media through op-ed pieces, letters to the editor, and news releases provides the means to reach vast numbers of people with information in a trusted and reliable format. This type of free advertising is valuable to everyone dealing with critical conservation issues and tight budgets. Guidelines of the structure and format for producing these materials are presented to ensure success in the competitive mass media arena. Fact sheets, flyers, brochures, and guidebooks are used by conservation organizations to build audience awareness, increase knowledge, and foster new conservation skills. The tips presented for clear writing and attractive graphic design help guarantee the production of effective print materials.

Chapter 11. Taking advantage of educational technology

Educational technology, such as radio, television, and the Internet, can dramatically increase the number of people we reach with conservation messages. It also allows audiences to vicariously experience natural events and places they might never see in person. Videos, Web sites, computer simulations, and distance learning allow conservation agencies and organizations to go beyond traditional face-to-face programming and consider ways to effectively reach their audience. Whether these technologies are successful depends in part on the degree to which they incorporate relevant learning theories. Strategies for evaluating Web sites, videos, and distance learning courses help ensure quality programs.

Chapter 12. Designing on-site activities

On-site activities can enhance first-hand experiences with natural areas by orienting, informing, and stimulating visitors. The development of on-site activities considers the visitor experience, resources of the site, and education and outreach objectives of the organization. An initial planning process at a site paves the way for developing trails, exhibits, demonstrations, nature awareness activities, and visitor centers. Guidelines provided in this chapter for implementing and evaluating these techniques help ensure achievement of conservation and education goals.

The challenge

Abundant examples illustrate the techniques described throughout the book. They represent the hard work and wisdom shared by conservation educators around the world. We hope these will inspire creative thinking and new ideas to fit the needs of the students, educators, environmentalists, program designers, resource managers, conservation biologists, and policy-makers reading this book. We hope the book will help people speak and act more effectively for wildlife and the environment. We hope it will help conservation programs amplify their results to bring about a more sustainable future for us all.

Designing successful conservation education and outreach

"Plan the work; work the plan."
"Proper planning prevents poor performance."
"Failing to plan is planning to fail."

We all recognize that careful planning is critical for success. As the need for conservation education and outreach grows, systematic planning will ensure its success. This chapter describes the development of education and outreach programs following guidelines for the Planning–Implementation–Evaluation (PIE) process. It provides a systematic design for identifying education and outreach goals, targeting specific audiences, selecting appropriate media and content, and evaluating the results. These guidelines will help ensure your effectiveness at using the many education and outreach techniques needed to safeguard the environment. Whether teaching a class, giving a public presentation, or designing a brochure or Web site, you can follow this systematic framework for success.

1.1 The need for conservation education and outreach

The need for improved education and outreach about the environment grows as conflicts over natural resources increase (Figure 1.1). The public affects the success or failure of environmental management efforts. Public opposition is the major constraint to implementing ecosystem management plans in the United States (Yaffee *et al.* 1996). Reintroducing endangered panthers in Florida (US) for example, is a social as well as a biological challenge. The public decides whether to allocate the needed funds, tolerate additional animals, or preserve enough land to sustain the large carnivores. The panther's fate depends in part on how well natural resource managers communicate with public groups and policy makers to raise concern and support for panther conservation. In essence, researchers could spend years designing plans or studying biological processes, but fail to achieve conservation goals without adequate public support. Failure to accurately assess and target public knowledge and concerns can result in opposition to conservation initiatives and costly political battles. When the public is informed and involved, however, conservation goals can be achieved. Successful outreach programs have helped increase sea bird populations in Canada and endangered primates in Brazil, and have improved farmland management in Australia.

Fig. 1.1 Human conflicts with wildlife range from concerns about diseases and safety to opportunities for ecotourism. Effective outreach is needed to better manage wildlife and people (photo by S. Jacobson).

Programs such as the Global Rivers Environmental Education Network (GREEN) demonstrate the power of education and outreach (Stapp *et al.* 1995). GREEN is a water quality monitoring program. Through participation in the program, students explore their local rivers, present their findings to government officials, and exchange data and insights with students in other cultures throughout the world. GREEN has developed a global communication network that includes countries such as Bangladesh, Argentina, Australia, Italy, Kenya, and the United States. While the countries differ culturally, many face similar conservation issues concerning the pollution and degradation of aquatic systems. Through our common need for healthy rivers and watersheds, GREEN has helped learners take action to address local watershed issues in more than 80 countries.

The public is exposed to conservation issues through print media, radio, television, and the Internet, through interpersonal activities, such as demonstrations and workshops, and through the school system, informal clubs, and other groups. Public opinion polls have found that 61% of survey respondents in the United States recognize that humans are the main cause of species extinctions (National Opinion Research Center 1995); 30% say they have heard about the loss of biological diversity, and the majority of these say maintaining biodiversity is somewhat to very important (Biodiversity Project 2002).

The majority of citizens in Eastern and Central Europe and Latin America believe that the quality of their environment has worsened in the past 10 years

(Gallup International 2002). Seven out of 10 Europeans believe that environmental protection and fighting pollution represent "an immediate and urgent problem," yet, many of them felt "fairly badly informed" about environmental problems and the steps taken by their governments to protect the environment (European Commission 1999).

Although people report that they care about the environment, public knowledge about conservation is minimal. Concern for wildlife is largely confined to attractive and emotionally appealing species. Among people surveyed in the United States, 89% believed endangered bald eagles (*Haliaeetus leucocephalus*) should be protected; however, only 24% believed the endangered Kauai wolf spider (*Adelocosa anops*) deserved protection (Kellert 1996). Most people view invertebrates with indifference or dislike, despite their crucial role in maintaining ecosystems.

Many school curricula include environmental topics, but too few offer comprehensive programs or focus on achieving conservation goals (Chapter 4). Although innovative extracurricular materials have been developed, teachers may not find the time, training, or motivation to use them. Yet, the need for conservation education continues to increase as problems become more complex. From cumulative impacts on wetlands restoration to declines in biodiversity, a knowledgeable public is needed to effectively protect the environment and address the goals of sustainable development.

Many conservation agencies and organizations have education and outreach programs. These programs focus on nature and natural resources and emphasize capabilities to solve environmental problems. Conservation education shares many goals with the broader field of environmental education. These include providing learners with an opportunity to gain (UNESCO 1978):

- Awareness—to acquire an awareness of and sensitivity to the environment and its associated problems.
- Knowledge—to gain a variety of experiences in and acquire a basic understanding of the environment and its associated problems.
- Attitudes—to acquire a set of values and feelings of concern for the environment and the motivation for actively participating in environmental improvement and protection.
- Skills—to acquire the skills for identifying and solving environmental problems.
- Participation—to encourage citizens to use their knowledge to become actively involved at all levels in working toward resolution of environmental problems.

Conservation education and outreach programs also share goals with newer programs, such as education for sustainable development. These are an outgrowth of the United Nations Conference on Environment and Development held in Rio de Janeiro, Brazil, in 1992. Education for sustainability emphasizes the need to solve problems to address three goals: protect the environmental systems that sustain life, enhance social justice for all people, and ensure appropriate economic development (Pigozzi 2003). Some conservation education programs, particularly those that focus on community resources, address issues of sustainability, and explore local concerns to develop solutions, promote this broad mission.

Education is often defined as a process of imparting or acquiring general knowledge, developing the powers of reasoning and judgment, and preparing oneself or others intellectually for mature life. Conservation education techniques include issue investigation, citizen science, professional development, interactive Web sites, and local school–community action projects. In classrooms, teachers use discovery learning, experiments, simulations, debates, and other techniques that employ the environment as a classroom. First-hand experiences help students understand natural systems, their community, and environmental issues. Infusion of environmental themes into the curriculum stimulates effective creative writing and language arts studies. The outdoors can serve as a laboratory to study math and science.

Conservation outreach programs encompass communication and information approaches. Unlike organized, formal education programs, outreach programs often target a non-captive audience in social and novel settings. Outreach programs are designed to increase understanding of conservation issues, exchange opinions and experiences, and establish a dialogue among sectors of the community (Fien *et al.* 2002). Outreach programs use techniques, such as publications, presentations, posters, and exhibits to improve awareness and knowledge of conservation problems. Outreach also includes techniques from the field of public relations. This involves marketing conservation products or services to meet the needs and interests of identified audiences. Techniques include newsletters, public meetings, advertisements, television, billboards, and the Internet targeting youth and adults.

Researchers have shown that appropriate education and outreach can foster sustainable behavior, improve public support for conservation, reduce vandalism and poaching in protected areas, improve compliance with environmental regulations, increase recreation-carrying capacities, and influence policies and decisions that affect the environment and natural resources (e.g. Knudson *et al.* 1995; Jacobson 1999; Day and Monroe 2000). The goal of this book is to present the many techniques available for creating effective education and outreach programs.

1.2 Designing education and outreach programs

Education and outreach programs help conservation organizations and agencies to address public needs and solve environmental problems. Some programs target a broad audience with an awareness campaign, such as providing information about recycling to homeowners in a city, or introducing wetlands ecology themes into a curriculum. Other programs target groups practicing specific behaviors the organization wishes to change, such as providing information about harvest regulations to deer hunters. Some programs teach problem-solving skills or issue investigation to youth. To succeed, organizations and agencies must respond to their audiences' existing needs, interests, and behaviors.

Program development should follow a systematic framework to be successful. Planning involves identifying goals and objectives, audiences, and educational strategies. Implementation concerns the operation of activities. Monitoring and evaluation of the results help identify successful activities as well as components in need of improvement. This iterative process—PIE—leads to an education and outreach program that avoids common problems, such as targeting the wrong audience or using an inappropriate message or medium. It is as easy as pie!

The expression, "as easy as pie," alludes to sitting around at one's leisure eating dessert. Yet, if you have ever baked a pie, you know that careful measuring, baking, and monitoring are required for success. During the planning process decisions must be made about the ingredients to be included and measured. What kind of pie is required—a meat or vegetable pie for supper or a berry pie for dessert? Who will eat the pie and what are their dietary requirements or preferences? What kind of crust? Do you have time to make a topping or money to buy blueberries? It is a good idea to test a new recipe out on your family—a pilot test—before inviting many guests to a banquet. The implementation phase involves preheating the oven, baking the pie, serving the pie, maintaining the equipment, and other operational activities. Evaluation includes how the pie looks and tastes. Did your

---

Box 1.1 As easy as PIE—questions to guide the design of an education and outreach program.

Planning

- What is the conservation problem or issue you want to address?
- What are your goals and objectives?
- What audiences or stakeholders are involved in the issues to be communicated?
- What are their backgrounds, needs, interests, and actions?
- For each audience, what changes or actions are desired?
- How can audience members be involved in the planning process?
- What constraints and resources are there?
- What messages must be sent?
- What channels and activities will most efficiently result in the desired changes in knowledge, attitudes, or behaviors?

Implementation

- What modifications are indicated by pilot tests of activities and materials?
- Is scheduling, funding, and staffing adequate and efficient?

Evaluation

- How will you know if the strategy worked?
- What are the outputs and outcomes of the program?
- Have you assessed key indicators of success, such as changes in the environment or in audience knowledge levels, attitudes, or behaviors?

oven cook evenly? Do your guests eat a slice and ask for seconds? Even though the outcome appears easy, a lot of hard work goes into the production.

In a similar way, you might envision designing a conservation education or outreach program. Box 1.1 outlines an "easy as pie" plan for guiding just about any conservation education and outreach program. Most conservation concerns are urgent. These guidelines help avoid wasting time on ineffective practices or programs.

1.3 Planning

Planning starts with a review of the mission of your organization or agency and the goals for the education and outreach program. With this foundation you then can identify target audiences and develop objectives for each audience. Based on the audience and objectives, you devise specific activities to attain your goal. An inventory of your resources in terms of staff, materials, and funds shapes the design of realistic activities. The entire planning process assesses and then narrows the menu of activities, media channels, and messages for the education or outreach campaign. A number of approaches can be followed in the planning process, but all of them should include the development of goals, objectives, and actions in response to conservation issues important to your organization.

1.3.1 Review the mission

The mission of an organization, agency, or school is the reason it exists. A mission statement articulates the guiding principles for the institution's actions and long-term goals. It defines, "Who we are, what we do, and why." The answers will circumscribe the range of activities and opportunities to be pursued in an education and outreach program. The mission statement provides overall direction in light of audience needs, the organization's resources, and external constraints and opportunities. Like a beacon in the night, the mission illuminates your path so you can get where you want to go. Without a clear mission, you risk wandering in the darkness of an ineffective program.

The mission of The Nature Conservancy is "to preserve plants, animals and natural communities that represent the diversity of life on earth by protecting the lands and waters they need to survive." Their activities accordingly include land purchase, resource management, policy, and public outreach programs.

The Royal Society for the Protection of Birds is a UK charity "working to secure a healthy environment for birds and wildlife, helping to create a better world for us all (Figure 1.2)."

The Teton Science Schools in Wyoming (US) seek to "provide and encourage experiential education in natural science and ecology while fostering an appreciation for conservation ethics and practices. The Greater Yellowstone region serves as our outdoor classroom and model for year-round programs that offer academic, professional, and personal benefits to students of all ages."

These mission statements describe why each group exists. Their mission helps them define a path to success. If a proposed education or outreach activity does not help an organization achieve its mission, then it probably should not be implemented.

for birds
for people
for ever
RSPB

HEALTHY, WEALTHY AND WISE

Sustainable communities:
creating the right environment

Fig. 1.2 Attractive brochures and other outreach activities help promote the mission of the Royal Society for the Protection of Birds.

1.3.2 Identify goals and objectives

Goals of conservation organizations may be to protect endangered species, manage a community forest, increase energy conservation, or improve sustainable farming practices. Education and outreach goals generally address problems. Conversely, identifying specific conservation problems is a good way to formulate goals. One of the Save the Manatee Club's goals, for example, is the recovery of manatees (*Trichechus manatus*) in the wild. The more clearly the problem is stated, the more targeted a goal will be. The problem, "Manatees are an endangered species," is less helpful for identifying potential education and outreach-based solutions than: "Collisions with motorboats cause manatee deaths in Florida." This problem statement helps to identify a specific goal: "Reduce boat collisions with manatees." It also helps identify specific audiences, such as boat owners, marina operators, or coastal regulators.

With the goal clearly in mind, specific objectives can be identified for each audience. In a school setting, teachers may have goals such as to teach the scientific method or develop student skills using maps. Research on manatee ecology or distribution can provide a real-life example for students. While goals are broad, objectives are specific and measurable for each target audience. Thinking at both levels helps you form a bridge from the mission statement to the activities you eventually carry out. Later the objectives can serve as benchmarks for measuring a program's performance.

Education and outreach objectives may be related to changing a target audience's knowledge, attitudes, or behaviors. Commercial advertisers view objectives in the form of a staircase leading up to their goal of selling a product. The first step is building consumer awareness—the consumer's ability to recognize and remember the product. The next step piques the consumer's personal interest. This increases the consumer's desire to learn about some of the features of the product and to evaluate these attributes. The remaining steps lead to the consumer's first purchase. If all goes according to plan, the consumer will continue to purchase and use the product.

This same process can be duplicated in conservation education and outreach programs, where each objective focuses on one or several steps. An initial message may try only to increase awareness about a conservation issue or agency service. A further objective may focus on increasing concern or shifting an attitude, and a final objective may encourage conservation action. A program that only increases general awareness about a problem or product does not guarantee action. Chapters 2 and 3 offer a number of theories that can help move an audience from awareness to action.

To assess whether your education and outreach objectives are met, they must be SMART (Box 1.2). Often objectives specify the number of people that will display the desired concern or behavior and the dates by which these changes will be achieved (e.g., a 10% increase in visitors to a wildlife reserve per month or 80% passing scores on a classroom test). All objectives should identify the audience, the media and message, desired effect, and time frame to facilitate the implementation and evaluation of the strategy. When objectives are specific, program results can be compared with anticipated outcomes to judge success and make decisions about program continuation. Educators write objectives in terms of what they hope their intended audience will do. Consider the manatee example: "As a result of receiving a safe boating booklet in the mail, 90% of Florida boaters will obey voluntary speed zones by the end of the year." Objectives for an exhibit about the Great Bay Estuary in New Hampshire (US) described what the majority of visitors should be able to do after exposure to the exhibit. This included describing characteristics

Box 1.2 A SMART program objective meets the following criteria.

Specific	Describes a behavior or outcome that is observable
Measurable	Provides quantifiable indicators of progress toward achieving the objective
Audience-focused	Identifies the audience and describes outcomes in terms of what the audience will be able to do
Relevant	Details a meaningful and realistic task and impact for the audience
Time-limited	Gives a definite time frame for achieving the objective

of an estuary, and listing two reasons why the Great Bay is valuable (Heffernan 1998). Keep in mind that you may not anticipate all the positive results of your program. The evaluation phase also will help uncover unexpected outcomes.

1.3.3 Identify target audiences

Understanding and involving your target audience is vital in designing messages and selecting media. You must address the concerns of your audience, whether it is hunters using a resource, homeowners living at the wildland–urban interface, farmers testing integrated pest management, or students investigating environmental problems in their neighborhood. Audience research can help orient your conservation education and outreach program to meet your audiences' needs while promoting the "conservation products" of your organization. You must know how the audiences are connected to the issue, what actions you wish them to take, and their current knowledge of the topic. Understanding factors, such as audience demographics, lifestyles, and media use will help you select appropriate educational approaches and objectives.

Methods for identifying and targeting audiences include collecting data through public surveys, interviews, public meetings, and workshops. Networking with organizations that already serve the audience can offer additional insights. To help a wildlife agency tailor messages to the needs of their audiences, researchers might use demographic information, psychological profiles, consumer behaviors, geographic residence, and a host of other social variables (Figure 1.3). Unobtrusive measures, such as observation of your audience, content analysis of documents or Internet sources, and analysis of similar cases allow you to study the problem with minimal influence on the audience. In a classroom setting, understanding the background and cognitive development of students is critical.

These methods for collecting data can be used not only to better understand the target audience but also to test assumptions about techniques, media, materials, and messages for the audience. Audience research allows you to assess alternative strategies for education and outreach channels and messages. It provides a foundation for building support for a program or influencing audience behaviors. Audience research also provides baseline information to evaluate the results of your conservation education and outreach efforts.

The Canadian Yukon Department of Renewable Resources developed an education program to reduce the number of female grizzly bears (*Ursus arctos*) killed by hunters (Smith 1995). Their program targeted outfitters and hunting guides. Their baseline research about this audience suggested messages and media to be used. The most critical message for the campaign was the information that gave guides the ability to judge the age and sex of a grizzly bear. This information was delivered using a videotaped workshop by a respected Alaskan bear guide. He convincingly demonstrated that guides could judge bear age and sex, leading clients to kill only male bears. The videotape also took advantage of motivational factors for the outfitters by including the symbolic value of the bear as a lone, powerful, wild figure, an image the agency found greatly appealing to their audience.

Fig. 1.3 A researcher interviews boaters to obtain baseline information for an education program promoting safe boating with manatees (photo by S. Aipunjiguly).

In a similar manner, programs designed for school students must consider their needs and backgrounds. For example, park programs designed for inner-city students must first make students comfortable in a novel setting in order for learning to occur.

An understanding of your audience's baseline knowledge and beliefs is a prerequisite for creating effective conservation education. In traditional classroom settings this information is often readily available from student testing, but for addressing controversial subjects among adults, research is often needed. The controversy in the Pacific Northwest US surrounding conservation of old growth forests and endangered Spotted Owls (*Strix occidentalis*) offers an eye-opening example. Some residents framed the conservation debate as extremist groups that cared more about useless animals than human needs. Audience research revealed that only 8% of the people understood why habitat loss was a cause of species extinction (Beldon and Russonello 1995 cited in Jacobson 1999). Some people believed that Spotted Owls were simply being stubborn by refusing to move from old growth forests to other places where they would cause less trouble. Therefore, effective outreach programs here needed to address knowledge of the owls' habitat needs as well as the value of conserving the animal and its habitat in support of human needs. Obviously, messages framed for an uninformed audience must address their present knowledge and beliefs.

Audience research also reveals people's attitudes toward your conservation concern. Many factors influence the formation of environmental attitudes. Researchers have shown that socioeconomic characteristics, such as income or education levels, often correlate with certain attitudes toward the environment. A series of surveys conducted in the United States revealed that negative and utilitarian views of wildlife are most common among groups with the following characteristics: lower socioeconomic levels, elderly, rural, and engaged in natural resource-dependent occupations (Kellert 1996).

Studies such as these help you design specific messages to attract your audience and achieve your program objectives. An example is an effort to communicate with local hunters about the problems caused by feral hogs on a public reserve in Florida. Managers emphasized that hogs damaged regenerating forests and competed for food with other game animals (of particular concern for hunters!). The managers did not focus on their additional concern about hog damage to endangered plant species. These managers understood their audience and used a utilitarian message rather than an ecological one to target their rural hunting public. (Jacobson 2005). The outcome was a win–win situation—hunters showed more support for hog removal programs and endangered plants were spared further hog damage.

A leading environmental organization, WWF-International, conducted audience research to better understand the public in the many countries in which they have national members. They wanted to examine people's attitudes and behaviors with respect to the environment and their willingness to support WWF initiatives. They conducted a series of focus groups and broad public surveys in 25 countries. The results allowed WWF to understand the wide range of attitudes and support within and across nations. It helped them design a two-pronged approach for promoting sustainable environmental behaviors.

For countries where environmental awareness was relatively low (e.g. Greece, Italy, and Spain), WWF developed 30-s television advertisements with light-hearted messages that conveyed the need to curb consumption and wasteful habits, such as leaving the tap running or lights on. One ad depicts an elderly woman, in comical fast motion, knitting a sweater to stay warm, rather than turning up a thermostat. In contrast, for countries with high levels of environmental awareness (e.g. the Netherlands, New Zealand, and Sweden), WWF used advertising to convey a more sophisticated environmental message. They focused on issues, such as tropical forest protection, which they found to be of high public concern

Box 1.3 WWF designed a TV advertisement to increase public awareness about the need to conserve tropical rainforests. It showed a tropical hardwood table being sold at auction and itemized the values of tropical forests (from Klingeman and Rommele, p. 130 cited in Jacobson 2005).

Ladies and gentlemen, what am I bid for this 19th-century reproduction, a handsome table in the Regency manner? Note the rare tropical woods, the ebony, the mahogany. My friends, many gave up their lives for this classic piece (An affluent lady waves her handkerchief to signal interest) $5,000, may I have $5,000? Thanks, that may have held the cure for disease; people have died, all for this table. Ten species of birds—gone forever. I have $10,000 in the bank, $15,000, $20,000. Ladies and gentlemen, this table contributed to global warming. Surely that is worth something $40,000. I have $100,000. This is the last table to come out of the rainforest. Going once, going twice, gone!

The hardwood table was then depicted standing alone in an empty landscape of destroyed forest.

(Klingemann and Rommele 2002). In these countries, their research suggested that people already were aware that rainforest destruction contributes to global warming and reduction in biodiversity and that this audience would not require an explanation. Rather, they needed a reminder, or prompt, to examine their own consumer behavior to avoid buying furniture made from tropical hardwood that was unsustainably harvested (Jacobson 2005)(Box 1.3).

Audience research forms the backbone of a conservation education or outreach program. It guides the development of activities and techniques. Audience research helps you orient your conservation program to meet your audiences' needs and to market the products of your organization. Knowledge of your target audience can facilitate diffusion of environmental information to opinion leaders or critical members of your audience. It allows you to assess alternative communication channels and messages for building support for a program or influencing audience behaviors. This type of research is known as formative evaluation. It occurs during planning and implementation of a program to provide feedback for improvements in the "formation" of early stages of the program. The collection of baseline data about your audience also provides information for the later summative evaluation of your education efforts, assessing your end results. Research conducted before and after your program will reveal if your objectives have been achieved and if you had unanticipated or secondary impacts. This allows you to make decisions about whether your program should be modified, expanded, or cut.

1.3.4 Include audience members and potential partners

Effective programs involve audience members and program partners in all stages of a program to meet the mutual needs of the audience as well as those of the organization. Involvement of target audience members and other stakeholders in the planning process helps ensure relevancy, and promotes a commitment to the implementation and long-term sustainability of the program. Involving a diversity of people provides different perspectives and ideas to enhance the development of a program. Their involvement in the planning process also helps to create a sense of community participation, ownership, and interest in the program. It can generate additional resources to assist with labor or funding. For example, involvement of landowners in planning a native plant workshop helps ensure that content and format will be appropriate for other landowners who participate in the program. Involvement by parents and community members in schools improves student learning. Teacher involvement in planning programs for schools ensures that the programs will address education standards and specific curriculum objectives.

Potential partners might include local businesses, resource users, teachers, parents, community leaders, landowners, and agency staff. Your planning process will help identify potential audience members and partners as you assess the needs of the program. To develop a group of people to serve in an advisory capacity, think about the ultimate users of the program. The following questions can help guide the process (Seng and Rushton 2003):

- In what specific ways can audience members or potential partners work with you?
- What do you or your program have to offer them?
- Who are the end users for your program?
- Who has special expertise to contribute to the design or content?
- What groups and individuals should be represented on your advisory committee?
- What are the specific roles and responsibilities of the committee members?

A program in Fez, Morocco, designed to improve solid waste collection through better outreach and technology involved many stakeholders in the planning process. These included policy-makers, neighborhood residents, community leaders, municipal truck drivers, sanitation engineers, health professionals, and solid waste experts. During the 2-week planning meeting, the group researched the solid waste system as well as the social structure of the community. The end result was a program that represented a broad consensus and investment from all groups in implementing the program (Grieser 2000).

1.3.5 Inventory resources and constraints

A realistic view of the strengths and constraints of time, personnel, money, and other resources provides a basis for selecting activities. Constraints limit the activities you can consider. For example, television ads may be beyond your budget, periodic newsletters may be too sluggish for your scheduling needs, or the lack of enough trained staff may turn a wilderness field trip into a nightmare.

Timing of your activities may correspond to the annual calendar, seasonal occurrences, natural phenomena, school schedules, or special days. Other activities may be required at a moment's notice to respond to an unexpected opportunity or problem. No matter what the timing, the schedule for any activity requires careful planning to provide sufficient days to successfully complete each task.

Staffing requirements must be realistically gauged. Who will do what? If many people are needed to help with events, where will they come from? Empty niches can be filled by volunteers from the community, scouts, or club members, interns from local colleges, or temporary hires. Each alternative has advantages, but may require additional resources for training, or stipends.

Accurate assessment of the funds needed versus the funds available is critical for planning an education and outreach program. Resources must be allocated efficiently, focusing on priority activities. How much money should be spent on your public programs? Budgeting involves determining your educational objectives, identifying tasks to accomplish the objectives, and determining the costs of these activities. If your budget is too small, consider substituting alternate activities, redefining the objectives, or finding partners to share the costs. If funds are unavailable, you may be able to get grants or donations of money, goods, or services in support of the program. Often fundraising may be an objective in itself.

1.3.6 Select activities and messages

The final stage of the planning process identifies appropriate activities to meet the objectives for each target audience. This book provides a selection of techniques from personal interactions to mass media. In developing materials and choosing delivery systems for your program, consider the background, needs, and interests of your learners in order to determine what teaching or outreach techniques will work best under your conditions. For example, abstract ideas are fine for older teens, yet, nonsensical for young children (Chapter 2). Activities must be attuned to the situation and should depend on the results of your inventory of budgets, personnel, time frames, and other constraints and opportunities. Your own experience and observation of similar effective programs can provide a wellspring of ideas for specific activities.

How do you choose techniques that will effectively reach the target audience? Using a variety of techniques may be the answer to sufficiently reach audiences segmented by factors, such as age, education, occupation, recreation activities, or geographic location. Different audience characteristics may call for techniques, such as speeches, demonstrations, e-mailings, or placement of messages in specialized newsletters. Knowing the media habits of your audience is one obvious way to target a message. For example, an article in the *London Times* may effectively reach policy-makers' eyes, but few farmers may ever read it. An article in an agricultural newsletter may better target farmers. The use of more than one technique increases the likelihood of reaching a greater audience; repetition reinforces the message. When selecting techniques to teach about conservation, evaluate factors, such as potential impact, production and dissemination expense, and audience size (Box 1.4). The following chapters describe the strengths of various techniques.

Box 1.4 Factors that influence decisions regarding appropriate messages and media (Jacobson 1999).

Factors	Questions to ask
Background and habits of the audience	What are the interests and media sources of your target audiences?
Attributes of the message	Does it require background knowledge, maps, graphics, color, or sound?
Urgency of the message	Do you need a response today or next month?
Complexity of the message	Is a 30-s sound bite adequate for the message or is a lengthy educational publication necessary?
Frequency of the message	Is repetition needed regularly, seasonally, infrequently? Do new people keep joining the target audience?
Personnel required	Is staff time available for personal contact, developing materials, providing outreach activities, interfacing with media, or training volunteers?
Cost	How many in your target audience can be reached, for how long, with what detail, at what price?

As you are identifying techniques and activities to meet your objectives, research other programs that use similar techniques. Do not waste time reinventing the wheel if quality materials are already available or could be adapted to your use. Just as you involved your target audience in planning, invite key stakeholders and experts to participate in program development, such as gathering resources, writing lesson plans, reviewing drafts, and creating graphics. Experts, such as wildlife extension or agency staff can provide quality data and feedback for program development. Teachers can help develop curriculum supplements. Offering free materials, stipends, or recognition helps provide stakeholders with incentives to participate in the development of your program. You may need to raise funds to pay for outside experts or materials to complete your program development.

1.4 Implementation

1.4.1 Pilot test activities

Once your activities are developed, you are ready to pilot test and implement your program. Just like a rehearsal before a theater production, practicing your activities with an audience before implementing them allows you to fine tune your performance or postpone opening day while changes are made. This is called pilot or pretesting the activities. Pilot tests are a form of evaluation conducted as a program is being developed, also part of formative evaluation. Pilot tests involve subjecting a small group of the target audience to your planned activity or materials. Pilot tests answer questions such as:

- How do members of the audience react to the activity?
- Which alternative versions of the activities are most successful?
- Does the activity communicate the appropriate message to achieve the objective?
- What feedback do people give about the activity?

Pilot tests are conducted before a program is implemented fully. They help ensure that the activities are effective and allow you to make needed changes before incurring the expense of implementing the entire program.

Methods for collecting data about an activity include focus groups or surveys with members of the target audience as they are exposed to your program. For written materials, you can use simple approaches, such as a portfolio review. This entails placing the new materials in a folder alone or with alternative materials and asking the test subjects from the target audience for their reactions. The audience can relay their impressions of the materials verbally, or they can complete a questionnaire using ratings or scales (e.g. *very attractive–attractive–neutral–unattractive–very unattractive*). Impressions about the main message, level of detail, appropriate audience, illustrations, graphics, and relevance guide decisions about the need for modification. Once feedback is obtained, the activity

Fig. 1.4 During pilot testing, a visitor provides feedback on the design for conservation messages embedded on park picnic tables (photo by S. Jacobson).

or materials can be revised, if needed, and pilot tested again before the campaign is executed. Pilot testing ensures that the best materials and activities are implemented (Figure 1.4).

Pilot tests were critical for the Florida Wildlife Federation. Before the Federation embarked on a new magazine for its members, they wisely pilot tested it. They sent a trial magazine issue with a written survey to a sample of 240 members. The survey asked the members what they liked most and least about the publication, and how much they might pay to receive such a magazine. Based on the negative responses to the survey and the high cost of the project, the Federation rejected the entire activity of publishing a magazine (Jacobson 1999). Teachers participating in a pilot test of a resource guide about pine forests in the Bahamas helped identify unclear directions and several student activities that would not work in large classrooms. The resource book was completely revised before distribution to schools throughout the islands.

1.4.2 Program operations

During implementation, develop a final schedule and budget and formalize project staff responsibilities. The establishment of a system for oversight of program operations ensures that the activities you select will continue to achieve your objectives. Monitoring tells you when your program activities are running smoothly or if modifications are needed.

The remainder of this book provides examples and inspirational models of the implementation of more than 50 techniques for education and outreach programs. Experiential approaches such as field trips and hands-on activities engage participants through direct experiences (Chapter 5). Using the arts for conservation can elicit emotional responses and attract new audiences (Chapter 6). Project-based learning connects schools and communities and leads to finding creative solutions to local problems (Chapter 7). Forming networks with other organizations and audiences will increase the success of your conservation efforts (Chapter 8). Social marketing approaches are useful for targeting specific audience behaviors. The adoption of new behaviors, such as recycling or composting, frequently occurs as a result of friends, family members, or colleagues introducing people to them (Chapters 3 and 9). Using mass media and educational technologies have a powerful role in setting the public agenda and reinforcing opinions (Chapters 10 and 11). More detailed publications or interpersonal methods are generally needed to change an audience's fundamental knowledge or shift their opinions, as described in a number of chapters. Activities conducted in natural areas and community centers can provide profound experiences in nature or foster social interaction to effectively reach your audience (Chapter 12). The many examples in each chapter help demonstrate effective implementation of the exciting techniques available to support conservation goals.

1.5 Evaluation

During and after any education and outreach program, conservation educators must ask if their objectives are being met:

- Did the target audience receive the message?
- Were the materials recognized; was the message remembered?
- Was the audience interested, satisfied, or engaged with the activity?
- Are there changes in audience knowledge, attitudes, or skills?
- Are there long-term changes in resource use or participant behavior, or increased concern for environmental management or conservation legislation?
- Are there improvements in the wildlife, ecosystem, or other conservation target?

Evaluation reveals if the program is worth the time, money, and resources. Is it cost-effective? Does it need modification? Should it be continued or cut? Are there unanticipated results? Evaluation is the only systematic means of finding the answers to these questions. It is key to providing feedback for improving conservation education and outreach programs and providing accountability to funding agencies, staff, and audiences. Without evaluation, ineffective outreach can harm the natural resources that your efforts seek to protect. A program may fail to reach the target audience, the message may be misunderstood, or the wrong behavior may be targeted. Without evaluation, hope of recognizing or improving a failed program is small.

Program outcomes may be immediate or long term and expected or unanticipated. Methods to evaluate education and outreach include a wide range of techniques from formal before-and-after surveys to direct observations of the target audience or their impacts on the environment. To measure the effectiveness of an education and outreach program to conserve a rare wildlife species, you might count new members joining your organization, funds donated to purchase key habitat, legislators' votes to pass protective measures, increases in public awareness after your campaign, and ultimately, the status of a wildlife population or conservation target after a certain time period. You also can design controlled experiments where, for example, radio ads are used in one market and newspaper ads are used in another. Results can then be compared to test the efficacy of different media. Continuous assessment allows you to modify activities based on timely feedback and new information. Results often are evaluated in terms of changes in public awareness, attitudes, or behaviors, and cost-effectiveness and efficiency of the program. Evaluation of outcomes tells whether your educational strategy worked. It allows you to make decisions about the fate of the program—whether it should be continued, cut, or expanded.

1.5.1 Designing an evaluation

A popular method for planning an evaluation uses a logic model to guide program design and assessment (Box 1.5). Program inputs are the materials, staff, funds, facilities, and environmental and community resources identified during the

Box 1.5 A logic model diagrams elements of program planning, implementation and evaluation (adapted from Seng and Rushton 2003).

PLANNING	IMPLEMENTATION			EVALUATION		
Inputs	*Throughputs*		*Outputs*	*Outcomes*		
Resources	Activities	Participation	Counts/Feedback	Short-term	Medium-term	Long-term
				Learning	Action	Conditions
Teachers	Curriculum	Participants	Numbers reached	Awareness	Practice	Social
Staff	design	Customers	Experiences	Motivations	Decisions	Political
Volunteers	Product	Stakeholders	Surveys	Knowledge	Action	Civic
Curricula clinics	development	Volunteers	Feedback	Values	Behavior	Environmental
Donors	Citizens	Trainers	Service units	Attitudes	Stewardship	Public relations
Time	Workshops	Teachers	Cost per unit	Opinions	Policies	
Money	Meetings	Youth	Service quality	Skills		
Materials	Counseling	Families	Aspirations			
Equipment	Facilitation	Students				
Technology	Assessments					
Partners	Media work					
	Training					

INFLUENTIAL ENVIRONMENTAL FACTORS AND ASSUMPTIONS

planning process. Program throughputs include the techniques and activities that were implemented and the participants involved. Program objectives target specific outputs and outcomes. All facets of the logic model are influenced by external environmental and social factors, such as political, economic, or institutional constraints.

The model requires that *logic* be used to align program activities with the desired outcomes stated in your objectives. If your program focuses on factual information, you would expect participants to display a short-term outcome of learning. If the goal is action, your program may need to provide similar information, but also incorporate experiential activities that develop skills needed for participants to take action. Your evaluation will measure whether these outputs and outcomes are achieved.

Planning an evaluation requires making choices regarding the evaluation design, data collection, analysis, and use of the evaluation results. The answers to the questions in Box 1.6 are derived from negotiation among the evaluator,

---

Box 1.6 Key questions for evaluation (Jacobson 1999).

Questions for Evaluation

Design of the evaluation

What are the evaluation questions or objectives?
What are the evaluation criteria or indicators of success?
Who will be involved in the evaluation? If an experimental design is selected, how will random assignment to control and treatment groups be done?
How will the results from the evaluation be used?

Data collection

What are the information sources?
What data collection methods are most appropriate for the evaluation questions?
How large is the sample? What sampling methods, if any, are needed?
How will the quality of the data be ensured?
Will data collection instruments need to be pilot tested?
When will the data be collected?

Data analysis

How should the data be analyzed?
Are computers, software, and expertise available for statistical analyses, if needed?
What is the most useful format for the data?

Reporting and use of evaluation results

What do the findings reveal for making programmatic decisions and recommendations?
How should the evaluation findings be presented for different groups?
What follow-up measures can assess implementation of recommendations?

program staff, audiences, and administrators. In schools, standardized tests dictated by educational institutions may set benchmarks that should be included in the evaluation of your program. Other large organizations may have existing benchmarks or metrics to help measure success.

1.5.2 Evaluation objectives

Evaluation methods range from experimental and quantitative methods, such as before-and-after tests, to informal qualitative measurement indices, such as focus groups, observation, and direct interview techniques. The type of evaluation questions you are asking will guide your data collection. For example, the level of detail in your evaluation will be based on your educational objectives. You will probably want to assess both outputs and outcomes of your program as detailed in your logic model.

Outputs

The easiest level of impact to evaluate simply measures the activities and products of the education or outreach campaign. What did you produce and how much exposure did your activities receive? How many people attended your program? Who participated and who did not? Was content suitable for the audience? How many events, press releases, and publications were produced? Were they likely to have reached your target audience? Outputs include the number of students participating in a schoolyard ecosystem program, the number of workshops on biological corridors provided to journalists, or the number of brochures on integrated pest management provided to farmers. Thorough records of activities, schedules, and participants are used to measure these outputs.

Short-term learning outcomes

Measurements of outcomes are often more complex. Measurements attempt to determine whether or not the target audience received, paid attention to, understood, and retained your conservation message, and in what form. You can conduct public opinion surveys to find out if your target audience has been exposed to your conservation message. Data collection tools, such as focus groups, in-depth interviews, and polling of target audiences are necessary to answer these types of evaluation questions. Before-and-after tests and portfolio reviews are used in school contexts. Students' knowledge about ecological principles is evaluated on a science test, or their field journals are reviewed for comprehension of a class on pond ecology.

Medium-term action outcomes

The next level of impact seeks to measure changes in target audiences' opinions, attitudes, and behaviors, and improvements in conservation targets. Are audiences using new knowledge and skills to help protect the environment? Are more people removing exotic plants from their landscapes or participating in recycling programs? To determine these outcomes, you may use techniques such as surveys and experimental designs, and unobtrusive methods such as observation and role-playing. Behavioral changes also may be measured through secondary indicators. For example, targeted homeowners may be expected to use less fertilizer to protect local water bodies, which could be measured as reduced sales at local garden centers.

Long-term condition outcomes

Finally, the last level of impact measures changes in the environment that are due to changes in the target audiences' behaviors. Have populations of threatened species recovered? Is the environment cleaner? Are resources being used sustainably? On-the-ground measurement of environmental indicators is necessary to monitor impacts in the field and to identify additional programs needed beyond education and outreach activities.

1.5.3 Data collection for an evaluation

The following section describes useful tools for collecting data to answer evaluation questions (adapted from Jacobson 1999). The advantages and disadvantages of the methods are outlined. These techniques also are useful for collecting baseline data to guide program planning.

Production of activities and participant counts

Simple measurements of outputs of your program are useful for evaluating impacts. These include monitoring the number of products and activities produced, as well as number of participants. Such documentation provides a simple and inexpensive means of recording activities. It is useful for long-term monitoring of activities and audiences. Baseline data from such indicators can provide comparison for measuring increases in program products, audience size, and new activities. Other monitoring methods, such as monitoring mass media to determine coverage of a press release, are used to document outputs. Keep in mind, however, that measurement of outputs does not measure actual impacts and may sometimes be misleading. Thirty people may attend your workshop on sustainable forest management, but none may implement new practices.

Tests of attitudes, knowledge, skills, and behavior

This method measures a sample of the target population with attitude questionnaires, or knowledge or skills tests that address the objectives of the conservation education program. The evaluator tests the sample before implementation and then after program completion to measure specific impacts of the education intervention. Tests are easy to administer, particularly in a school setting. They also make it easy to obtain a large sample and to score. Tests may involve an experimental or time series design, in which the evaluator measures subjects on some indicator before a program and at several data collection points during and after the program to assess changes over a period of time. In an experimental design, subjects are randomly assigned to a treatment (i.e. educational activity) or a control group. The results of their tests can be compared statistically to determine any changes resulting from the education program (Figure 1.5).

Fig. 1.5 Malaysian students take a test after visiting a medicinal plant demonstration at Kinabalu Park in Sabah (photo by S. Jacobson).

A disadvantage of tests is that they may be threatening to participants, and are often not appropriate outside of school. They also miss unexpected program outcomes.

Surveys

A survey is a systematic way of collecting information from a sample of people. Results are used to make generalizations about a target population. Survey research is used commonly during the planning stage of a program to measure public attitudes and knowledge, and to characterize a target audience, where they live, and what are their concerns. Surveys of voters' attitudes about conservation legislation can tell promoters what types of benefits to emphasize, for example, clean drinking water rather than recreational opportunities for legislation about wetlands protection. Surveys can be conducted on site, through the mail, or via telephone or the Internet. Insight provided about your audience's knowledge levels, opinions, and attitudes toward conservation issues can be used to successfully reach

them with appropriate programs. Surveys are useful both for collecting information on broad societal issues, and for detailed site- or program-specific problems.
Specific objectives of a survey might be to describe:

- Social and political make-up of your target audience.
- Sources of information for audience members.
- Audience opinion about current issues, such as new ecosystem management
  practices.
- Priorities of conservation problems for the audience.
- Compromises people may be willing to make on environmental or economic issues.
- Current levels of knowledge among the audience; for example, which specific
  aspects of ecosystem management do park neighbors not understand.
- Possible audience reactions to a new policy or communications campaign message.
- Audience willingness to support or pay for new initiatives.

Many books are available for designing valid and reliable surveys (e.g. Dillman
2000). Results of systematic surveys conducted before and after a program help
measure changes in audience knowledge or attitudes that can be associated with
the program.

Surveys are cost effective for collecting factual data from large groups. Results
from random samples surveyed can be generalized to the larger population.
Problems with surveys are created by bias from small numbers of respondents, or
inability to probe the target audience for complex information.

Interviews with participants

In an evaluation, interviews provide immediate data regarding program effectiveness. The personal contact between the evaluator and the respondent allows for
collection of more detailed information than a written survey. Participants in interviews can provide data on the strengths and weaknesses of a program and suggestions
for improvement. Questions can ask for opinions about specific learning or skill
outcomes. Common interview questions you might ask participants include:

- What did you like most about the program?
- What did you like least about the program?
- How would you improve the program for the future?

Interviews also may focus on other stakeholders besides the target audience,
such as donors, program staff, parents, government officials, and community
members. Interviews can be structured or unstructured, either following a preset
interview guide or allowing the interviewee to digress from the initial questions.
Interviews with visitors to an environmental art exhibit may reveal that the
program provided new perspectives about a conservation issue or, alternatively,
that the material was too obscure and the objectives were not achieved.

One specialized interview technique makes use of photographic documents.
Photo interviewing involves taking photographs of participants engaged in an
activity, and then using the photographs as interview prompts with participants.
Showing a photograph of a farmer planting a native shade tree may stimulate a rich
discussion about the effectiveness of a reforestation program. Photographs help

participants remember their attitudes and receptivity toward an activity at various points in the program.

Interviews can provide in-depth information, in contrast to written surveys and other techniques that limit feedback. On the other hand, personal interviews can be costly and time-consuming to conduct. Responses can be difficult to analyze, and interviewer or response bias is hard to eliminate.

Focus groups and meetings

Focus groups, public meetings, and advisory committees can provide information for designing new programs and fostering audience support and participation. Focus groups are a useful data collection tool to explore general attitudes, motivations, and behaviors of your audience. Focus groups are often used during the planning (formative) stage of a program to receive feedback from a specific target audience about ideas and educational approaches. Educators use focus groups to:

- Obtain background information about people's perceptions of a specific topic.
- Generate ideas or effective approaches for introducing a new service, product, program, or organization.
- Stimulate new research or interpret previous results from quantitative research.

A focus group consists of an interview with a group of 7–12 individuals who share some key characteristics (e.g. age, occupation, or interests), such as urban high school students or community leaders in a rural town. A moderator facilitates the focus group meeting using an interview guide. Respondents interact with each other as well as with the interviewer, potentially stimulating new ideas or insights. The session is videotaped for review and analysis. A focus group engaged in a discussion about recreation and other uses of a national park can reveal a range of options for park management planning (Figure 1.6).

Other group interview techniques involve drawing and mapping exercises. These are most appropriate when evaluating community or individual perceptions of a program with visually oriented cultures or children. Drawings produced by individuals or a group can facilitate discussion, as well as provide comparison for later drawings. Maps produced by community members are useful tools for monitoring land use and perceptions of land use change (Davis-Case 1990). Maps drawn by different groups of people can serve as baseline and comparison data before and after a conservation outreach program is conducted.

Focus groups and other group techniques are socially oriented and allow moderators to probe into a complex issue. This method is inexpensive and provides quick results. Conversely, group methods can be difficult to control and to analyze. They require a trained moderator.

Observation

Observational techniques provide a useful evaluation tool, especially for assessing behavior. The method involves observing participants before, during, and after a conservation education or outreach program. Some evaluations use behavioral checklists for recording presence, absence, and frequency of behaviors. Observers

Fig. 1.6 Researcher Browne-Nunez conducts a focus group with Maasai women about their experiences with wildlife around Amboseli National Park, Kenya (photo by R. Nunez).

must be properly trained to ensure reliable results. If it is impossible to observe the actual program, evaluators can record and analyze audience behaviors from the observation of videotapes. Photographs also can be used to document changes before and after a program. For example, photographs of a natural area can document a reduction in the amount of trash following a public awareness campaign on littering.

Other forms of systematic observation of behavior involve role-playing or simulation games, in which participants enact 'theoretical' situations of interest to the evaluator. Such dramatics can provide insight into the impact of a program, as participants often express attitudes that would be concealed in an interview or survey. Participants in a workshop on conflict negotiation may be asked to role-play a dispute over fisheries issues to test whether new skills were learned.

In general observational techniques are nonthreatening to participants and offer an effective way to measure behavior changes. Keep in mind that the results of observation can be unreliable and trained observers are required.

Content analysis and document review

Content analysis represents an efficient and objective tool for evaluating the text of documents or other materials. To conduct a content analysis, you must first determine which documents to evaluate and define the unit of analysis. For example, you may wish to review all Ugandan Wildlife Authority brochures from 1990–2005 for Queen Elizabeth National Park. Your variable of interest might be references to lion conservation messages. After coding the references to lions from each brochure, the evaluator calculates the frequency and type of lion-related information over the years. To evaluate coverage of conservation issues in the mass media, content analysis can be used to quantify frequency of press releases on a specific topic, placement within the newspaper, number of people potentially reached, messages expressed, and attitude conveyed toward the topic or organization. A content analysis of newspaper coverage of wildfires and prescribed burning in Florida found that the media emphasize the benefits of prescribed fire for preventing wildfires, but ignored the benefits of maintaining wildlife habitat or controlling pests. This suggested new topics for an outreach campaign about Florida's fire-adapted ecosystems (Jacobson *et al.* 2001).

Case study

A case study is a qualitative evaluation tool that focuses on an in-depth portrayal of one program, institution, or community. The results of a case study provide detailed insight into a particular program and its outcomes. The specific context and impacts are analyzed in detail. Based on the results of a case study, evaluators can make recommendations concerning methods, tools, audiences, and impacts of a program. A case study of the public outreach program to support reintroduction of wolves into Yellowstone National Park provided recommendations for modifying the program (Jacobson 1999). The case study also provided examples to other agencies faced with the public outreach challenge of reintroducing carnivores into a region. Although the results of a case study cannot be generalized and are dependent on data availability, case studies provide in-depth understanding of a situation. Multiple case studies of similar programs might be used to make comparisons or draw broader conclusions.

Portfolios and projects

Portfolios and projects are useful tools for measuring audience growth in developing new knowledge and skills. Portfolios are collections of participants' work that demonstrate increasing capabilities. Portfolios may include journals, reports, displays, illustrations, videotapes, or other media selected by the participant and evaluator. Scoring guidelines, called rubrics, are often used to evaluate individual projects. Students are graded on factors such as their participation in the project, accuracy of information, presence or absences of specific facts or skills, or clarity of presentation or visual aids. Group projects are often prescribed as the outcome for cooperative learning assignments in schools. Products might include group presentations or interaction with a community issue, such as the restoration of a city park. These alternatives to traditional testing may be difficult to use in some school settings.

Using multiple evaluation techniques

Using more than one technique for evaluation will strengthen your analysis by capitalizing on the strengths of different methods and minimizing the weaknesses. Using several methods combines multiple data sources, perspectives, and investigators or participants in data gathering. This provides crosschecks to validate your findings. Accurate program assessments result in more effective education and outreach programs.

1.6 Summary

Effective education and outreach are essential for influencing conservation policy, involving more people in conservation initiatives, improving people's knowledge and behaviors, garnering funds, and sharing scientific advances. The fate of our environment depends on effective communication with a great variety of audiences.

Systematic planning, implementation, and evaluation are the foundation of effective education and outreach programs. This chapter provided the Planning–Implementation–Evaluation guidelines for developing programs. Planning activities begin by identifying education and outreach needs, objectives, and audiences. Possible strategies are selected based on your resources and constraints of time, money, and staff. The planning process provides information for making decisions about the nature and scope of your actions. Implementation involves pilot testing strategies and monitoring ongoing operations. Evaluations should be conducted both during and after implementation of your program. Formative evaluation occurs during planning and implementation of a program and provides feedback for design and improvement in the early stages of a program. Summative evaluation is conducted at the end of a program and addresses questions about the program's worth. Program staff, consumers, participants, and funding agencies use information from summative evaluations to answer questions regarding program impacts and continuation.

Evaluation of products and outcomes tells if your tactics worked. A variety of quantitative and qualitative techniques provide you with data to make decisions about the fate of the program—if it should be continued, cut, or expanded. Following a systematic approach helps avoid problems, such as targeting the wrong audience or using an inappropriate message or medium. Most conservation concerns are urgent. Applying these guidelines as you implement the techniques described in the rest of this book will ensure your success.

Further reading

Environmental Education and Communication for a Sustainable World: Handbook for International Practitioners, by Brian Day and Martha Monroe (eds.) (2000). Academy for Educational Development, Washington, DC.

Communications Skills for Conservation Professionals, by Susan Jacobson (1999). Island Press, Washington, DC.

Interpretation of Cultural and Natural Resources, by Douglas Knudson, Ted Cable, and Larry Beck (1995). Venture Publishing Inc., State College, PA.

Fostering Sustainable Behavior: An Introduction to Community Based Social Marketing, by Doug McKenzie-Mohr and William Smith (1999). New Society Publishers, Gabriola Island, British Columbia, Canada.

The Practice of Public Relations, by Fraser Seitel (1995). Prentice Hall, Englewood Cliffs, NJ.

Handbook of Practical Program Evaluation, edited by Joseph Wholey, Harry Hatry, and Kathryn Newcomer (1994). Jossey-Bass Publisher, San Francisco, CA.

2

Learning and teaching with adults and youth

Why do we develop conservation education and outreach programs? We develop them because we want people, our audiences, to know more about their environment, care about their environment, and make choices that support conservation. We believe educational programs will help achieve this important goal. Examples abound of people who make the wrong choice because they lack information or who fail to act because they just do not care. Clearly, offering conservation education programs that address knowledge and attitudes ought to help improve conservation practices. This chapter offers an introduction to the educational theories behind effective programs for adults and youth. These theories are also useful for program planners who interact with school personnel. Chapter 3 offers additional theories that can be used in education and outreach programs to motivate people toward appropriate conservation behaviors.

2.1 Pedagogy for youth and adults

Common sense tells us that adults and youth may prefer to learn in different ways and therefore should have their own educational programs and strategies. Jean Piaget, a Swiss psychologist, noted that as children grow, they develop mentally as well as physically. The process of cognitive development begins to override instinctive reflexes even in infants, who develop the ability to use sensory input to understand their world (Box 2.1). Toddlers begin to use language to symbolize physical items, and usually between the ages 7–11 develop the ability to generalize from concrete experiences, but are less adept at situations they have not experienced. Beyond the age of 11 children are better able to deal with abstractions, solve problems, and manipulate their mental models to hypothesize about events they have not witnessed. Age does not determine development, however, since some adults have yet to develop some of these abilities (Huitt and Hummel, 2003).

Cognitive development theory suggests that young children should be able to learn from concrete objects and experiences they can manipulate and sense. This would suggest that global issues like climate change and biodiversity, or abstract concepts like air pollution and energy conservation may not be well understood by youngsters under age 10. Conservation education programs should respect the abilities of their audience and make sure their program is "developmentally

Box 2.1 Piaget's stages of cognitive development (adapted from Huitt and Hummel 2003).

Sensorimotor stage (Infancy). Knowledge of the world is limited (but developing) because it is based on physical interactions or experiences. Physical development (mobility) allows the child to begin developing new intellectual abilities.

Preoperational stage (Toddler and early childhood). Intelligence is demonstrated through the use of symbols, language use matures, and memory and imagination are developed, but thinking is done in a nonlogical, nonreversable manner. Egocentric thinking predominates.

Concrete operational stage (Elementary and early adolescence). In this stage intelligence is demonstrated through logical and systematic manipulation of symbols related to concrete objects. Operational thinking develops (mental actions that are reversible).

Formal operational stage (Adolescence and adulthood). In this stage, intelligence is demonstrated through the logical use of symbols related to abstract concepts. Only 35% of high school graduates in industrialized countries obtain formal operations; many people do not think formally during adulthood.

appropriate," both cognitively and emotionally. Because youngsters (age 4–7) explore a world that is fairly close to home, educational programs should be similarly limited to concrete, home-based concepts. As older youth expand their territory to include community centers, nearby parks, and more abstract elements, educational programs can accommodate their interests by focusing on complex, global, or abstract issues (Sobel 1996).

Malcolm Knowles extended Piaget's concepts of cognitive development by focusing on how the adult differs from a youngster. Maturity generally brings a different set of priorities for learning, especially a greater sense of self-direction, more experiences from which to draw upon, and a desire to learn things that can immediately be useful in work or social situations. When the characteristics of adult education are contrasted with youth education, the assumption seems to be a comparison of traditional school-based settings for young children and professional training for adult workers. When youth are taken out of the traditional school setting, however, some of the characteristics of youth learning may disappear (Box 2.2).

These distinctions may not always be helpful for adult groups, either. There may be times when adults are self-motivated and times when they prefer to be told what to do. A needs assessment will help the educator determine how to develop a successful adult education program. Questions such as the following could help organizers better meet the needs of adult learners (Newstrom and Lengnick-Hall 1991):

- How concerned are you with the immediate applicability of this material?
- To what degree do you question sources of information?
- How fearful are you of future change?

Box 2.2 Adult and youth learning characteristics.

The outside columns are frequently used to distinguish adult from youth learning. The assumption for youth, however, is based on the typical school experience not what is ideal for youth. The center column suggests there could be some common characteristics for learning at any age, depending on the nonformal education opportunities.

Adult education	Nonformal audiences	Youth in school
Voluntary participation	Voluntary participation	Compulsory participation
Learner oriented	Learner and leader oriented	Teacher oriented
Need it now	May or may not need it	Probably need it later
Share information	Explore information	Receive information
Build on past experience	Build on past and current experience	Not much experience yet
Intrinsic motives (self-esteem, competence)	Intrinsic and extrinsic motives	Extrinsic motives (grades, teacher approval)

- Do you have a relatively long attention span?
- How interested are you in these topics?
- Describe other successful educational programs you have attended.
- How much organizational support do you think you will need to implement these changes in your job?

In the likely event of a diverse group of adult learners, it is important to use a variety of learning strategies to appeal to the widest possible set of learners, allow adults to choose their subgroups or task, and permit a variety of responses and outcomes. As a result, most adult training workshops include icebreakers to introduce people, large group information-delivery, small group discussions, engaging activities, and opportunities for participants to share experiences and ideas. As long as adults believe the learning experience will meet their needs, respect their differences, and provide useful information, they often are willing to engage in a variety of learning strategies. If there is a risk of embarrassment, however, or the slightest indication that the activity is a waste of time, an adult audience may rebel.

The assumption that adults have more relevant experience to build on as they learn new information may be true in some circumstances and not in others. As the world becomes more complex it is common for adults to need totally new information and skills for which they may not have relevant experience. Conversely, some youngsters have a great deal of expertise with selected topics, even though they are children. Consider the child who keeps a pet snake. He or she knows a great deal about that snake's habits and nuances. Information about the snake's smooth scales or how a snake sheds will not be of interest to this child, but could be fascinating to the average adult. Constructivism, another learning theory, suggests that everyone—even a child—brings ideas, information, and preconceived notions to every learning opportunity; the job of deciphering what learners already know and believe falls to every educator, not just to those who work with adults.

At any age, new information should be relevant and interesting to the learner. A presenter should acknowledge what the learner already knows, and engage him or her in expanding that knowledge base. In this way adults and youth are not so different. Our assumptions about what is new and interesting, however, may be different for adults and youth. Youth are interested in concrete aspects of the world around them—insects, stars, predators, and edible plants, for example. Because organized educational programs for adults bring together a group with some commonality—a Kiwanis club meeting, a religious study group, an environmental organization, or a professional development workshop—we tend to use those elements to make our programs interesting and relevant. Professional development programs make connections to workplace successes, while civic groups may focus on parenting challenges or community well-being. The more we know about our audience, the better we are able to generate an educational program that meets their needs, builds on what they already know, and prepares them for new challenges.

2.1.1 Youth-based environmental education

In many nations, formalized public and private education have institutionalized opportunities for environmental education. A national curriculum may include environmental objectives in science or social studies, as well as skills for communicating and solving problems in language arts, social studies, and other disciplines. Teaching supplements are produced as resources for educators, and in-service workshops help them use new techniques and materials. Annual tests and assessments that include questions about the environment motivate educators to cover this information in school.

A variety of agencies and organizations external to the Ministry of Education may support and contribute to local educational efforts, offering field trips, guest speakers, residential opportunities, and after-school clubs (Chapter 4). Apart from the formal school-based curriculum, nonformal educational opportunities exist for youth. Scout, Guide, Y, and other organizations offer programs for groups of youth, particularly those in urbanized communities. In some countries, rural youth have nonformal program options that can be conducted by individuals. The North American Association for Environmental Education publishes Guidelines for Excellence that describe how to develop high quality programs and materials for formal and nonformal audiences (NAAEE 2005).

As long as conservation educators identify ways they can help these organizations and institutions meet their goals, they should be able to identify opportunities to work with youth in a variety of situations.

2.1.2 Environmental adult education

In contrast to youth-based education, adults have fewer organized, institutional educational programs. The programs that do exist are usually attended by choice, as opposed to the required school attendance for youth. The conservation educator wishing to work with adults could look for guest speaking opportunities at service clubs or community groups with regular meetings. Many professions offer technical training for their staff that may be relevant to conservation issues, like waste reduction programs in industries or carpooling options in office buildings. Some faith-based organizations have adult education programs that deal with steward- ship of resources and self-improvement. Several agencies employ adult educators to develop community outreach efforts, like agricultural extension or a reading literacy program. In general, developing an adult program will require identifying an adult audience that you can access or advertising your program to attract an audience.

Some educators have successfully created their own environmental adult educa- tion group around an issue of concern that engages learners in gaining awareness of the political process and the power they could mobilize in their own community. Communities wrestling with landfills or hazardous waste dumps often generate interest and power to make change. The plight of a red-tailed hawk nest on an apartment building in New York City has involved a book and TV documentary, citizen protests, and candlelight vigils (Lueck and Lee 2004). While it is difficult to know when a protest becomes an educational program, those engaged in the activ- ity are learning, often teaching themselves, as they navigate the media and legal systems. Because adults are more responsible for their lives than youngsters and an adult brain is physically more capable of making judgments and weighing alternat- ives than a child's, it may be more appropriate for adults to engage in programs to change certain aspects of their lives.

If the presence of an adult educator helps to make advocacy efforts educational, adult education can include opportunities that may not be recognized by natural resource managers or conservation organizations. Adult education programs engage learners in basic skills, literacy, and training. These programs often use local problems and concerns to galvanize interest, build skills, facilitate discussion, and empower learners (Freire 1982). The reality of the interconnection between polit- ical, economic, social, cultural, and environmental strands in any issue makes adult education politically powerful. As a result, the distance between education and advo- cacy grows quite small, leading some to believe that eventually all good adult education efforts evolve into social action (Hill 2003; St Clair 2003).

So close is the connection between adult education and advocacy that some adult educators claim that examples of environmental action are education, like the group of women who took over the oil terminal in Escravos, Nigeria in 2002 and bargained with corporate officials for a town hall, schools, electricity, and a water system (Clover 2003). Villagers protesting dam construction, farm workers demanding better health protection, teach-ins at universities, faith-based commun- ities working toward environmental justice all become examples of environmental adult education. Paul Bélanger (2003, p. 86) insightfully summarized the link between environmental education and traditional adult education: "Precisely because environmental adult learning is directly grounded in immediate problems and the expressed needs of individuals living within a sociocultural–economic

Box 2.3 Learning theories at a glance.

Learning theories	Contribution to education
Brain-based learning	Construct learning opportunities to reinforce neural networks and build mental models
Experiential learning	Base learning opportunities on experiences
Learning cycle	Design learning opportunities to include an experience, then processing, generalizing, and applying that experience to better understand the concept
Learning styles	Recognize that learners might have different preferred learning styles hence develop programs to appeal to the diversity
Multiple intelligences	Recognize that intelligence comes in a variety of skills and abilities hence develop programs to appeal to the diversity
Constructivism	Engage learners in constructing their own knowledge, through inquiry, experiences, and questions
Cooperative learning	Develop learning opportunities to use groups of learners, each with a special role, to achieve a common goal. Help learners improve social and group communication skills
Creative thinking	Provide opportunities to generate and synthesize
Critical thinking	Provide opportunities to interpret, evaluate, and analyze
Systems thinking	Provide learning opportunities to understand the structure and function of complex systems.

context, educators must participate in real-world activities and work with local people to create concrete, constructive action."

Despite some obvious differences between youth and adult learners, there are many similarities that are simply common to the process of learning. The following theories help educators design effective learning situations and programs (Box 2.3).

2.2 Brain-based learning

Educational theories are based on our knowledge of how we learn, which is to say how our brains process information and build understanding. A human brain is a complex system of billions of neurons that store memory by generating links that connect neurons in a neural network. Neurons continue to grow throughout our lifetime though the learning period may be slower and more frustrating for older learners than young. There are limits to what youngsters are capable of, however.

Recent research suggests that the area of the brain associated with judgment, weighing risks, and considering consequences (i.e. the dorsal lateral prefrontal cortex) is among the last to develop and may not mature until the age of 20 (Davis 2005). New research taking advantage of technologies that track neurons as they become activated is helping us know even more about brains, memory, and the learning process.

The goal of an educational event is usually for the learner to gain and be able to use knowledge. Our measure of having gained that knowledge is the ability to repeat or show what we have learned. This makes memory the key to learning. The more we know about how memory works, the better we can devise educational opportunities to enhance learning.

One form of memory functions below the level of consciousness. It enables us to automatically perform routines out of habit like getting dressed, harvesting food, or going to work. Because these behaviors are not performed with conscious recall, they are difficult to change, though skill development can help modify them.

Within the area of consciousness, memory enables us to learn facts and figures by applying effort, and also to learn from repeated experiences without any effort at all. It is a powerful tool that allows us to abstract common features from many events and store them collectively. This means we do not remember every distinct event, but instead form a general impression that is quite serviceable. When educators offer slightly different learning experiences about one concept, they help learners build a cognitive map of common features of the concept. When learners practice skills, repeat concepts, or write summaries, they are reinforcing the neural networks that include this new information and making it more likely that the proper associations will be made if it is triggered by a cue. Reading about a forest, visiting a forest, drawing a picture of a forest, and listing the plants and animals that live in a forest are different learning activities that employ a range of skills but repeatedly reinforce the concept of a forest. The more frequently the memory is activated, the more quickly it is recalled. Educators can assist learners in the recall process by helping them identify the cues that will stimulate the appropriate information (Bauer 2005).

Our brains are able to retain information from a variety of sources simultaneously—smells, tastes, sight, and sounds, as well as context and time. Things that are experienced together become connected by neural links. If one piece of information is triggered, the rest are also activated. This enables us to learn vast amounts of information and maintain an organizational system so that appropriate information can be retrieved when needed. Emotion can modulate memory so that interesting, challenging, or relevant information is more easily remembered (Bauer 2005). Educators who work to connect new information to what learners care about or who ask learners to help resolve community problems are using the learners' emotional resources to increase learning.

The way we learn new information is also coded into our memories along with the information itself. Reading about fishing regulations, nets, and lures will give people information they can repeat, but may not give them skills they can use. If an educator wants learners to perform skills, the teaching process should include practice with those skills. Being clear about the learning objectives will help educators decide what concepts and skills should be experienced and reinforced.

Our brains are not foolproof. Memory shifts over time and is susceptible to suggestion. We can observe our own mental mistakes and realize when an inappropriate memory is triggered or when information we knew is not recalled. An observant educator will notice when a learner's neurons are activating the wrong network—as when a tolling church bell near a nature center prompted a child to ask what kind of bird made that sound! Not knowing how developed the area around the nature center was, this child assumed everything was natural. Since churches were not expected, that memory was not activated when the bell rang.

Familiarity is the key to building conceptual frameworks and the basis of understanding. Being familiar means having a mental model that allows for comfortable exploration of new ideas around the edges of what is already known. Building that familiarity requires a significant process of contact or experience with the topic. Research suggests that even adults learn unfamiliar concepts best by experience or by examples so concrete and real that they mimic experience (Kaplan and Kaplan 1982). Although television and motion pictures seem to be a reasonable substitute for experience, listening to and reading stories can be a more effective strategy to prepare people for the world they will experience. Listening and reading enables learners to conjure up their own mental images. Because these learners have done the hard work of creating those mental images, they may be more memorable than a televised show.

Mental models can be used to enable us to function in unfamiliar places. Landing at an airport for the first time is not incapacitating because we expect gates to be sequentially numbered, restrooms to be labeled for males and females, and baggage to be near the exit. This skill of determining where to go in an unfamiliar place is not unlike problem solving—figuring out what to do in a new situation. The skills of problem solving can be practiced in conservation education programs, which can improve the chances that those skills will be used if activated in the future. Young brains may need to practice the skills of problem solving through exercises that enable them to collect and analyze information, compare and contrast alternative solutions, listen to different opinions, and reach a considered conclusion. A variety of techniques to achieve these goals are described in Chapter 7.

2.3 Experiential learning

During the 1800s several educators broke away from the established norms of textbook learning and encouraged students to study the real world. Louis Agassiz, the Swiss scientist noted for contributions as diverse as fish and glaciers insisted his graduate students "study nature not books." Liberty Hyde Bailey, Anna Botsford Comstock, and Wilbur Jackson made significant contributions to elementary science education through their nature study movement (Disinger and Monroe 1994). Children are naturally curious about their world, they reasoned, so by encouraging their observations teachers can use weeds, rocks, or caddisfly larvae to

Fig. 2.1 Youth enjoy exploring aquatic organisms at the Smithsonian Environmental Research Center on Chesapeake Bay (photo by G. Traymar).

convey concepts about life and earth science (Figure 2.1). John Dewey, the American philosopher whose work greatly influenced American and British education, was a proponent of learning by doing rather than the commonly practiced rote memorization and authoritarian teaching style. His work and that of others in the progressive education movement, Jean Jacques Rousseau, Johann Pestalozzi, and Friedrich Froebel, promoted practical education—industrial training and agricultural education—to make education relevant to living experiences. Schools, Dewey maintained, should reflect society. These philosophers further help justify citizen science initiatives, service-learning, and project-based education (Chapter 7) that engage learners in community projects and problem solving.

Brain research gives educators a firm footing in explaining why "hands-on" or experiential learning is effective. Our senses help us gather information about the world. By grasping, moving, counting, and manipulating, children learn more about the world around them. As they grow older, their increased familiarity with the world enables them to use concepts that represent ideas rather than things. The "learn by doing" philosophy is firmly embedded in elementary education through field investigations, but begins to fade in secondary education, where students are expected to learn from books. This may account for the fact that many secondary students find school less relevant and meaningful (US Department of Education 2002). By giving older youth an opportunity to explore community issues and develop plans for action (Chapters 4 and 7), schools are realizing important benefits of experiential learning.

It is important to note that even though experiential learning has an element of "doing," the experience alone is not sufficient to build understanding. Understanding requires the mental effort of comprehending what the experience entailed, connecting prior knowledge to new information, and accommodating that new information into the mental model. Dewey called this "critical reflection" and suggests that it is an important part of an educational experience for youth and adults. Brain researchers call this re-activation of the memory. Opportunities to write poetry, draw pictures, complete evaluations, design an action plan, present information to others, and create a visual presentation enable learners to reflect on and learn from information and experiences.

2.3.1 Experiential learning cycle

Recent theorists have developed systems to help educators guide learners through a process of learning: the learning cycle. While there are many different versions of learning cycles, they all have two things in common: an experience or active endeavor and a thoughtful, reflective analysis of that experience.

Some popular versions of the learning cycle have four steps (Figure 2.2). These steps can be identified in most environmental education materials like Project WET's Activity Guide (The Watercourse and the Council for Environmental Education 1995). The first step is the *experience* itself, while the following three steps are strategies for helping learners understand and accommodate the information. Step 2 emphasizes *processing*. Questions about what happened, what data were collected, and what was observed help learners pull the experience into their mental framework. The next step involves *generalizing* the information from the experience. Learners might hypothesize why one group's responses were different than their own. Worksheets and journals are two strategies that help educators guide learners through these two stages by reporting their findings, comparing and contrasting, and making sense of their experiences and feelings. Finally, the fourth step makes the concept more usable and flexible in their mental model. Learners are asked to *apply* the concept to a new situation (Braus and Monroe 1994). What would happen in another case?

Science education in the United States was transformed by the launch of Sputnik in 1957 with greater emphasis on teaching the process of science and

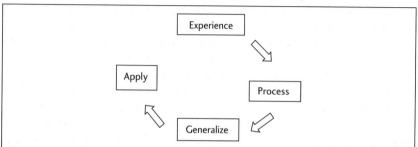

Fig. 2.2 The Experiential Learning Cycle guides educators to use four different approaches in teaching a concept.

inquiry, not just scientific information. In the decades that followed, a five-step learning cycle developed by Bob Karplus at UC-Berkeley was adapted by science educators at BSCS. Today it is known as the 5E's (Saguaro 2004):

- *Engage.* Learners are introduced to the task and make connections to things they already know. Motivating questions and problems are often used to spark curiosity and discussion.
- *Explore.* Learners are directly involved in activities, field observations, data manipulation, etc. Questions and interaction with others help them build concepts about the new information.
- *Explain.* An instructor provides information about the concepts through lecture, readings, and discussion. Abstract concepts are introduced and explained.
- *Elaborate.* Learners apply these concepts in a new situation, testing patterns and ideas and verifying their understanding.
- *Evaluate.* Instructors employ assessment techniques to monitor and quantify learners' progress.

In the exploration step, the experience can be almost anything that enables learners to realize new information or skills—a game, simulation, hike, video, puzzle, experiment, observation, or even a memory of a previous experience. An educator can help learners share their ideas with others, confirming similarities or discovering unique perspectives. Questions that generalize from this experience to a broader concept and promote elaboration would ask the learners to explain what this experience means, what it is related to, or what they would expect would happen if one thing changed. In schools, the application or elaboration process often occurs in a homework assignment. In adult programs it could be an opportunity to conduct a similar experiment in a different circumstance, develop a poster that explains their new ideas to others, or create a plan that uses the new information.

Building a lesson on a common experience helps to erase differences among learners due to poverty, family, culture, or extracurricular activities. Beginning a lesson by asking youth how an animal moves or what birds eat would enable the most talkative and those who watch televised nature shows to answer with confidence. Not all youth have the same skills and opportunities. An experiential lesson would have them observe a dog and a rabbit and duplicate the movement, or simulate bird beaks with a variety of tools and attempt to eat an array of foods. After such an exercise, everyone has ideas to share about how animals move and eat, which will enable to generalize, elaborate, and apply.

A study in Colorado, US, tested three different experiential science activities to see if student knowledge increased in each case. One activity was from Project WILD, a popular environmental education supplement that included only the first three steps of the experiential learning model. The second group was exposed to a modification of this activity that included all four stages. The third group received instruction from a similar activity designed to follow Kolb's four-step learning cycle. After instruction, there were no significant differences in student knowledge gain among the three groups and all groups increased post scores; all three activities successfully achieved increases in knowledge (Powell and Wells 2002). From this study it is clear that learning occurs when students are engaged in an experience, have an opportunity to reflect on it, and use that information to build new ideas. The specifics of which model and how many steps are much less important.

The learning cycle also makes obvious where educational programs fail to achieve all they should. Some educators are tempted to stop an activity after the "fun" stage and then move on to another. While youth may enjoy the game or skit, it is not likely that they will have learned something without a structured opportunity to think and reflect on their experience. It is also possible for educators to design exercises that do not have a relevant concept behind them. The too-popular crossword puzzles and labyrinth mazes are examples of seatwork that occupy youth and may teach them to spell or draw, but not accomplish the conservation objectives.

The learning cycle also is useful for adults in a learning program. A few questions could remind them of a common experience they have had recently, or a small group activity around a case study could serve as an experience to introduce a concept. Process questions would ask a few representatives to describe what happened to them. Generalizing questions would ask them to comment on the underlying constructs. Opportunities to apply that information to their workplace would complete the cycle. When educators neglect to use a learning cycle, they tend to throw a lot of information at the audience. Some of it may stick, but much will be forgotten. It is critical for educators to think about what they want their learners to gain, to identify the main concepts, to design an experience that will enable them to realize those concepts, and to guide them through a process to understand and use these concepts.

2.4 Learning styles

When educators vary the way they teach, they might observe that some learners are more satisfied with one approach. People have learning preferences and an inventory can help determine if people prefer concrete experiences, abstract conceptualization, active experimentation, or reflective observation (Kolb and Kolb 2004). While these learning styles are not forever fixed (they can be shaped by experience and may change across subject matter) they are fairly stable over the short-term. Because practice in nonpreferred styles can help build new skills, educators should use the entire learning cycle to develop lessons and opportunities that will appeal to all learners (Kolb and Kolb 2004). People can become more effective learners if they can use more learning styles.

Another version of learning styles suggests that people have sensory preferences for learning. Auditory learners readily grasp spoken information. They may read aloud or listen to the radio for news. Visual learners get information from books, charts, maps, and notes. They visualize what they read in their "minds-eye." Tactile

Box 2.4 Learning styles inventories.

Several Web sites offer assessment forms to help learners identify their preferred learning style. Sites include 30–70 questions about learning, such as: I enjoy learning about a new or difficult topic by reading about them, I best learn something new or difficult by working with my hands and making things related to the topic, or about abilities, such as: I am excellent at finding my way around even in unfamiliar surroundings, or I am good at putting jigsaw puzzles together (LdPride 2005). One site offers learning tips for each type of learner (Abiator 2005):

- Visual learners should write down things to remember, look at the person speaking to you, work in a quiet place, and work alone.
- Auditory learners often learn better when working in groups so they can talk about the information and hear it, recite material and read out loud, make flashcards and use them while talking out loud.
- Tactile-kinesthetic learners often learn best when moving around, they tend to fidget or snack while studying, they may not prefer to sit at a desk and may enjoy having music in the background.

For more information, use a Web search engine for "learning style inventory."

learners prefer to build, manipulate, trace, and take notes. They are physically engaged in the learning process. Several Web sites offer short assessments to help users determine their sensory learning style (Box 2.4). Again, most people do not fall into only one category but have a strength or preference for certain types of learning situations.

Because people are different we should not expect everyone to learn the same way. By varying teaching styles and approaches, conservation educators can reach more people, more effectively. Information about hunting regulations that is presented on a sign is best suited to those who learn by reading. By keeping the written information short, educators can hope that other learners will tolerate this presentation. Hunters with auditory preferences could be accommodated by oral information or an audiotaped presentation.

2.5 Multiple intelligences

Our common definition of intelligence comes from the IQ test, which evolved from French efforts in the early 1900s to identify first-graders in need of remedial assistance. This primarily verbal and numeric test gives us a narrow view of intelligence. The Theory of Multiple Intelligences suggests expanding the concept of intelligence by including seven skills or attributes (Armstrong 1994) (Box 2.5).

Box 2.5 Howard Gardner's seven types of intelligence, how they are used, and educational strategies that promote them (adapted from Armstrong 1994).

Linguistic Intelligence is the ability to use words effectively, either orally or in writing. Writers, storytellers, politicians, poets, and journalists use this intelligence effectively. Activities: Small group discussions, book reports and discussions, speeches, reports, journal keeping, publishing, and storytelling.

Logical–mathematical Intelligence is the capacity to use numbers and reason effectively. This expertise is displayed by tax accountants, computer programmers, and statisticians. Activities: classifying, categorizing, puzzles and games, calculations and problem solving, measuring, quantifying, sequencing.

Spatial Intelligence is the ability to perceive the visual world accurately and transform those perceptions. Hunters and detectives do the former while architects and inventors do the latter. Activities and tools: visualization, photography, 3D displays or dioramas, painting, drawing, mind-maps, future wheels, decision trees, microscopes, and binoculars.

Bodily-kinesthetic intelligence is using one's body to express ideas and feelings or specifically using one's hands. Actors, athletes, dancers, artists, mechanics, surgeons express this intelligence. Activities and tools: creative movement, field trips, theatre, drama, mime, physical activity, running games, crafts, and making products.

Musical intelligence is the capacity to perceive, transform, and express music. A critic, a performer, and even an appreciator of music show this type of intelligence. Music can help set the mood during a break. Activities: singing, humming, whistling, rap and chants, putting new words to old tunes, and music software.

Interpersonal intelligence is the ability to perceive moods, motivations, and feelings of other people. People with this intelligence may become counselors, social workers, or political leaders. Activities: cooperative groups, conflict mediation, peer teaching, community involvement, simulations, clubs and social gatherings.

Intrapersonal intelligence is the ability to access one's own feelings and emotions, strengths and weaknesses. Religious leaders and psychotherapists may use this intelligence. Activities: independent study, reflection, options and choices for projects and styles, self-esteem building activities, goal-setting.

An attribute qualifies as an intelligence if it can meet a number of criteria. Each of these abilities is valued in some way by human society, can be traced to early human culture, can be measured by a test or cognitive task, and is located in a discreet part of the brain.

Of importance to educators is the proposition that each person possesses all seven intelligences, though some will be stronger or weaker than others. Having an intelligence does not guarantee that it will flourish, as environmental, social, and cultural constraints might limit the options (e.g. lack of funding for special programs, lack of resources or tools, needing to care for other siblings). Many people, however, find that their stronger intelligences are the easiest way for them to understand the world and to express themselves.

Thus, conservation educators should try to develop programs that engage learners by covering all seven intelligences. That way everyone will get an opportunity to succeed and achieve feelings of success and accomplishment that help promote learning. Outdoor activities engage learners in different intelligences and learning styles than classroom activities, which may enable typically boisterous and difficult students to stay focused and engaged during outdoor lessons.

Utilizing these strategies does not mean that only one intelligence is employed. Journal-keeping uses both linguistic skill and intrapersonal reflection. Running games may be primarily bodily kinesthetic activity, but also involve understanding rules (linguistic), staying in bounds (spatial), and keeping score (logical–mathematical).

Conservation programs that follow a theme, like animal migration or wetlands habitat may intuitively use multiple intelligences by presenting the same concept to learners in a variety of ways—draw it, map it, write about it, build it, talk to your neighbor about it, measure it, and reflect on it. Considering multiple intelligences helps conservation educators develop well-rounded programs that are more interesting, as well as more accommodating to all learners.

2.6 Inquiry learning

Inquiry learning is a teaching method that uses and cultivates our natural sense of curiosity. Most commonly used in science education, inquiry learning also fosters an understanding of the scientific process. In inquiry learning, learners actively question, engage in an experience, seek information, and use that information to make sense of the world. Educators who practice inquiry learning emphasize the process of questioning and help students learn to ask questions effectively. Because youngsters do a better job of asking questions when they have learned something about the topic, many inquiry lessons begin with "guided inquiry" where the educator asks questions that lead students toward a discovery. "Pure inquiry" encourages learners to ask their own questions. For some, answering the question is less important than asking, predicting, and figuring out how to test a hypothesis.

Of course the topic of inquiry matters. Learners may not be motivated to invest time in questions that others have already answered. Instructors may use questions like, "what is this butterfly?" or "why is the sky blue?" to guide learners to the Internet or library to learn to find answers. Other questions are not so easily answered. "What is wrong with this tree?" could lead learners and teachers on an exploration of soil fertility, insect damage, or tree diseases. A teacher using inquiry methods would help guide learners as they continue to make sense of their questions.

Educators who use the inquiry method acknowledge that the world is changing. Simply teaching students information will be limiting, because the information may become obsolete. In an uncertain and changing environment, a better strategy is learning to question, to make sense of the world, to appreciate the inquiry process, and to develop habits that will continue to promote inquiry (Matsuoka 2004).

Biologists use inquiry when setting up an experiment to determine if burrowing owls use dung to attract their food supply—dung beetles. Community planners also use an inquiry process when using traffic data and growth projections to determine which roads should be widened. Inquiry learning is the process we use to answer questions we raise in all types of situations.

Inquiry is not merely using questions to engage learners in discussion. Likewise, it is not simply allowing learners to ask questions, and then answering them! Inquiry learning focuses on using questions and curiosity to guide the learner's efforts to figure out an answer.

"Why do you suppose there is more water in this pond in November than in September?" a naturalist might ask a group. If she responds to the guesses with "no" or "you are getting close," this is just a questioning technique to motivate the audience. It would be guided inquiry if the subsequent conversation went more like: "How could you find out?" When a group member suggests it might have rained a lot in October, the naturalist could distribute rainfall data and ask the group to determine which month had more rainfall. If someone suggests a creek was dammed, a topographic map could be pulled out and the group could compare the existing landforms to the map. If the group needs help, a clue might be, "Instead of thinking about more water entering the pond, consider what might change in the watershed to make the water level higher." An aerial photograph of the watershed might help the learners realize that it is mostly forested, and that between September and November the leaves fell from the trees, removing the tree's ability to pump gallons of water into the air, making more groundwater available to increase the pond's depth.

Inquiry is more than asking questions, hands-on learning, or experiential education. It is structuring a learning opportunity to engage learners in the process of pursuing their questions. Because experiences with single objects or problems do not usually generate understanding, it is important that learners continue a line of inquiry with several different scenarios, tools, or hypotheses. Our naturalist may wish to identify another question or event related to seasonal change or groundwater movement (depending on the program topic) that allows the learners to build on what they experienced at the pond. This type of "rehearsal teaching" allows learners to use a concept in a similar but not identical way, using what they just learned. It helps build connections in the brain and more firmly place this new information or skill in an appropriate mental model (Lowery 1998). Because this process takes time, it may be better suited for classroom teachers than nonformal field trip programs.

Some critics complain that inquiry learning takes too much time, and takes time away from studying the content that will be on standardized tests. In areas where learners are required to take such tests, educators must prepare learners for them. For tests that try to assess student ability, not just knowledge, teachers may better prepare students with an inquiry-based, problem-solving approach to learning.

Other critics recognize that learners who are accustomed to learning content and receiving rewards for getting the right answer could be terribly frustrated when a teacher changes the rules of the learning game. Some youth will not easily switch from traditional to inquiry learning programs. Poorly orchestrated inquiry could encourage students to ask the wrong questions or lead them to incorrect conclusions.

Inquiry does play a role, however, in community-based projects that encourage participants to actively collect data, consider multiple perspectives, design a project, and implement it. Conservation educators may find themselves on the receiving end of inquiry learning if they serve as a resource for groups engaged in community problem solving. If so, student questions could be answered with more questions that refocus students and encourage their continued exploration. Citizen science programs that engage adults in research projects may also use the inquiry process (Chapter 4).

2.7 Constructivism

A relatively new concept in education is constructivism, which became popular in the 1990s. This theory acknowledges the role that experience and reflection have in the learning process, but explains that people construct their own understanding from those experiences (Brooks and Brooks 1993). Therefore, two people may have the same experience but arrive at different understandings. The strength of political debates in recent years underscores the possibility that people can develop very different ideas from the same set of circumstances.

When people are confronted with information or an experience that does not fit their concept or understandings, they have several choices. They can ignore the information that does not fit, or they can actively construct a new understanding that encompasses the new information. Creating new cognitive structures, what Piaget called accommodation, is a personal effort. Teachers or supervisors cannot do it for anyone, mandate it, or prevent it. Those who choose to ignore the new information present a different problem for educators. When prior mental models do not match the new information, the learner may be holding on to a misconception. Misconceptions (also called naïve theories or misperceptions) are personal understandings that explain observations differently than the accepted theory. It may not be easy to discover exactly what the misconception entails except through discussion and careful listening (Box 2.6). To help someone construct a new understanding, the educator must first understand and acknowledge their existing concept and try to show how that belief is not always valid. Then the educator can present a new theory and an example of its explanatory power (Dunwoody 2003).

If the educator comes from a different culture or environment, however, it may be the educator who holds a misconception. Conversation, explanation, and inquiry may be helpful in sorting out which ideas are useful and which are not. Although most research does not find cultural differences in learning, in many parts of the world it is important for educational programs to be culturally sensitive (Box 2.7).

Box 2.6 Misconceptions about fire ecology can lead to poor communication (Monroe 2005).

The concept of defensible space is often used to encourage residents to reduce vegetation to increase the survivability of their home in case of wildfire. Resource managers often suggest that residents, "Clear 10 m of defensible space to protect their home from fire." Unfortunately, residents who have experienced wildfire know that burning embers travel well ahead of the fire front and believe that 10 m is not sufficient to prevent their home from burning. They discount the concept of defensible space because it obviously will not work.

The misconception buried in this example has to do with what constitutes protecting a home. The resource manager knows that if vegetation is at least 10 m from the home, the heat released when those shrubs burn is not likely to ignite the home. The fire will move on to more accessible fuel. The residents do not realize that defensible space could protect their home even if a fire burns over it. It is important to engage in a discussion with a resident to understand what they think and clarify where a different perception may be helpful.

Box 2.7 Native educators in Alaska created a set of standards for culturally responsive education that incorporates a number of strategies that nurture the cultural well-being of youngsters, a few of which are reproduced here (Alaska Native Knowledge Network 1998).

- Culturally knowledgeable students are able to build on the knowledge and skills of the local cultural community as a foundation from which to achieve personal and academic success throughout life.
- Culturally responsive educators incorporate local ways of knowing and teaching in their work.
- Culturally responsive educators participate in community events and activities in an appropriate and supportive way.
- A culturally responsive curriculum uses the local language and cultural knowledge as a foundation for the rest of the curriculum.
- A culturally responsive school fosters the on-going participation of Elders in all aspects of the schooling process.
- A culturally responsive school has a high level of involvement of professional staff who are of the same cultural background as the students with whom they are working.
- A culturally supportive community incorporates the practice of local cultural traditions in its everyday affairs.

Five guiding principles of constructivism are helpful for educators of both youth and adults (Brooks and Brooks 1993):

- *Posing problems of emerging relevance to learners.* If they do not find the topic inherently interesting, an educator must find a way to make it meaningful. Prompting questions, changing perspective, asking for a testable prediction, and offering immediate feedback are some suggested strategies for increasing relevance.
- *Structuring learning around primary concepts.* Constructivist teachers organize their teaching around major themes or essences. If these themes are problems, they engage learners in solving them. By relying on the major theme to connect the lessons, they focus on the whole, the system. It is much easier to break concepts into understandable pieces than to assemble the pieces into a whole system.
- *Seeking and valuing students' points of view.* Understanding learners' perspectives enables educators to challenge learners, reframe questions, and make information meaningful. It takes time to understand and value learners' views, but it also is key to engaging them and providing appropriate reinforcement. Workshop leaders who ask for participants' expectations should return to the list and confirm they have been addressed. Teachers who receive a confusing homework assignment should ask the student what he or she was trying to convey. Questions like, "what do you mean?" and "how do you know that?" and "what do you think?" are helpful in building understanding and encouraging learners to take responsibility for their own thinking. The more educators work to understand a learner's viewpoint, the more culturally and developmentally appropriate the lessons will be.
- *Adapting the lesson to address students' suppositions.* Once an educator understands how a learner perceives a situation, he or she should redesign the lesson appropriately. Just as the resource manager uses an explanation that address a misconception, a teacher can fit a learning opportunity to a student's cognitive ability. Examples of this cognitive "mismatch" occur when young children do not understand (Box 2.8). Misconceptions can be identified in adult audiences through surveys about issues such as the sources of water pollution or the causes of global warming.
- *Assessing student learning in the context of teaching.* The act of providing feedback is part of the learning process. A constructivist teacher uses it as an opportunity to continue teaching, not a chance to quell creativity or launch a guessing game for the "right answer." The process of providing nonjudgmental feedback often involves, again, asking the learner questions like, "what do you like about this?" or "why do you think that is the correct answer?"

Constructivism is an opportunity for a skilled educator and learners to engage in the learning process together. It uses inquiry and experiential learning to guide the process of building knowledge and making sense of the world. Like most skills, it requires patience and practice to do well. And like many concepts in education, it

> **Box 2.8.** How educators look at the world may not be the same as their students (Brooks and Brooks 1993).
>
> Two second-grade teachers developed a unit on changes to trees as cold weather sets in. Some of the topical questions were: What makes a tree in the autumn look different? Are all leaves the same? Are all trees of the same color? Do you see patterns? They engaged the youth in a nature walk, collecting leaves, rubbing leaves, making observations, watching a filmstrip, and classifying leaves. As an assessment, they asked the children to paint a picture of trees in the fall. With a full complement of paints available to them, every child painted a green tree. The teachers decided that children have different understandings for leaves and trees; practice with one does not necessarily mean application to the other.

overlaps with other theories. The following example illustrates how an insightful educator can combine strategies to teach fifth-graders about photosynthesis (Roth 1989).

Teaching photosynthesis with a constructivist perspective begins by asking students, "What is food for plants?" and eliciting their initial understandings. Food was explained as something that contains energy. As students hypothesized about what plants use for food, the teacher helped them design experiments to allow them to discover answers. An experiment with germinating seeds helped students realize young plants use stored food since water was not required once the germination process started. The sprouts were then allowed to grow in dark and light environments. The class read about von Helmont's experiment in 1642 and predicted whether the tree growing in the tub would "eat" the soil. His result, that the soil did not change weight significantly after the tree grew, convinced the students that soil was not a food. The teacher continued to encourage their questions and focused them on how plants obtain food.

When the initial ideas about plant food had been ruled out or questioned, an explanation of photosynthesis was finally offered. Additional work with the concept continued as students applied the idea of photosynthesis to trees and plants in their environment, asked more questions, and conducted experiments about their evolving ideas.

These students thoroughly understood the difficult concept of photosynthesis after this unit. In their reflections, one wondered why only plants could make food. Another commented that plants are not just sitting there, they are really very busy on the inside.

2.8 Cooperative learning

Solving conservation problems is not something usually done by single individuals. It is a group effort. New legislation is only passed through concerted efforts by scientists, lobbyists, advocates, citizens, and legislators. Identifying where a new landfill should go requires that the hydrologist works with the geologist, traffic engineer, regional planner, county commission, and neighborhood residents. Restoring beach dune habitat may involve youth groups to plant grasses, as well as biologists and ecologists to identify which species are most important and where. Addressing environmental issues and moving communities toward sustainability require people to work together. It is helpful to engage learners in activities emphasizing teamwork to model and practice group skills.

Group tasks have other educational benefits as well. Because learners have different strengths, groups that function to take advantage of members' unique skills may achieve more than an individual could. Research on student achievement that compares three styles of teaching indicates that cooperative learning results in significantly higher achievement and retention than do competitive and individualistic learning. People learn to work together and care about each other when learning is designed to be cooperative. When people care about their collective work, absenteeism drops and satisfaction increases. Cooperative learning increases social skills and has been effective in teaching skills for conflict resolution (Johnson *et al*. 1994).

As cognitive scientists explore dimensions of learning and thinking, research reveals that much of the learning process is social. The period of reflection common to learning cycles can be a process of interacting with others—explaining what learners saw, discovering what that means, and justifying ideas to a group. Many adult educators use small group discussions to engage more people in talking about the topic at the same time. Learning is essentially a social activity; people tend to learn better when they talk through a concept (Leinhardt 1992).

Designing an educational program to include cooperative learning principles, however, requires more than small group discussions or assigning worksheets to teams. Cooperative learning should include the following essential elements (Johnson *et al*. 1994):

- *Positive interdependence*. Each member of the group must recognize that their efforts are required and indispensable for group success. This is often accomplished by assigning different roles to learners (timekeeper, notetaker, materials gatherer, question asker, facilitator).
- *Face-to-face positive interaction*. Learners facilitate each others' success. Trust-building, conflict management, and negotiation strategies can be practiced to help people learn to give each other encouragement and support (Stahl 1994).
- *Individual accountability and personal responsibility*. Groups fail when individuals slough off responsibility or "hitchhike" on the group goodwill. This can be avoided by structuring the cooperative learning task with small groups, asking individuals to demonstrate learning, observing the groups at work, and asking each member of the group to explain what they have learned to other groups.
- *Interpersonal and small-group skills*. Success requires that learners master not only the content but also the process of working together. There are several skills inherent in this process: (1) get to know and trust each other, (2) communicate accurately and unambiguously, (3) accept and support each

other, and (4) resolve conflicts constructively (Johnson *et al.* 1994).Young children (ages 5–8) may lack the social skills to succeed at cooperative activities; instead they should practice sharing and taking turns, in preparation for cooperative learning skills that blossom later.

- *Group processing.* Cooperative learning works when groups function well, and this is dependent upon whether groups reflect on what is working and what should change.

Conservation educators can use cooperative or collaborative learning to make learning more fun and effective for groups of adults or youth. It may be challenging if the program is short-term and the educator has limited contact with the learners. If cooperative learning is appropriate for your program, use the following tips to guide you:

- Restructure the lesson to engage groups in a task. Identify roles for each person in the group and provide the skills and resources for them to do their job.
- Clearly explain the objective of the lesson, the task, the roles, and the evaluation procedure.
- Enable the groups to function collaboratively by focusing on the problem and the big picture. This may engage creative and critical thinking skills (see next section).

Educators can use a variety of methods when designing a cooperative task. A jigsaw is an effective way for learners to review information and share it with others (Figure 2.3). Reading alone, interviewing a partner, and sharing information with others is another procedure that can build cooperative skills. A process called "Think, Pair, Share," where individuals contemplate their response to a question and share their ideas with a neighbor, can be used during lectures to informally take advantage of the social aspect of cooperative learning without the formal process.

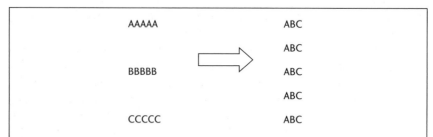

Fig. 2.3 A jigsaw is a cooperative learning activity that engages all learners by making them responsible for teaching something.

A group of 15 learners is first divided into 3 groups. Each small group is responsible for mastering a different aspect of the information. (In a lesson on aquatic insects, Group A identifies dragonfly and damselfly nymphs, Group B distinguishes stonefly from caddisfly larvae, and Group C works with waterbugs and beetles.) When everyone is sure of their information, the 15 learners "jigsaw" into 5 groups with 1 member from each of the 3 original groups. Each person is responsible for sharing their expertise with their two colleagues. This achieves the cooperative learning goals of individual accountability and positive interdependence.

2.9 Creative, critical, and systems thinking

When pressed to identify what conservation educators really want to achieve in educational programs, some might say their goal is for audiences to "understand" the ecology of the area or the implications of issues. What does it mean to understand something? A constructivist would say that understanding is built by the learner, and that the beams and boards of that mental structure are experiences, investigations, reflections, and discussions. A cognitive scientist might add that the more an individual experiences and learns, the denser that mental structure becomes. As we add richness, detail, and examples to the framework, we become more confident and more familiar with the concepts. That familiarity enables us to use that structure in novel ways, allowing us to solve problems. After all, if our brains were only capable of repeating the structures we have built, we would not be capable of new ideas. To have a functional and flexible mental structure capable of solving problems is to understand (Kaplan and Kaplan 1982).

Thinking is the activity that builds mental structures. It includes processing and generalizing, reflection, accommodation, and assimilation. Long heralded as the goal of a good education, thinking has become a skill that educators aim to improve. Conservation educators who work with schools may need to develop programs that enhance thinking skills. Other conservation educators may find these techniques worth using because they build understanding and environmental literacy.

2.9.1 Bloom's taxonomy

Educators who wish to enhance learners' ability to think often use Bloom's Taxonomy of the Cognitive Domain (Trowbridge and Bybee 1986) to help structure and evaluate their lessons. Bloom suggests that the first step in cognitive growth is *knowledge*—recall, recognition, and memorization. Questions that test this level of thinking are easy to develop and score, so they are frequently used. Unfortunately this level of instruction and evaluation does not equal understanding.

Comprehension is the next step in the hierarchy, the ability to interpret information, describe it in the learner's own words, and go beyond the information to explain what it means or demonstrate its use when asked. This is the first level of understanding. The next level, *application*, requires that the learner spontaneously apply the appropriate information without prompting. These first three levels of Bloom's taxonomy are often referred to as "lower-order thinking skills." They are also the thinking processes stressed in the learning cycle, leading toward understanding and mastery of a concept.

The three "higher-order thinking skills" are often grouped together because they complement each other in well-designed activities. *Analysis* is the ability to break down material into separate parts and relationships. *Synthesis* is the ability to put those pieces together into a functional organization. Seeing patterns, applying previous experience to new knowledge, and creating a whole are strategies that reflect synthesis skills. Finally, *evaluation* is the process of judging ideas, solutions,

Box 2.9 Bloom's Taxonomy of Cognitive Domain suggests six levels of thinking skills that can be practiced with a variety of educational exercises (adapted from Learning Skills Program 2003).

Competence	Skills	Activities
Knowledge	Observe events and processes Know species and concepts	List, define, tell, describe, identify, show, label, collect, examine
Comprehension	Understand information Translate and interpret Order, infer, predict outcomes	Summarize, describe, interpret, contrast, predict, distinguish, estimate discuss, extend
Application	Use methods, theories and information in new situations Solve problems using skills or knowledge	Apply, demonstrate, complete, illustrate, show, solve, examine, modify, relate, classify, experiment
Analysis	See patterns and organization Identify components Recognize meaning	Analyze, separate, order, connect, classify, arrange, divide, compare, select, infer
Synthesis	Generalize from facts Relate ideas from separate areas Draw conclusions	Combine, integrate, modify, rearrange, plan, create, design, compose, formulate, generalize, rewrite
Evaluation	Compare and discriminate between ideas Make choices based on reasoned arguments Verify evidence	Assess, decide, measure, recommend, convince, judge, support, conclude, summarize

methods, and strategies to ascertain their worth and effectiveness. Setting standards or criteria is a key aspect of evaluation.

Assessment tests in many states encourage teachers to help students master higher-order thinking skills. Box 2.9 suggests some strategies that a conservation educator might use to convey each of these levels of cognition. Educators often use questions to guide learners' thinking, and some of the verbs that trigger each thinking skill are included.

2.9.2 Creative thinking

Creative and critical thinking are both needed to solve problems. They are represented in Bloom's Taxonomy (synthesis uses creative skills, analysis and evaluation use critical skills). Creative thinking helps generate new solutions to problems and also builds learners' confidence.

Creativity is an ability, an attitude, and a process (Harris 1998). It is an ability to imagine something new, often by changing an assumption or rearranging information. Educators believe everyone has the ability to create, but some educational programs and cultures suppress it. This ability can be reawakened and encouraged.

As an attitude, creativity is a willingness to play with possibilities, to try something different, or to experiment. It may take nerve to strike out against the norm and be creative. Creative works are not usually produced on the first attempt. Creativity usually takes work and patience, trying new strategies, and tweaking to make improvements.

Educators can encourage creativity by designing open-ended assignments and activities that engage learners in generating their own ideas. Several decades ago, the "Anti-Coloring Coloring Book" was developed, providing just the background and a sentence to prompt a creative drawing. A caption, such as "You just found the most bizarre insect in your backyard" below an illustration of a grassy foreground, a picket fence and a big space, invited children to use their imagination rather than coloring inside predetermined lines.

Brainstorming is a technique that encourages creative thinking first, and critical thinking second. Educators can prompt creativity by pushing learners to generate several different answers to every question. Not only might they realize that there is more than one right answer, but that it is OK to think divergently.

The skill of generating new ideas is called "lateral thinking"—a way of thinking that seeks a solution to an intractable problem through unorthodox methods or elements that would normally be ignored by logical thinking (APTT 2004). Using specific tools can improve creativity. Tools and techniques include searching for alternatives, challenging traditional thinking patterns, generating provocative statements, and reshaping ideas.

2.9.3 Critical thinking

Logically critiquing complex information, uncovering and understanding bias, and having the motivation to use these skills to shape future actions are ideal traits for learners. Sometimes called decision-making skills, problem-solving skills, or responsible citizenship skills, educators are converging on definitions of critical thinking. There has been some debate over what critical thinking is, how to encourage it, and how to assess it, mostly centered around whether it is dependent upon contextual learning or an independent skill. The American Philosophical Association gathered input from over 35 specialists in an effort to define critical thinking. They concluded that it is the process of purposeful, self-regulatory judgment that drives problem solving and decision making (American Philosophical Association 1990). A core set of skills can be taught to learners to help promote critical thinking (Facione 1998):

- *Interpretation*. The ability to understand and analyze information (or comprehension in Bloom's language).
- *Analysis*. Identifying relationships between ideas, main points, assumptions, and bias.
- *Evaluation*. Judging the credibility and value of an argument, based on logic and evidence.
- *Inference*. Understanding the consequences of an action and the ability to decide what to do.

- *Explanation*. Communicating one's reasoning process to others.
- *Self-regulation*. Monitoring one's own thinking and correcting flaws.

Teaching critical thinking can be done explicitly by practicing these skills, or can be done in context by using the skills. Of course different situations make it possible to employ different types of thinking. It may be most useful for educators to mimic the real world and provide problems that have missing, uncertain, or extraneous data. The processes of sifting and filtering what is provided and determining how to obtain additional needed information help build important critical thinking skills.

Some standardized tests have been devised to measure these skills. Their use with environment-based education programs indicates that project-based programs contribute to a gain in critical thinking skills among high school students (Ernst and Monroe 2004).

2.9.4 Systems thinking

Systems thinking is a term from cybernetics and science. The concepts are versatile enough to be frequently applied to ecosystems, astronomy, business, social systems, and computer technology. As the world becomes more complex, problem solving is necessarily more complex. Viewing problems and their solutions as if they are simple—with one cause and one effect—is not realistic or useful. We need to help people see the big picture of relationships and interactions (Hough and Day 2000). Changes in fishing technology could mean changes in the menus at urban restaurants as well as in target fish populations, ocean floor geology, and populations of predator and prey species. As we develop the abilities to collect and synthesize data, we learn more about pieces of the system that may have a bearing on other elements, such as the effects of temperature and ocean currents on lobster larvae distribution. Unfortunately, we must train our minds to look for these relationships, to ask questions about connections, and to think in terms of system interactions. Our brains, while being magnificently capable of many things, are not very good at compiling data from multiple sources, analyzing, looking for relationships and feedback loops, and predicting how a change will affect the system. But we can train our brains to do just that.

The tools of systems thinking help us understand complex systems, like national debt or climate change. The tools help us keep track of all the different relationships between interacting components and how changes in one affect another. Systems thinking, such as understanding how resources flow from one process to another and how feedback loops change responses, enhances our ability to solve environmental problems.

Donella Meadows, one of the authors of *Limits to Growth* and *Beyond the Limits*, took her knowledge of systems dynamics to the public in a weekly newspaper column called The Global Citizen (Meadows 1991). She believed that the public could understand how systems worked, if properly introduced through analogies and stories. She used the task of weeding plants in her garden to discuss population issues and the cost of city services and annual wages to quantify a billion dollars.

Her collection of essays is a good example of how systems thinking can affect the way we look at environmental problems and how we can communicate complex systems to the public.

Conservation educators can assist in the communication of environmental issues by using a systems framework to identify objects and their relationships. Educators can also help develop a citizenry that can effectively tackle environmental issues by enabling them to think about systems and ask questions about connections and unanticipated outcomes.

2.10 Summary

What makes an effective conservation education program? Effective programs often incorporate several learning theories to increase understanding, such as engaging learners in an experience or activity and then using the experiential learning cycle to help them process and apply that information to new situations. Learners of all ages tend to be more interested in information that is meaningful and relevant, which is usually easy for environmental themes. Learning also happens in a social context, so it is valuable to have people share experiences, explain the results of an activity to each other, and work in cooperative groups. Because people bring different experiences and skills to an educational program, it is helpful to use a variety of teaching styles and learning methods to appeal to everyone's strengths and intelligences.

Additionally, there are specific things educators can do to maximize learning opportunities. Identifying the main concept and developing an interactive experience to reveal or stress that concept will help learners rehearse the main ideas as they process the information. Talking to learners to find out what they already know about a topic and how they explain evidence enables educators to provide experiences that will help learners construct their own understanding. Guided questions can help stimulate curiosity and prompt learner-generated questions and theories about how the world works.

Much of our work in conservation education helps build thinking skills in addition to knowledge so that learners can actively engage in solving problems in their community. Skills in creative thinking, critical thinking, and systems thinking are all needed to make sense out of environmental issues and identify solutions.

Many tips for working with older learners are appropriate to learners of every age (Spigner-Littles and Anderson 1999). Constructivism, experiential learning, cooperative learning, and inquiry learning contribute to our understanding of what makes learning effective:

- Learners do well when allowed to have some control over their learning environment. This is often achieved by offering choices of topics, situations, and exercises to learners. It is appropriate when an educational goal is to empower learners. By choosing what they are more interested in, learners of every age are likely to learn more about it.

- Learning is most effective when new information is connected to and built upon the learner's prior knowledge and real-life experiences. A brain-based educator and a constructivist would say that even young learners have prior knowledge that must be understood and addressed to make learning effective.
- Learning occurs when learners reflect on, recall, and rework information they have experienced and stored into memories. Having different experiences, using the information in different contexts, and delivering this information through different forms of communication can strengthen learning.
- Learners respond best to collaborative learning environments where they can share information. Sharing personal experiences tends to be motivating for many people, and is easier the more experiences one has to draw on. Learners of every age increase their learning in social contexts, because they can learn from each other. Group discussions allow learners to examine and investigate a problem from multiple perspectives.
- Learners do better when instructors ask probing, thoughtful, open-ended questions. These questions tend to encourage higher thinking skills, and promote generalizing and application. Learners may need practice to be comfortable with open-ended assignments if they have perfected the ability to get the "right" answer.
- Learners are encouraged and stimulated to share responsibility for learning with active, rather than passive, instructional methods. Active, experiential instruction creates more engaging and meaningful learning at any age.

It matters less which educational theory is used to build understanding in learners, and more that at least one of these theories is put into practice to improve conservation education and outreach programs. The interest of an organization's staff or volunteers, the age and interest of the audience, and the resource itself should help educators select the most useful theories and apply them.

Further reading

A Piaget Primer: How A Child Thinks, by Dorothy G. Singer and Tracey A. Revenson (1997). International Universities Press, Madison, CT.

Essentials of Learning for Instruction, by Robert M. Gagné (1975). Holt, Rinehart and Winston, New York, NY.

How People Learn: Brain, Mind, Experience, and School, by the National Research Council (2000). National Academy Press, Washington, DC.

The New Circles of Learning: Cooperation in the Classroom and school by David W. Johnson, Roger T. Johnson, and Edythe Johnson Holubec (1994). Association for Supervision and Curriculum Development, Alexandria, VA.

The Adult Learner: A Neglected Species, by Malcolm S. Knowles (1990). Gulf Publishing, Houston, TX.

Changing conservation behaviors

Despite our dedication to developing conservation education programs to help people understand the issues and develop an ethic that will support a host of conservation behaviors, it is difficult to measure program success by actual behavior changes. Moving people from "awareness to action" is not a simple task. Knowing about something does not guarantee caring or doing anything about it. Programs that simply provide information often may not lead to the hoped for changes, except where the lack of information is a significant barrier to conservation behavior (Schultz 2002). The good news is that educational programs can provide more than information; they can help learners consider values, build skills, and offer opportunities to practice new behaviors in a supportive environment. They can empower learners to take their own actions and can build community support for social change. They can help people decide how to improve the quality of their life and provide opportunities to build the skills to do so. Chapters 4–12 offer techniques that help educators and communicators address these broad goals and create successful conservation programs. This chapter reviews what we know about designing education and outreach programs to influence conservation behavior.

3.1 What do we mean by behavior?

Although it is easy to talk about "behavior" as a single concept, it means different things to different people. There are at least three different and common uses of this simple term. Educators often define any learning as a change in behavior; behavioral objectives are designed to enhance the measurement of this change. For them behavior could mean the ability to recite the stages of the water cycle. Performing this behavior would indicate having knowledge about the water cycle. The term behavior is broadly used in educational programs that measure cognitive learning in everything from art to technology. There is often no presumption that these behaviors will directly lead to changes in anything else.

Conservation behaviors are referred to quite differently. Some conservation educators design activities to create "responsible behavior." These activities provide the background people need in knowledge and skills, plus the motivation and commitment to make appropriate choices. This is also called environmental literacy. These programs have learning objectives that guide the interaction between the learner and the educator. Behavior in this second case means a sense of stewardship or an environmental ethic, but not a specific action associated with a single conservation issue. These programs are often seen as an investment in the future. They are quintessential environmental education. While some educators shun controversial issues to avoid the appearance of being one-sided, others thrive on presenting all sides of an issue and helping people sort through the consequences of various options.

Still other educators use psychology, sociology, and marketing principles to encourage specific changes in behavior. Their programs usually are designed for adult audiences and tend to promote socially acceptable behaviors. Messages about energy conservation or protecting habitat for endangered species can often be used without generating backlash, but messages that promote purchasing certain items or that suggest lifestyle changes that are not yet an acceptable norm can lead to complaints about inappropriate educational practices.

These last two strategies, education to promote an ethic of conservation that leads to environmentally responsible behavior and communication campaigns to change specific behaviors, are two avenues frequently used to achieve conservation goals (Monroe 2003).

3.2 Building environmentally responsible behavior

Five different models are explained here that help educators design programs to build environmentally responsible behavior. There is no single best model, but different situations may suggest one model over another. Since the models are somewhat similar, any of them could help conservation educators explain why their program is

---

Box 3.1 Theories at a glance.

Theories to build environmentally responsible behavior

Environmental citizenship behavior model	Three broad categories of major and minor variables lead to to responsible environmental citizenship.
Value–Belief–Norm model	Behaviors arise from personal values, beliefs about the world, and an individual's sense of appropriate actions.
Reasonable person model	People are motivated to understand the world around them; to learn, discover, and explore at their own pace; and to participate in solving problems. Creating opportunities to obtain needed information and to make a difference could lead to increased conservation behavior.
Systems thinking	Understanding the interactions and complexities between elements in ecologic and social systems could improve our ability to affect these systems.

Significant life experiences	Positive experiences in nature, adult role models, environmental organizations, education, environmental degradation, media, and work experience are common dimensions that environmentally active adults report helped shape their own attitudes and behaviors.

Theories to change specific behaviors

Theory of planned behavior	Target behaviors are determined by intention which is a product of beliefs and attitudes about the behavior, the opinions of important others, and the perception of control over the behavior and its outcomes.
Elaboration likelihood model of persuasion	Attitudes about a behavior are more likely to be permanently changed by communication activities that engage people in thinking about the behavior.
Motivational theories	Behaviors are a product of motives and needs. Understanding those motivations will help us predict and manage behaviors.
Stages of change	People go through a process of deciding, committing to, and achieving a change in behavior that affect their personal health, such as smoking or exercise.
Diffusion of innovation	New ideas spread through a community in a predictable manner, accepted first by innovators and last by laggards. The speed of acceptance can be accelerated by using opinion leaders to convey ideas to target audiences.
Social learning theory	People can learn from observing and modeling other people as well as vicarious experiences.

worth using. In some cases, these models may suggest additional strategies that educators could incorporate to strengthen already existing programs (Box 3.1).

3.3 Environmental citizenship behavior model

In the Environmental Citizenship Behavior Model, Hungerford and Volk (1990) suggest that three broad categories of variables work together to make it more likely that someone will act to protect the environment (Figure 3.1):

1. Entry-level variables are a prerequisite to environmental interest and knowledge. They include environmental sensitivity, an empathetic perspective toward the environment.
2. Ownership variables help personalize environmental issues, such as in-depth knowledge of environmental issues and their consequences and a personal investment built out of prior experience or knowledge.

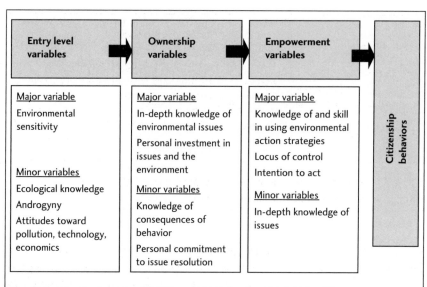

Fig. 3.1 The Environmental Citizenship Behavior Model identifies determinants of behavior that can be addressed in educational settings (Hungerford and Volk 1990).

3. Empowerment variables give people a sense that their actions can help resolve a problem. They must perceive that they have the skill and knowledge to act effectively, as well as the belief that they can and should.

These variables can be addressed through age-appropriate educational programs that increase environmental sensitivity and knowledge, engage people in direct experiences, such as backpacking to build a personal relationship with nature or data collection activities to explore a local issue in great depth, and practice with skills to improve proficiency with action taking. An issue investigation and analysis process that emphasizes the ownership and empowerment variables was developed for secondary teachers (Hungerford *et al.* 1985). A variety of other environment-based projects are also available that achieve similar goals (NEETF 2000 and Chapter 7). Research supports the development of very specific issue investigation and analysis skills leading to action-taking skills that help learners engage in community environmental problem solving, such as persuasive communication and political action (Marcinkowski 2004).

3.4 Value–Belief–Norm model

The Value–Belief–Norm model helps explain the process and the variables that seem to create and drive environmental behavior (Stern 2000). In a nutshell, behaviors are an outgrowth of personal values, beliefs about the world, and an individual's sense of appropriate actions. The model suggests that people are motivated by three different values: egoistic, altruistic, and biospheric. Each value is a type of predisposition that makes certain beliefs and attitudes more likely. A nurse, social

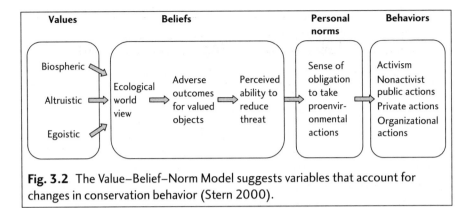

Fig. 3.2 The Value–Belief–Norm Model suggests variables that account for changes in conservation behavior (Stern 2000).

worker, or teacher may have a stronger altruistic value, which makes them more likely to believe they can relieve suffering, empower individuals, and improve society by helping others. An environmentalist would build on their biospheric value to hold beliefs about the role of people on the planet, the severity of ecosystem threats, and the promise of actions to reduce those threats. These beliefs influence the personal norms a person lives by, which in turn influence actions (Figure 3.2). Researchers often use values and value orientations to predict beliefs, attitudes, and behaviors (Decker *et al.* 2001).

Environmental beliefs and behaviors may also stem from egoistic values, since helping the environment helps provide clean air or recreational opportunities that benefit oneself, and from altruistic values since helping the environment helps guarantee a future for children (De Young 2000; Kaplan 2000). This complicates matters for educators looking to design a program to elevate the values that will enhance environmental behavior; however, it simplifies matters because most environmental behaviors have many good reasons that people use to justify their actions. We need not be limited in our belief that environmental action must stem from caring about the environment. Indeed, it can also come from caring about family, caring about engaging in community activities, and caring about future generations.

The Norm Activation Model helps explain the final element of the Value-Belief-Norm Model (Schwartz 1977; Stern *et al.* 1993; Vining and Ebreo 2002). It suggests that the motivation for environmental behavior is an individual's personal sense of the morally right choice, or obligation. The factors that activate that personal norm have been an interesting source of research. A study in Germany indicated that three factors contribute to a personal norm to use public transportation: subjective norm (how others perceive the behavior), awareness of consequences, and feelings of guilt (Hunecke *et al.* 2001). Giving participants a free subway ticket also increased ridership, indicating that in certain contexts, reducing personal cost complements personal norm and influences behavior. Other studies have suggested that acceptance of responsibility and perceived efficacy is a key factor that triggers personal norms. There is similarity between these factors and those mentioned in other models. An awareness of the negative consequences, for example, is an important element in the development of the ownership variables in the Environmental Citizenship Behavior Model. Subjective norm and perceived efficacy are key factors in the Theory of Planned Behavior (see below).

Many studies have analyzed aspects of the Value–Belief–Norm Model. One, in Sweden, looked at the influence of values, awareness of the environmental problem, and personal norm of cooperation in the context of reducing personal car use (Nordlund and Garvill 2003). The choice was between self-interest (using a car) and the interest of the collective (using more environmentally appropriate transportation choices). Values (e.g. biospheric, egoistic), beliefs (awareness of the problem and possibilities to reduce the problem), and norms (responsibility and willingness to cooperate) were hypothesized to play a role in this dilemma. After analyzing the responses from over 1500 drivers in five cities, researchers concluded that environmental values help increase awareness of an environmental problem and provide a foundation for a personal norm for environmental action. Beliefs about the specific problem of automobile traffic and air quality also helped activate the personal norm for cooperation. Programs that emphasize the value of such action to others and the negative consequences of using a car may help people reduce the use of their personal car (Nordlund and Garvill 2003).

This model helps conservation educators justify their programs by suggesting that activities that strengthen and support biospheric values or that build connections between conservation behaviors and altruistic or egoistic values will help lay the foundation for people to develop norms of environmentally responsible behavior. An educational program that increases awareness of the consequences of our behaviors and builds a sense of personal responsibility for our actions also could be justified with this model.

3.5 Reasonable person model

The Reasonable Person Model (Kaplan 2000) was derived from a consideration of human evolution, cognition, and motivation. It considers how humans make decisions that govern behavior and in which circumstances people are more likely to be able to consider changes. It suggests that people are motivated to understand the world around them (people do not enjoy being confused); to learn, discover, and explore at their own pace; and to participate in solving problems (they do not like being incompetent or helpless). These elements, of course, interact. In some cases the more information people have about environmental issues, the more helpless they feel (Levin 1993). Reducing helplessness may be the key to motivating people toward environmentally responsible behavior.

Participatory problem-solving activities are recommended to help people focus on a problem that needs to be solved. In the ideal situation, people would have an opportunity to work creatively to develop solutions that provide a variety of benefits and do not require heroic suffering. Being able to make this type of contribution requires that groups of people understand the issues and have among them a variety of resources and skills. The most productive participatory groups often work with experts, of whom they ask questions, get information, and check reality.

One critical element of a successful program is to start solving small problems. Large problems can easily overwhelm the most dedicated activists, whereas smaller problems often lead to concrete and measurable change in a reasonable period of time (Weick 1984).

In some cases people make appropriate shifts in environmental behavior when they are presented with a reasonable opportunity. What makes an opportunity reasonable will likely vary from person to person. Would-be car-poolers may need comfortable seating, the ability to fetch a sick child mid-day, or stops at a grocery store, for example.

The challenge for conservation educators is to create opportunities where people can be empowered with information and have the opportunity to particip-ate in solving a problem. Too often sponsoring agencies define the problem and the solution, only using public participation when complications arise. In Florida, prescribed burning permits are carefully managed and controlled; burns on large properties are planned years in advance and implemented if weather conditions are appropriate. One would think there is little opportunity to engage the public in planning a burn. However, a group of unhappy hunters who did not want certain areas burned during deer season helped the forest agency rethink their approach. Now hunting season and the location of deer stands are additional variables used when planning burns.

When the solutions to environmental problems must be site-specific it is diffi-cult for educators to develop programs that tell people what to do. In these cases it is easier to provide guidelines for how groups can figure out the most effective solu-tions to their problems. The Building Sustainable Societies Program of the Harmony Foundation in Canada provides problem-solving guidelines for communities (http://www.harmonyfdn.ca/programs.html). The EcoRecycle program, Waste Wise, in Australia, provides a toolkit for businesses and industry to explore source reduction strategies (EcoRecycle Victoria 2003–2005). The first step includes establishing a group to conduct the waste assessment and engage employees in the process of defining goals and identifying ways to reduce waste. After exploring the production process and quantifying the inputs, the outputs, and waste, the assessment team brainstorms ideas for reducing waste. Before mak-ing any recommendations the economic, environmental, and technical feasibility of each strategy is assessed. When recommendations are implemented, monitoring should occur to document whether waste is reduced. Only by conducting these experiments, unique to each industry, will people have an opportunity to know if their suggestions are effective.

Real participation in protecting the environment could be a powerful motiva-tor. These programs may be successful also because they build community, overcome helplessness, and engage people in exploring and understanding their world.

3.6 Systems thinking

Another approach to developing conservation education and outreach programs to build responsible behavior comes from work on systems thinking (Chapter 2). To create sustainable communities, people need to understand how communities and ecosystems are organized and function (Capra 2000). This is not accomplished by teasing apart an ecosystem and identifying pieces but by understanding the whole system and seeing patterns of cycles and energy flow at a variety of levels (e.g. cell, organism, population, community). The Center for Ecoliteracy, based in California, US, has been established to help educators and school communities apply systems thinking as they reform their approach to education (Figure 3.3). They suggest using a project-based approach to identify a local, meaningful, real problem that interests students. By exploring a problem in the context of the local natural and social environment, students will learn from landowners, resource users, resource agencies, government decision makers, and others (Barlow 2000). These varied perspectives help build connections and deepen the awareness of the problem and consequences, and the specificity of the problem helps ground the understanding of the system in something concrete. The real world may be the best source of complex systems—quite suitable for learners to practice on!

Systems thinking will become a necessity if citizens are to make wise decisions about sustaining life on this planet. The Millennium Ecosystem Assessment (WRI 2003) makes clear the multitude of links between four categories of environmental

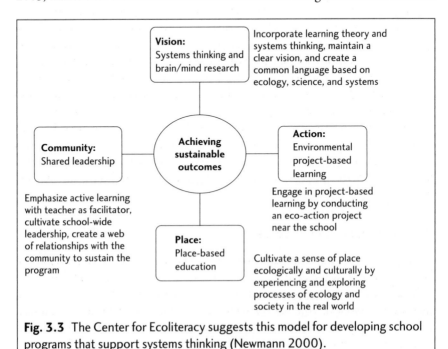

Fig. 3.3 The Center for Ecoliteracy suggests this model for developing school programs that support systems thinking (Newmann 2000).

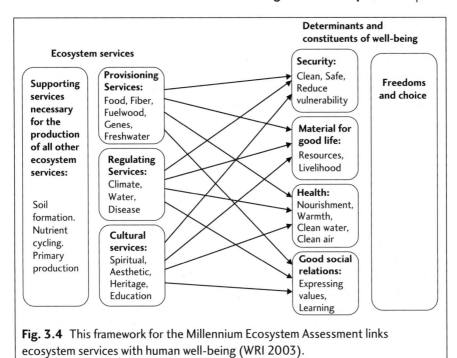

Fig. 3.4 This framework for the Millennium Ecosystem Assessment links ecosystem services with human well-being (WRI 2003).

services and five aspects of quality of human life (Figure 3.4). Mayors, builders, presidents, generals, chiefs, and landowners may be far more likely to care about the environment if we make clear to them the many connections between the environment and their health and happiness. Seeing and understanding these connections and feedback loops may be improved with practice in systems thinking. In this model, the influence that people have on the environment is not a direct link, but is felt through a number of indirect and direct factors. Our conservation education programs must help make clear the links between demographics, economics, politics, technology, and culture and human well-being as well as with factors that directly impact the environment, such as land cover changes, species changes, harvesting practices, and climate change. How else but through systems thinking can we begin to understand our many and complex impacts on the planet?

3.7 Significant life experiences

Many people who work in the conservation field attribute their career choice to childhood experiences in nature. Because they see links in their own life between experiences in nature, an appreciation of nature, and working to protect nature, they want everyone to experience and love the natural world. Conservation education programs developed in this vein usually emphasize outdoor adventures, nature walks, and field trips for youth. No doubt these experiences help youth learn more about the natural world, but they may fail to replicate earlier generations' childhood nature experiences, which were based on the freedom to explore, build forts, and play outdoors and independently (Nabhan and Trimble 1994; Sobel 1996). If this element of independence is critical, then a walk in the woods with a class of 30 other fifth graders will not build the same connection to nature.

Researchers have investigated the memories of significant life experiences by interviewing environmentally active adults. Two reviews of research on significant life experiences (Chawla 1998, 1999) indicate that the following factors are common to several studies:

- positive experiences in natural areas
- adult role models
- environmental organizations
- education
- negative experiences with environmental degradation
- books and other media
- on-the-job experience

Unfortunately, this research does not help explain why others who experienced these factors did not become environmentally active as adults. Was it the quality of the experiences that mattered? Might the experiences only matter if they complement a fertile bed of values and beliefs?

While several items on this list (i.e. positive experiences, education, and adult role models) are typically childhood experiences, many of these factors can influence adults. It may be never too late to experience the wonder of nature, to understand the threats of environmental contamination, or to gain experience in taking environmental action. Agencies and organizations that develop programs to guide and support youth and adults in local action-taking experiences may be making an important contribution toward environmental literacy by helping to create significant life experiences.

3.8 Building environmental literacy

Conservation educators should not abandon their desire to help others identify and appreciate the wonder of such things as hawks, water striders, and thunderstorms in their quest to influence conservation behaviors. At younger age levels and with introductory programs, this approach is quite appropriate to build awareness and knowledge. But the five models discussed above suggest that conservation behavior is not likely to evolve from nature study alone. We need to also teach how the parts are connected, how the system functions, what threatens the system, and what we can do about those threats. We need to empower learners with skills to solve problems, work with others, imagine new possibilities, and build structure around hope. We need to offer groups of people a supportive environment to develop their own solutions to environmental challenges. Practice in problem solving and systems thinking can help develop a host of skills that will ideally allow these learners to contribute in the future. Environment-based projects that involve real issues in the school or community are certainly instrumental to achieving these goals (Chapter7).

3.9 Changing specific behaviors

Psychology, sociology, anthropology, and their applications to communication, education, and marketing, form the basis of many theories used to change specific conservation behaviors. These theories developed from work on questions about what determines and predicts behaviors, how people change behaviors, how communities accept new ideas, and what influenced people to change; six are described below (Box 3.1). The role of attitudes, values, and motives are key, as they help shape how the individual perceives situations and opportunities for change. In this section, the term behavior refers to a specific action. When developing a campaign to encourage that action, researchers pay careful attention to who conducts the action, where the action occurs, and what barriers prevent that action. Specifying who, what, and where helps them better identify and influence a behavior. Applications of the theories described below are discussed in Chapter 9.

3.10 Theory of planned behavior

How we think and feel about something, our attitude, plays an important part in what we plan to do about it. Attitudes are the combination of knowledge (cognitive element) and a positive or negative evaluation (affective component). Although attitudes have a role in determining behavior, behavior is best predicted by measuring intention, which is a combination of several types of attitudes. Knowledge about the consequences of the behavior and the degree to which one cares about those consequences are one type of attitude. The other is a function of the importance we place on doing what "important others" would approve of. This Theory of Reasoned Action (Fishbein and Ajzen 1975) has been well researched and was later revised into the Theory of Planned Behavior (Ajzen 1985) (Figure 3.5). The latter includes a third element, perceived control, which makes it more applicable to conservation. In this theory, intention is based on:

1. The attitude an individual holds toward the behavior. This attitude is derived from beliefs about the behavior's consequences and a desire to achieve that outcome. There are usually more than one relevant belief about the behavior; this element is the sum of all those beliefs. In studies of boater behavior near manatees (*Trichechus manatus*), for instance, attitudes about boat speed, manatee injury, and manatee population status are suggested as influences on behavior (Aipanjiguly *et al.* 2003).

2. The perception of social pressure to conduct the behavior. This attitude is a combination of the beliefs attributed to people who are important to the individual and the motivation to comply. It measures attitudes like: if I think

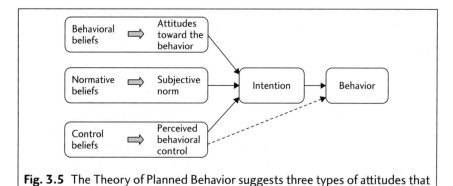

Fig. 3.5 The Theory of Planned Behavior suggests three types of attitudes that contribute to the intention to conduct conservation behaviors (Ajzen 2002).

my father would think well of the people who do this, I am more likely to do it.

3. The perception that one has the ability to conduct the behavior. This attitude is a function of beliefs about personal control and actual control. Someone may be able to legally hunt, but if he has run out of ammunition, he will not be able to do so.

Specificity is an important dimension of this theory. The attitudes people have about general things—such as wildlife or endangered animals—may not be predictive of the attitudes they have about a particular animal in their neighborhood. Consequently, educational programs involving nature study, while interesting, may not affect how people address specific environmental issues.

The third element—perceived personal control—is closely related to self-efficacy, a belief about the capability one has to exert control over his or her life (Bandura 1997). Self-efficacy is such a broad and powerful concept that it is often credited when people rise above adversity or approach challenges with resolve. Bandura suggests that self-efficacy can be enhanced through three types of strategies. The most effective strategy is to personally master a new skill, which can be promoted by educational programs that engage learners in practice or project-based education and issue investigation. A second strategy is vicarious learning—watching others succeed and fail. In the third, persuasion can be used to convince people that they have the ability to succeed (Bandura 1997). Persuasive materials may be more effective, however, if they use examples and models of success, blurring the second and third strategies.

Researchers use the Theories of Reasoned Action and Planned Behavior to help understand the strength of the factors that lead toward intention. In an analysis of voting behavior in California (US) for example, the Theory of Reasoned Action's specificity (i.e. evaluation of that policy becoming law as a result of voting) enabled researchers to connect ecological concern with voting intention (Gill *et al.* 2001). The generalized environmental attitudes, however, were not influential in voting intention.

The Theory of Reasoned Action was used to develop a rapid assessment tool to identify the social and attitudinal barriers to using alternatives to a depleted supply of fuelwood in northern Ghana (Batchelor *et al.* 1999). Researchers asked open-ended questions about the beliefs of the outcomes of conducting three different behaviors associated with fuelwood (firewood collection, improved stove adoption, and wood lot planting), as well as about their perception of social pressures. The project team found that the theory helped develop useful tools to understand the internal motivations of the audience. In addition, they developed ideas about the types of messages that could be used to change behavior. For refugee groups and for the behavior of woodlot planting, they discovered social norms are more important than attitudes, perhaps reflecting their tenuous and insecure status. Although all three behaviors involve fuelwood, decisions about collection and stove adoption are household bound and therefore, may be less influenced by others.

Many behaviors are a function of the interaction of knowledge, attitudes, social pressure, and perceived control. Unfortunately, different behaviors and contexts change the relative importance of these variables. In cases where the behavior is conducted in public, for example, social pressure may be greater than when the behavior is conducted in private. Understanding the motives of the audience helps educators and communicators develop a program that will be more likely to reduce barriers and support motives to change.

3.11 Motivation theories

One way of understanding why people behave as they do is to explore the motives that underlie that behavior. Usually a variety of reasons provide the motivation to prompt a behavior. For example, hunters often say they look forward to spend time in the woods, test their skills in nature, relax, or be with friends as motives for hunting. Like hunting, multiple satisfactions may be behind the majority of conservation behaviors.

Some motives are based on needs. Maslow's Hierarchy of Needs suggests that people are motivated by five different levels of needs. They are hierarchical because the lowest level needs must be met before the next level becomes recognized as needs. Basic needs such as food, shelter, and safety must be satisfied before higher-level needs such as belonging, love, self-esteem, and self-actualization (Decker *et al.* 2001).

In the field of outdoor recreation, several theories explain why people choose certain forms of recreation. One perspective suggests that people obtain eight broad psychological benefits through leisure activities (Driver *et al.* 1991):

- Self-expression—the need to use one's talents to express oneself
- Companionship—the need to sustain relationships
- Power—the need to be in control
- Compensation—the need to experience something new
- Security—the need to make a safe commitment

> **Box 3.2** Recreation experiences involve a variety of perceived preferences (Driver *et al.* 1991).
>
> | | | |
> | --- | --- | --- |
> | Enjoy nature | Physical fitness | Escape physical stressors |
> | Outdoor learning | Share similar values | Independence |
> | Family relations | Introspection | Be with considerate people |
> | Achievement/stimulation | Physical rest | Teach/lead others |
> | Risk taking | Risk reduction | Meet new people |
> | Creativity | Nostalgia | Agreeable temperatures |

- Service—the need to be of assistance
- Intellectual Aestheticism—the need for intellectual stimulation
- Solitude—the need to be alone

Another perspective (the Recreation Experience Preference Scales) quantifies benefits that people fulfill through recreation. According to this expectancy-value theory of motivation, people expect that certain behaviors will lead to desired events (i.e. if I drive to this trailhead, I can go for a hike) and that those events will lead to valued psychological outcomes (i.e. if I go on this hike I will enjoy seeing wildlife and being with my family). Researchers have identified 19 different motivations that tend to drive recreation behavior (Box 3.2). Of course different types of people have different motivations. By understanding the population of users and what they wish to achieve, managers can provide a more appropriate suite of opportunities. In the United States, for example, Hispanic families tend to seek large picnic areas for extended family gatherings, while other visitors prefer single, isolated tables. Picnic tables that can be moved together or shelters that accommodate large numbers of people would be a reasonable way to accommodate both groups of visitors.

These internal motives tend to be successful as drivers of behavior because they lead to satisfactions or benefits. Bird-watchers may be satisfied with their outing if their experience matched their expectations for pleasant weather, interesting group interactions, the number of birds seen, or seeing one favorite bird. Recyclers report their satisfactions include participation in something that benefits the community and resource frugality (De Young 1988–89).

Conservation behaviors are sometimes motivated by external motivations, such as incentives or disincentives. If external incentives are used, however, they may be best as tools to merely launch a behavior. Once someone begins a new behavior, a satisfactory experience should help develop the internal motives that could sustain the behavior (De Young 1993).

3.12 Elaboration likelihood model of persuasion

People are inundated with information every day. Many messages are ignored, but some win our attention. Marketers conduct research to understand the characteristics of messages that increase our attention, become memorable, and lead to action. It is challenging research because behavior is difficult to measure, attitudes may or may not correlate with behavior, and people react differently to the same message. Most of us can probably remember situations where we chose a behavior that was not based on relevant information or attitudes, but instead on peer approval, and other situations where meaningful information made a difference. Such distinctions led to the development of the Elaboration Likelihood Model of Persuasion (Petty and Cacioppo 1981), which suggests that there are two different routes of persuasion that could lead to a change in behavior. One provides information that engages people in thinking about the circumstance, which changes attitudes, and ultimately influences behavior. The other changes attitudes by association or reference, not by thoughtful reflection on information.

Long-lasting and durable attitude changes are usually the result of thoughtful consideration and cognitive information processing, called elaboration. The elaboration route to persuasion involves mental effort, prompted by the message. Communicators are more likely to engage the elaboration route if they use strong arguments, credible sources, relevant topics, clear messages, few distractions, and favorable comparisons (Petty and Priester 1994). The receiver of the message actively considers whether this information is sufficiently interesting to think about and if the argument has merit. Everyone does not react the same way to a message—some people find religious arguments persuasive while others rely on legal information when considering an issue such as hunting. For whatever reason, if they think about the message they have engaged the elaboration route.

The alternate or peripheral route may generate a short-term behavior change (such as brand-switching) through less relevant cues and motivators. Especially when people are not inclined to engage in the mental effort needed to elaborate on an issue, their attitudes may be influenced by messages that make them feel good or that trigger agreement. Many advertisements for luxury items use status as a motivator, which has little to do with the quality of the item. Similarly, propaganda techniques like "Four out of five doctors agree..." could be effective with people who are momentarily cognitively lazy. If the alternate route generates a behavior change, people may find themselves seeking information and elaborating on their behavior. When people eventually learn more about the product or behavior, or when moods change, the reasons for agreeing with the message may change to more durable support or the behavior may disappear.

Many social marketing messages and public service announcements strive toward elaboration by using humor to attract attention and then adding vivid, concrete examples that make information personally relevant. They provide clear information that is easy to remember and act on, and emphasize what is lost by not acting instead of what is gained by acting (McKenzie-Mohr and Smith 1999). Knowing what people care about can certainly increase the possibility that a communicator can encourage elaboration. For those people who are too busy to elaborate, communicators can use messages that appeal to a sense of responsibility or pride—"Leave the world a better place," and "Keep our community beautiful."

Televised commercials provide examples of advertisements that are designed for both routes. A perceptive listener can identify those that prompt thinking and those that hope listeners will not!

Educators can use the Elaboration Likelihood Model (ELM) to develop programs that emphasize cognitive engagement. A workshop on assertativeness training was developed with this model and compared with a standard workshop (Ernst and Heesacker 1993). Volunteer participants were divided into two groups, one receiving the ELM workshop and the other receiving the typical workshop. Both workshops included the basic definitions and appropriate procedural information. Both groups received the same ten-page handout and engaged in role plays to practice the behavior. The ELM workshop emphasized the development of strong, favorable attitudes toward the behavior by asking participants to think about a situation when they could have used this skill (motivation), asking participants to recall their thoughts (memory consolidation), and presenting arguments that tested well in generating favorable thoughts about the behavior. (A previous study assessed a number of arguments about assertive behavior, and the four perceived as most favorable were used in the workshop.) Two weeks later, surveys of workshops participants and their roommates revealed the ELM workshop participants had developed more favorable attitudes toward the behavior and used the behavior more often. Presumably the ELM workshop helped participants elaborate on the value of assertive behavior by organizing activities that promoted the development of supportive attitudes through introspection (Ernst and Heesacker 1993).

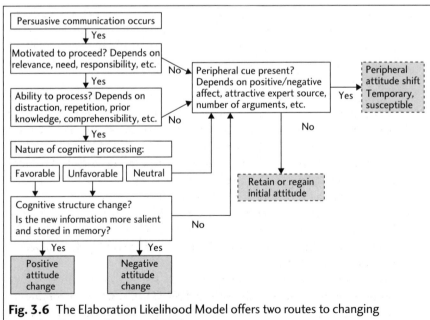

Fig. 3.6 The Elaboration Likelihood Model offers two routes to changing attitudes (Petty and Priester 1994).

Activities and experiences that encourage "yes" responses to the questions presented in the model (Figure 3.6) will help move individuals toward the central route. Enhancing motivation with vivid or personal stories, structuring opportunities to remember facts and tidbits, and providing strong arguments in favor of the new behavior can be effective.

3.13 Stages of change

Another approach to understanding the process of behavior change came from interviewing people who changed a health habit. The intentional approaches used by these self-changers had several common dimensions, which were used to create the model known as Stages of Change (Prochaska *et al.* 1992). This model has been extremely influential in the health field and with addictive behaviors. Educators first determine what stage someone is in, and then prescribe an educational intervention to help them move forward (Aschwanden 2000). Improving health has a direct and personal benefit, which may make this model more relevant to conservation scenarios when there is an equally strong element of self-interest involved. It may be possible to use other motives (such as social norms) to prompt change. The five stages include (Prochaska *et al.* 1992):

1. *Precontemplation.* In this stage people may be unaware of the problem or the benefits of change. They do not have enough information about the benefits to outweigh the costs and consider it for themselves. More information and more reflection on goals can be helpful at this stage.

2. *Contemplation.* At this point people have evaluated their situation and seriously consider doing something, but have not made a commitment to do so. People often get stuck here, having decided to act but not being ready to begin. Collecting information, evaluating options, and building a sense of commitment is important.

3. *Preparation.* In this stage people make plans to make change. They gathered the necessary equipment and form a mental plan for what will happen. They may have made initial forays into the new behavior, but have not yet completed the transition. Barriers have been reduced. A detailed example of a plan may be helpful at this stage, including contingencies for bad weather or sick children, and periodic rewards for sticking to the plan. This is the time to obtain written commitment to launch people into action.

4. *Action.* People are doing the behavior they desired and planned for. This may be the most obvious stage, but the process is not yet complete. Reminder notes, support from friends, rewards, and lower levels of guilt help people continue the new behavior.

5. *Maintenance.* Some behaviors require some work to sustain. In this stage people continue to reduce obstacles and reward themselves for continuing the new action. The fact that they have gotten this far helps people realize this change is no longer hypothetical. It is real. In this stage people can observe

and track the benefits they are beginning to accrue. This direct feedback can help motivate people to continue the behavior. Some changes require constant maintenance; others are more readily incorporated into the individual's new conception of self.

Conservation educators have used these stages of change with zoo exhibits to assess the likelihood of changing conservation behavior with exhibit messages. In a study at Disney's Animal Kingdom in Florida (US), significant changes were noted in visitors' interest and involvement with conservation over time. The impacts were not equal across all visitors, suggesting that variability among people allow some to be more influenced by messages than others (Dierking *et al.* 2001–2002). Due to the large numbers of influences associated with conservation behaviors, it is difficult to know what leads toward a predisposition for change. Nevertheless, we can be aware of the various needs for information, support, and assistance as people go through the process of changing conservation behaviors.

3.14 Diffusion of innovation

Taking the notion of change to the community level, the Theory of Diffusion of Innovation suggests that new ideas and changes move through a community toward acceptance in a predictable manner (Rogers 1995). Innovators will be the first to risk trying a new action, followed by early adopters, early majority, late majority and finally laggards. The characteristics of each group, however, influence acceptance. Since few innovators are respected trend-setters, the change could stall if only innovators try it. It often takes the involvement of the early adopters to launch an innovation in a community. Because the categories of adopters have certain characteristics, a communication campaign can be developed that speaks to each category (Box 3.3). Innovators, for example, are more likely to have networks and friends over huge distances and substantial financial resources to cushion bad risks than other groups. Innovators read national newspapers and magazines to keep up on the news that they care about. Early adopters are more integrated into the local social system; local newspaper articles are more likely to reach them than the innovators or laggards. If opinion leaders are among those who understand the innovation and believe in its benefits, their testimony and leadership could help diffuse the new idea more quickly.

Linking diffusion theory with stages of change makes it easy to see that some forms of education and communication are more effective at raising awareness among early adopters and early majority. Newspaper articles, television stories, and other mass media techniques can reach large number of people with basic information. But when it comes to actually doing something, such as voting, planting native flowers, or donating to a wildlife refuge, it is the personal contact we have with friends, neighbors, opinion leaders, and educators that makes the difference. The techniques of demonstrations, workshops, meetings, and events where people interact with each other can help give people the information they need to make a

Box 3.3 The Adopter Categories in a community suggests that any innovation is accepted by different groups at different rates (Rogers 1995).

Innovator:	2.5% of a group	Venturesome, risk takers, cosmopolitan set of relationships. Communicate among other innovators, even if they live in other cities.
Early Adopter:	13.5% of a group	Respected member of the local community. Often leaders; others look to early adopters for advice and information.
Early Majority:	34% of a group	Deliberate members of the community. Interact with peers, but usually not the leaders. Interconnected with networks.
Late Majority:	34% of a group	Skeptical elements of society. Not likely to change until the majority have already done so. They respond to peer pressure.
Laggard:	16% of a group	The last group to adopt an innovation. Suspicious of new. Traditional norms and resistance to change keep them isolated.

commitment and launch a new behavior (Figure 3.7). This two-step model, which uses mass media to reach community leaders and then those leaders to influence the public is simplistic, but it has helped educators realize that different techniques have different strengths and can be used strategically at different times (Rogers 1995). In order to promote the equitable distribution of the benefits of an innovation, Rogers suggests that educators target the opinion leaders who are well-respected by the late adopters and laggards, thus speeding their acceptance (Dearing 2005).

The suggested techniques for achieving diffusion at the second step may be effective because they rely on the social aspect of learning and may even develop a critical mass of people who are launching a new community-accepted social norm for that behavior. While some people cringe at the thought of being the first in their neighborhood to do something new, others may get a thrill from being part of the team that developed a new strategy. Finding the right group of people and orchestrating the appropriate educational program for them is a skill of matching audiences, needs, and opportunities. The norms, or standards of behavior, that a community agrees to follow are very important in molding behavior because people who want to be accepted by the community will probably conform to the accepted norms.

Certain aspects of a change make it easy or difficult to spread through a community, such as the degree to which it can be observed, the importance of the benefits, and the extent to which it conflicts with current culture (Rogers 1995).

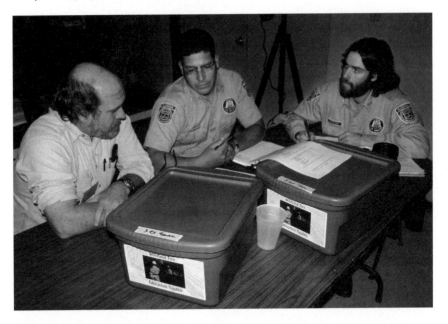

Fig. 3.7 Forestry field staff and extension agents plan new programs during a training workshop (photo by M. Monroe).

Conservation educators can overcome these barriers by advertising the benefits of conservation behaviors that may be less observable, providing feedback about the early successes, establishing demonstrations so others can see an example, and obtaining testimonials from community leaders so people recognize social approval (Chapter 9).

Extension workers in Eastern India used a range of media strategies to provide information about developing and sustaining community forests, including radio, television, newspapers, posters, leaflets, dance-drama, training events, field visits, and interpersonal contact by forestry supervisors (Glendinning *et al.* 2001). In an evaluation of the 10-year effort, researchers applied ideas about the diffusion of innovation through a community to understand the results of this agroforestry program. Low levels of education made the written materials of limited value. Although the majority of farmers report receiving information from the forestry supervisors, they tended to use other farmers for information about seedling availability or the impact of trees on other crops. More advantaged farmers (socially and economically) were better able to access information through formal and mass media channels than other farmers. To extend the program to reach small farmers this study recommends using direct interpersonal contact with extension workers and enhanced interactions with neighboring farmers (Glendinning *et al.* 2001).

3.15 Social learning theory

People can learn from observing someone and following their lead, not just from verbal interchange. The Social Cognitive Theory is the foundation for modeling (Chapter 9), also called observational learning. It recognizes that humans live in social environments—with other people—and the relationships we have with these people are sources of learning and change (Bandura 1997). School teachers know well the social pressures on them to be model citizens—their actions help establish the next generation's moral compass. Parents are concerned that television shows could become so powerful that children model the televised behavior rather than reality. This is problematic when the broadcast show does not match the child's cultural and social norms. Both social learning and diffusion theories recognize that individuals may not mimic the behavior exactly, but adopt the essence of the action and re-invent the innovation to be appropriate for their use (Rogers 1995).

In addition to providing models or demonstrations, educators use social cognition theory when they employ examples and case studies to help participants understand the possibilities involved with a change in behavior. In this instance the learners are experiencing the model vicariously.

3.16 Summary

These theories and models are but a few of the many that help explain human behavior. They are, however, the ones most commonly used to support a variety of conservation education and communication techniques, from social marketing to community education.

Many conservation education programs reflect the exuberant but naïve approach of wanting to tell people how to "save the world." Few people enjoy that sort of dictatorial treatment, and some react against it (Brehm and Brehm 1981). Borrowing a few theories from the social sciences will help conservation educators think strategically about a conservation message:

- How can I get people to pay attention to my message? Your best strategy is to design a message that appeals to them. It is important to understand their concerns, motivations, and values to match your message to their needs. Several theories of motivation offer perspectives on how to organize and appeal to different motives.
- If I can get their attention, how can I impress upon them the need to change behaviors? The Elaboration Likelihood Model provides suggestions for designing messages that will engage the listeners in more careful deliberation, which might lead to a shift in attitudes.
- Of all the good reasons to change a behavior, which are most influential to this audience? Theory of Planned Behavior suggests that knowledge about consequences, the influence of significant others, and the perception of effectiveness contribute to forming an intention to act. The Reasonable

Person Model reminds us that people change behaviors for many different motives; it may be possible to develop new opportunities that attract a number of followers for different reasons.

- Which media channels are most useful for reaching my audience and what should the message be? Stages of Change can help identify the messages that are most appropriate as people move through the process of accepting a new behavior; Diffusion of Innovations suggests which segment of the audience to target with each message.
- Is it useful to try to affect values and attitudes? The Value–Belief–Norm theory suggests that programs which address and clarify deep-seated values and show their connection to conservation could help reinforce those foundational values that underlie responsible environmental behavior. Values alone are not likely to lead to behaviors, however, but could move people in that direction if programs also include opportunities to gain understanding and knowledge about environmental issues and consequences, and build a sense of personal responsibility for taking actions that can lead to improvements.
- How can I justify a new program of matching mentors with youth for outdoor adventures? This type of program could create Significant Life Experiences that could help form values and attitudes that continue to shape these individuals for their career. Even if these experiences are not significant, they are direct, hands-on learning opportunities that help build knowledge (Chapter 2).
- What benefits could result from engaging learners in community problem-solving activities? These programs often increase self-efficacy (Bandura), help teach issue investigation skills and action-taking skills (Environmental Citizenship Behavior Model); and reinforce systems thinking, problem-solving skills, and community leadership skills.

A program can use several of these theories to introduce the benefits of conservation behaviors, build peer pressure for change, empower individuals to gain skills, and help people understand the complexity of the multiple systems in which we exist. A better understanding of what people think, what motivates their behavior, how they perceive the change, and who else in their world cares, can provide educators with ideas and resources to build a conservation education program that leads to appropriate conservation behavior.

Further reading

Understanding Attitudes and Predicting Social Behavior by Icek Ajzen and Martin Fishbein. (1980). Prentice-Hall, Englewood Cliffs, NJ.

Diffusion of Innovations by Everett M. Rogers. (1995). Free Press, New York, NY.

Cognition and Environment: Functioning in an Uncertain World by Stephen Kaplan and Rachel Kaplan. (1982). Praeger, New York, NY.

Environmental Problem Solving: Theory, Practice, and Possibilities in Environmental Education edited by Lisa V. Bardwell, Martha C. Monroe, and Margaret T. Tudor. North American Association for Environmental Education, Troy, OH.

4

Conservation education in the schools

Conservation education in the schools takes many forms—from schools that use occasional supplemental materials funded by wildlife agencies to schools that focus on the environment throughout their curriculum. Conservation education in the schools can range from a 1-day field trip to a farm to campus greening initiatives that construct sustainable buildings on campuses.

In New South Wales, Australia, the Department of Education and Training supports the development of "landscapes," features on school grounds such as wildlife corridors or edible gardens that promote learning in the local environment. At the Minnesota Zoo, an innovative program allows 400 students to pursue their high school studies in a school located on the zoo grounds. Students in the Zoo School have scored higher on college admissions tests than peers in their same district (NEEP 2002).

International documents such as Agenda 21 from the 1992 Rio Earth Summit and the Millenium Development Goals from the 2002 World Summit on Sustainable Development suggest that governments across the globe enhance education for sustainability. Several state and national governments have taken a lead in developing environmental education policies and practices in the schools, such as New South Wales, Australia, the Netherlands, and Wisconsin, US (New Zealand Ministry of Education 2004).

Yet environmental education often remains marginalized from the formal school setting. There is no consistency in how conservation education is administered worldwide. In some countries, environmental education is housed within a ministry or department of the environment, while in others the ministry of education oversees its programming. Many countries have no formal policy or supporting institution for conservation education.

In this absence, nongovernmental organizations (NGOs) and agencies can provide informal, though less coordinated support. For example, international initiatives such as the Asia-Pacific EE Network have supported professional development for teachers in environmental education. State-level professional organizations in the United States, like the Environmental Education Alliance of Georgia, also provide training and conferences for both teachers and nonformal educators.

The plethora of approaches to conservation education in the schools highlights the importance of both understanding and using the school environment in your community to maximize benefits for students and teachers. Conservation education in the schools can be a powerful approach for students, teachers, administrators, conservation organizations, and community members. This chapter highlights techniques for working with the formal sector, from communicating with schools to integrating conservation education into legislation and policy (Box 4.1). Three case studies illustrate a sample of successful strategies for practicing conservation education in schools.

This chapter targets conservation educators aiming to work with schools, but also teachers or school administrators who wish to enhance their curriculum

Box 4.1 A variety of techniques and approaches can help integrate conservation education into the schools.

Technique	Purpose
Communicating with schools	Focuses first on understanding the school environment and then choosing appropriate communication strategies for teachers, parents, administrators, students, and education agencies.
Acting as a resource to the schools	Identifies strategies for meeting the needs of school systems, such as serving as a science fair judge, developing an outdoor lab, and providing professional development for teachers
Connecting to academic standards	Focuses on correlating conservation education programs with academic standards or a curriculum framework to increase the use of conservation education resources and improve student achievement
Integrating conservation education into legislation and educational policy	Strives for systematic inclusion of conservation education into government funding, training for teachers, educational standards, and legislation.
A sample of approaches to conservation education in the schools:	
• Environment-based education	Uses the local environment, either natural or social, as a framework for the students' educational experience, with the goal of increased student achievement.
• Education for sustainability	Integrates the study of social, economic, and environmental systems to enable learners to take responsibility for creating a more sustainable future.
• Action projects	Engage students in selecting, planning, implementing, and evaluating a real-world environmental project and making informed choices for action in the community.

through conservation education. We use the term nonformal educators to refer to conservation educators from organizations such as agencies, parks, and NGOs, while formal educators are teachers in a classroom setting. In this chapter, both types of educators are working with schools and students. For all readers developing conservation education programs in the schools, we also suggest reviewing Chapter 1 for general tips on developing goals, selecting strategies, developing and pilot testing materials, implementing, and evaluating your program. Chapter 2 provides details on matching conservation education programs with appropriate developmental levels of students.

4.1 Communicating with schools

If the last time you entered a classroom was as a student yourself, the classroom environment can seem like a foreign country. Even if you work as a teacher in a suburban high school, walking the halls of an urban elementary school in another school district can feel like a cross-cultural experience. The first step in planning your communication with schools is taking the time to understand the classroom experience and the school environment. As many teachers will admit, the language of schools—from preschools to universities—is one filled with nuances, acronyms, and terms such as standards, scope and sequence, outcomes, and accountability. But teachers and conservation educators share a common interest in one word: education. This connection makes conservation education in schools not only viable, but also valuable to all partners!

4.1.1 Planning

Like learning a new language, understanding the classroom environment requires learning the basics such as the structure of the administrative system, as well as complexities like the pressures of testing. Begin investigating the school by talking with and listening to teachers, students, and community members, exploring newspaper articles and Web sites of school systems, and volunteering in the schools. Consider again the analogy of planning a visit to another country. Savvy travelers talk to friends who have visited that country to learn about customs and study a phrasebook to master simple greetings before arriving in a new environment. In the schools, this understanding about factors such as time, administrative interests, academic standards, testing, backgrounds of students and local community, and environmental issues will promote effective communication about conservation education programming.

Time

A reality of most school systems is time constraints. Find out how the schools structure the day for various grade levels and subjects. For most primary schools, teachers are with their students the entire day, which provides flexibility in terms of field experiences and outdoor activities. In upper grades, teachers usually specialize in one subject, taught in blocks of time, often ranging from 50 to 90 min. In the

United States, middle schools often operate through "team teaching" where a team constitutes a math, science, English, and social studies teachers. This can allow integration of conservation education across the curriculum as teams of teachers meet to plan their curriculum and discuss student needs. Topics such as music, art, and physical education are often rotated between classrooms throughout the week. This rotation could provide a place for conservation education activities within a subject such as art for multiple classrooms during 1 week.

Administrative interests and structure

Whether you are a conservation educator with a wildlife agency or a teacher, you need the support of administrators such as the principal, curriculum coordinator, school board members, and even department of education officials. If you can identify the interests of administrators in a school system, you can dovetail the strengths of conservation education to match administrative priorities. The achievement gains shown by schools using the environment to integrate learning across the curriculum convinced administrators in schools such as Central Middle School in Grand Forks, Minnesota (US) to give teachers more control over curriculum development (Lieberman and Hoody 1998). Addressing administrative interests through conservation education can lead to greater cooperation, increased funding, materials for outdoor projects, release time for teachers to attend workshops, and the subsequent sustainability of your efforts.

Explore questions such as: Who makes decisions at the individual school level? Can teachers and department heads plan their own projects with or without approval from their supervisors? How influential and active are groups such as the Parent–Teacher Association? What kind of support does the school system receive from the national ministry or department of education? The answers to these questions place you in a better position to communicate with administrators. In fact, your planning may reveal that a colleague or neighbor has a connection to a key administrator to facilitate your first communications!

Academic standards

Schools are required to follow national, regional, or state academic standards, which include the general knowledge and skills students should gain from their education. Most national or state departments of education post their academic standards on their Web sites or in official documents. Different countries and states use different verbiage to describe academic standards. Terms such as curriculum framework, curriculum guidelines, and academic standards all refer to the same thing—broad goals for student learning. Benchmarks or objectives are the more specific achievement targets, often organized by grade level. The word "curriculum" is also used to refer to lesson plan guides and units used by educators, such as the "Project Learning Tree" curriculum guide. A review of the academic standards will reveal what information is taught and which subjects could be enhanced by conservation education. The standards will also show if conservation education is a part of the academic course of study, either as a separate course such as Environmental Science or infused throughout the curriculum. The section, "Connecting to Academic Standards," later in this chapter provides specific guidelines for correlating conservation education to educational standards.

By reviewing the academic standards and spending time in the classroom, you can get a feel for what content teachers emphasize and why. Many kindergarten classrooms, for example, spend more time on reading than math and science. The rationale is that reading forms a foundation for learning other subjects. If your conservation education program addresses gaps in the curriculum or supports a priority area, you will have an easier time gaining support.

Testing and accountability

Some school systems feed their students a hot breakfast the morning of testing to enhance their performance on standardized tests. The pressures of testing in schools are real, as outcomes on these tests can be linked to school funding and student advancement. Arm yourself with knowledge about how and when these tests occur before you propose conservation education activities in the schools. Is there a national test and when do students take it? What is the pass rate in your area? Do the schools use alternative assessments such as portfolios? Do standardized tests include knowledge about the environment and conservation? The standardized tests in Kenya, for example, includes environmental content. This gives a good opportunity to conservation educators to assist teachers in preparing students for the exam, through field experiences and hands-on activities that reinforce these concepts.

In contrast, many states in the United States do not test for science in primary grades such as third grade. Many elementary teachers do not have a strong science background and may be intimidated to integrate science into their curriculum. The EcoTeam program in the United States works in conjunction with Jane Goodall's Roots and Shoots program to provide a series of seven environmental education lessons correlated to national science standards for third graders. With the pressures of testing in other subjects, teachers do not have time to teach science to their classes, so they have been eager for college students with the EcoTeam program to provide this science instruction (Figure 4.1).

Background and developmental levels of students

Building support for conservation education in the schools requires knowing your target audience—the students. A conservation education program that only emphasizes conservation of elephants may not find an appropriate audience in a rural Central African region where elephants routinely destroy subsistence crops. An emphasis on ecosystem processes might be more successful. Conservation educators must understand the age, knowledge, attitudes, skills, ethnic background, special needs, and cultural norms of students and how these students might help address environmental issues in their community (Braus and Wood 1993).

As you consider the background of students, match your conservation education goals to their developmental level. Some conservation educators have created

Fig. 4.1 College students in the EcoTeam program facilitate conservation education lessons that correlate to science standards for third-grade students.

more harm than good by introducing complex concepts such as global warming to 7-year-old children. Using research from Swiss psychologist Jean Piaget, we can connect our exploration of conservation concepts to the cognitive development of students (Chapter 2). With young children (ages 8 and younger), for example, focus on sensory awareness and positive outdoor experiences, rather than more abstract concepts such as habitat loss or acid rain that can invoke a sense of power-lessness and fear (Braus and Wood 1993). Techniques such as games, hands-on activities, nature studies, and the arts are perfect for this younger age group. Older students in middle and secondary school can address complex topics and abstract terms. For these students, you can use role-plays, for example, to explore conflict-ing interests of stakeholders in a controversial environmental issue. Techniques such as issue investigation, community-based research, and citizen science also work well with these older groups.

Local environmental and community issues

Understanding the pressing issues in a community will help link your conservation education program to the local environment. What are the critical environmental issues for the students, teachers, and parents? Think about what knowledge and skills students need to address issues such as restoration of wildlife habitat or water quality. A conservation education program might involve students identifying and investigating the environmental issues in their community (Chapter 7).

As you assess environmental needs, also look at the dynamics of the local community. Do local businesses and organizations support the schools? Are there existing groups working in the schools with whom you could partner? If you already work in the schools as a teacher, you have an inside view of what issues matter to your students and your community.

4.1.2 Implementation

Your communication with the schools, whether with teachers, parents, or administrators, should focus on matching the needs of the schools with your resources and skills. Once school administrators or other teachers realize that you have an interest in addressing existing needs, from organizing field trips to judging science fairs, they will be more likely to try new or innovative programs. Often communication begins with a personal connection, such as a teacher or parent you already know. If administrators see teachers excited about a new possibility (and willing to help coordinate the efforts), they are more likely to get on board.

Start by asking what resources the school has for conservation education and what resources they need. Is there adequate money for field trips? In one South Carolina (US) district, the costs of transporting 60 kindergarten students and teachers for an afternoon field trip is $60. The schools typically ask students to contribute $1 each. As an agency or organization, you could raise those funds as a one-time donation, showcasing your program and building support for future collaborations.

Invite a teacher or group of teachers for a quick walk around the school grounds to assess physical needs. Does the school have any space for outdoor activities? Is there a garden, a nature study area, or a stream to study water quality? Do they have teaching materials related to conservation? Have teachers participated in any professional development opportunities connected to conservation education? Often you may find a few teachers within a school system enthusiastic about conservation education. Communication with these leaders can help promote your efforts. Also ask teachers to identify their most involved parents or those who may be interested in volunteering or supporting your project.

You may be working in a school district where basic needs have a priority over outdoor spaces (Braus and Wood 1993). As a conservation educator, you can help address needs such as supplies, chalkboards, whiteboards, books, paper, and audio-visual equipment. Does the school have information about available conservation education resources? Compile lists of nearby EE centers that offer free field trips for students or contacts of guest speakers from conservation agencies. Members of the school-based Wildlife Clubs of Kenya, for example, get free access to any national park in Kenya. Many clubs work in partnership with conservation agencies to help raise funds for transportation to the parks.

Remember that teachers do not sit in front of a computer or by the phone during the day. Instead they are interacting with students during school hours. Most schools do not have voice mail for individual teachers, and teachers cannot stop their lesson to pick up a phone call from an enthusiastic conservation educator. Ask teachers to tell you the best method to communicate with them. Some teachers prefer an e-mail message, while others respond better to a phone message or a note in their box in the main office offering a number where they can reach you after school. During individual meetings with teachers or administrators, take notes and summarize what you have heard. Keep meetings focused and brief to respect the time constraints.

In addition to communicating with individual teachers, find out how teachers communicate with each other. Do most teachers attend regular faculty meetings? Do they receive a school newsletter or e-mail messages from the school system? Are there informal gatherings of teachers at a local pub or meeting place to decompress after the week? As you begin to communicate with teachers, assess the substance of their own communications. How can your resources help address current stresses felt by teachers? Are they motivated by specific topics? How can you build on these interests? Do not be discouraged if there are only a few teachers who are open to a new conservation education program. Start small with those who are interested, and others will come to you as they see the benefits involved in the initiative.

Communications will identify key players—teachers, students, parents, administrators, community members, agencies, and organizations—who can help you in the next steps of conservation education programming. Even if you are planning a small-scale program, a steering committee can generate feedback for your ideas and delegate responsibilities. Invite a group of stakeholders to an initial steering committee meeting to identify ways you can serve as a resource and to generate support for conservation education in the schools.

Another important aspect of communications with a steering committee is identifying potential barriers, such as lack of planning time or administrative support, and strategies for overcoming those barriers (Box 4.2). With any meeting, come prepared with an agenda or list of topics to address. The more organized you are, the more effective your communications will be. Be sure to write down ideas on a flipchart, so everyone can see that you have documented their suggestions. Lastly, remember to send the minutes of the meeting in an e-mail or memo to all members of the committee.

4.1.3 Evaluation

A big part of good communications with any audience involves active listening. If people *feel* heard, they are more likely to hear what you have to say. Good communications involves paraphrasing and summarizing what you have heard. Assess if you are using these strategies, even in e-mail correspondence, to help evaluate the effectiveness of your communications with the schools. "From your last message, I understand that you have end-of-grade tests throughout May," wrote one conservation educator proposing a garden-based field trip in spring for an elementary school. "Would a date in mid-April work better for you?" Simple communication tips like paraphrasing help educators know that you have listened to their needs.

While you can evaluate the consistency and effectiveness of your own communications, ask teachers and other stakeholders to give feedback on the

Box 4.2 Overcoming barriers to conservation education in the schools. (Adapted from Pennock, M.T., Bardwell, L.V., and Britt, P. (1994). *Approaching Environmental Issues in the Schools EE Toolbox- Workshop Resource Manual.* Kendall Hunt Publishing, Dubuque, IA).

Structural Barriers

"The school year isn't long enough to cover the academic standards and address conservation education."
If conservation education activities and programs help teachers address academic standards, they will be more likely to integrate them into teaching.

"My class periods are too short, and there's no time to plan."
Administrative meetings and duties often consume teachers' planning period. Consider several alternatives: cover information in small steps over several months; ask students to explore a conservation project during after-school time; or restructure the school day through team teaching and occasional double periods.

"I don't have relevant materials for teaching about the environment."
Help teachers identify and adapt resources. Use local materials from environmental organizations and natural resource agencies.

Support Barriers

"My principal or headmaster doesn't support this kind of education."
Consider offering a workshop for administrators to learn about experiential methods for teaching and improvements in test scores from schools using environment-based education. Use research documents with real data on student improvements to support your case.

"My colleagues aren't supportive."
Sometimes just talking with colleagues about their concerns can build understanding. Invite teachers in your school to participate in a sample activity based on their needs or interests. Find one or two other teachers who share an interest in conservation education and pilot test an activity or program.

Personal Concerns

"I don't know enough about this subject and this kind of teaching."
Teachers are facilitators of learning. Do your part to provide resources that can help bridge knowledge gaps, such as conservation workshops, local experts, books, and articles. Provide training in experiential methods of teaching and learning.

"Parents and community members might not approve."
Emphasize the importance of teaching about multiple sides of any conservation issue and the role of conservation education in building responsible citizenship skills.

communications. At the end of the first steering committee meeting, facilitate a discussion on what has worked well so far in terms of communications and what needs improvement. Ask individual teachers and administrators for this same information.

4.2 Acting as a resource for the schools

Communication lays the groundwork for your role as a resource to the schools and will play an important role throughout any programming. Research on the effectiveness of government policies to promote environmental education in English and Welsh schools found only limited examples of success over the past 25 years (Scott and Reid 1998). This study argued that schools could not implement government environmental education (EE) policy without resources, support, and guidance for how to vision and implement EE in their individual schools. Ideally, your role as a resource can help provide that guidance. In turn, students, teachers, and parents can help achieve conservation goals in the community. Your role may start as small as serving as a judge at the science fair or providing materials for an Earth Day celebration. Starting small helps ensure successful experiences and establishes a good rapport with the schools.

4.2.1 Planning

Your initial meetings with teachers, administrators, or a steering committee should have helped identify strategies for meeting the needs of your school or school system. Now the goal of planning is to prioritize these strategies and decide which role to implement first. Start with a small task that has a high chance of success and then plan subsequent activities or programs to build the profile of conservation education. For example, a coastal education agency might start by bringing a hands-on display on estuaries into a few classrooms, then invite teachers for an afternoon field visit with meals and materials provided. Subsequently, work with teachers to design grade-appropriate field trips that correlate to the curriculum.

4.2.2 Implementation

The following ideas give some examples of resources needed by many school systems. Remember that with more extensive projects such as developing outdoor labs or resource materials, you will need to follow systematic steps of program design (Chapter 1).

Serving as a judge in fairs, exhibits, and celebrations

An age-old tradition in many schools is the annual science fair, and schools usually need outside judges from the community. Providing this resource to the schools allows you to spend a few hours with teachers, students, and administrators and get a feel for a school system. The projects also will give you insight into possible environmental issues on the minds of the students! Many schools have contests involving environmental projects for Earth Day, so consider helping with these events or providing an information booth at the school.

Creating scholarships and awards

Most professional organizations for conservation educators are eager to attract teachers to their conferences. Yet many schools do not provide funding for teachers to attend these conferences or pay for substitutes. Providing scholarships to teachers so they can attend conferences and network with conservation educators serves both the schools and conservation education. State and regional EE professional organizations often receive grants to offer scholarships to teachers to attend their annual conferences. Offering scholarships to pre-service teachers also provides a resource and a hook for future partnerships with schools.

Another resource to the schools is awards that promote environmentally responsible projects or actions. The Auckland Regional Council in New Zealand sponsors the Enviroschools campaign to encourage schools to incorporate best environmental practices into their school communities. Wildlife Clubs of Kenya sponsors a national art and essay competition for schoolchildren with a theme related to a pressing environmental issue such as marine conservation or forests. Chapter 5 and 9 includes additional details on organizing environmental contests and incentive systems.

Providing resources and field experiences

Students and teachers may lack specific resources to study environmental concepts or environmental issues in their community. You can collaborate with students to develop materials or design the materials yourself based on feedback from teachers. Often existing resources provide the needed information, but teachers lack the time to research these materials. Or they may have materials that are not relevant for their geographic region, so adapting resources to fit a local context often fills an acute need. Previewing the curriculum framework will have provided ideas for existing materials that could enhance the content teachers need to cover. If creating resources, consider hosting an afternoon workshop to get input from teachers into the design and focus of materials. If educators help develop materials, they are more likely to use them.

When selecting or developing EE materials, use established guidelines to help identify quality materials. The North American Association of Environmental Education (NAAEE), for example, involved more than 1000 practitioners in the development and review of the *Environmental Education Materials: Guidelines for Excellence* (NAAEE 2004). This publication identifies six key characteristics of quality EE materials, with accompanying indicators to help educators evaluate materials (Box 4.3). The National Project for Excellence in Environmental Education spearheaded the publication of other guidelines such as: *Guidelines for the Preparation and Professional Development of Environmental Educators* and *Nonformal Environmental Education Programs: Guidelines for Excellence* (NAAEE 2004).

Box 4.3 The *Environmental Education Materials: Guidelines for Excellence* includes six characteristics and indicators of quality EE materials. (Adapted from NAAEE (2004). *Environmental Education Materials: Guidelines for Excellence*. National Project for Excellence in Environmental Education. Retrieved July 2, 2005, from http://www.naaee.org/pages/npeee/index.html).

Key Characteristic 1: Fairness and Accuracy

Environmental education materials should be fair and accurate in describing environmental problems, issues, and conditions, and in reflecting the diversity of perspectives on them.

Indicators: Factual accuracy, balanced presentation of differing viewpoints and theories, openness to inquiry, reflection of diversity

Key Characteristic 2: Depth

Environmental education materials should foster awareness of the natural and built environment, an understanding of environmental concepts, conditions, and issues, and an awareness of the feelings, values, attitudes, and perceptions at the heart of environmental issues, as appropriate for different developmental levels.

Indicators: Awareness, focus on concepts, concepts in context, attention to different scales

Key Characteristic 3: Emphasis on Skills Building

Environmental education materials should build lifelong skills that enable learners to prevent and address environmental issues.

Indicators: Critical and creative thinking, applying skills to issues, action skills

Key Characteristic 4: Action Orientation

Environmental education materials should promote a civic responsibility, encouraging learners to use their knowledge, personal skills, and assessment of environmental issues as a basis for environmental problem solving and action.

Indicators: Sense of personal stake and responsibility, self-efficacy

Key Characteristic 5: Instructional Soundness

Environmental education materials should rely on instructional techniques that create an effective learning environment.

Indicators: Learner-centered instruction, different ways of learning, connection to learners' everyday lives, expanded learning environment, interdisciplinary, goals and objectives, appropriateness for specific learning settings, assessment

Key Characteristic 6: Usability
Environmental education materials should be well designed and easy to use.

Indicators: Clarity and logic, easy to use, long-lived, adaptable, accompanied by instruction and support, make substantiated claims, fit with national, state, or local requirements

Another strategy to provide resources is to create a network that links teachers to conservation educators. The Wisconsin Environmental Education Network is an association of schools and statewide agencies that promote environmental education in the state. Each school has a liaison to disseminate relevant EE information from quarterly mailings to other teachers in the school. In some cases, schools may need access to experts on specific environmental issues, such as extension agents, wildlife officers, or health officials for an environmental health issue such as high asthma rates caused by air pollution.

Lastly, nature centers and parks can provide a destination for field trips (Chapters 5 and 12), as well as pre-visit and post-visit lesson plans with supporting activities. Correlating these materials to the national or state academic standards again increases the likelihood of their use and effectiveness.

Providing professional development

Lack of training in environmental education is a barrier identified in repeated studies about environmental education in the schools (Ham and Sewing 1987–88; Monroe and Cappaert 1994). Many teachers perceive that they lack the experience and knowledge to teach about the environment. In other cases, teachers share a common misconception that conservation education connects only to the natural sciences. Agencies and organizations can provide teachers with well-organized workshops that include ready-to-use materials suitable for the classroom. Including a wealth of resources, quality facilitation, efficient use of time, incentives, and good food at these workshops will encourage teachers to spread the word about your work.

The state of Wisconsin (US) mandates that new teachers complete training in environmental education to receive their certification in elementary education, secondary science, social studies, and agriculture. The Center for Environmental Education in India provides an International EE Training program each summer that includes teachers as participants. Agencies such as the USDA Forest Service and US Fish and Wildlife Service sponsor workshops on curricula such as Project WILD, Project Learning Tree, and Project WILD Aquatic. Teachers attending these workshops generally receive free teaching materials, as well as credits for continuing education. In many cases, state agencies also have correlated these national curricula to their state academic standards to promote use of the lessons in the classroom.

Designing creative workshops that help teachers do their job more efficiently will draw the most participants. One popular workshop session, for example, involves showing teachers how to design a "literacy bed" using a school garden (Box 4.4). This workshop session helps teachers learn to integrate props from children's literature into an existing or new garden bed at a school. For the

---

Box 4.4 This lesson plan helps teachers use 'literacy beds' in the schools to integrate literature and conservation education. (Adapted from Jackson, E. (2005). *Literacy Beds*. Unpublished document by the Appalachian Sustainable Agriculture Project, Asheville, NC)

Literacy Beds

Objectives:
By the end of the lesson, students will be able to:

- Create and integrate props from the plot of a children's book into a garden bed.
- Plant produce described in a children's book in their school garden.
- Work in cooperative groups to depict the book in their school garden.

Participants:
Students in a classroom

Materials:

- Children's literature that integrates gardening or environmental themes.
- Recycled props to depict the plot of the book, such as a gardening hat, old buckets, stuffed animals, and gardening tools.
- Plyboard, paint, and paintbrushes.
- Seeds or seedlings of produce featured in the book.
- An existing garden bed or a new bed.

Procedure:

1. This activity creates a garden bed based on the information in a specific piece of children's literature, such as Beatrix Potter's *Peter Rabbit* or Dr. Seuss' *The Lorax*. Begin by reviewing the suggestions for books below and engage your students in choosing the book they want to use for the project. Explain to the students that the class is going to 'grow' their book by creating a garden bed based on the plants, characters, and plot line of the book.

2. Tell the students that they are going to include as many of the props that figure prominently in the book. Since you are using a garden, the class will also grow as many of the plants mentioned in the book as possible. If the book includes vegetables that do not grow together at the same time, such as peas and corn, improvise by using a laminated drawing or a piece of plywood painted like a corn stalk! Color a picture of the book's cover on plywood, so everyone in the schools can see the connection between the book and the garden bed. Divide

students into cooperative groups to accomplish the necessary tasks for this activity.

3. If you do not have gardening experience, model the learning process for your students by inviting a grandparent, local farmer, or extension specialist. Local nurseries often will donate seedlings. Use your imagination to integrate art, math, history, literature, nutrition, and literacy into this conservation education project. Invite parents, school administrators, and the local media to an official "reading" of the story by the students.

4. Below you will find a few suggestions of books and activities that correspond to a literacy bed:

   • *The Lorax* by Dr. Seuss—Consider planting a 'thneed' garden, reminding students that thneeds are things you don't need! Plant seeds and transplants in old buckets or shoes. Integrate a lesson on recycling into this activity. You could also build a strange tree that would be a Truffala Tree for a lesson on extinction.
   • *Growing Vegetable Soup* by Lois Ehlert—Use bright paint to color props such as old shovels, rakes, hoe, watering can, soup pot, ladle, and spading fork. The book includes numerous vegetables to plant.
   • *Scarecrow* by Cynthia Rylant—This book is appropriate for children or even adults. Props can include scarecrows, owls, rabbits, and worms. Have students research these characters for the project. Write stories from the perspective of the scarecrow.
   • *Other suggestions: The Ugly Vegetables, The Carrot Seed, Sunflower Sal . . . or any book you find that will work!*

Evaluation:

Have students assess their own effectiveness working in groups on this project by engaging them in discussion several times during the project. How well are they working together? What have been their successes? What have been their challenges and areas for improvement? Take pictures of the garden bed before and after the project for concrete assessment of the project's outcomes and achievement of lesson objectives.

children's book, *Growing Vegetable Soup* by Lois Ehlert (1987), conservation educators painted garden tools the color of tools in the book (Figure 4.2). Students could even paint the cover of the book on a piece of plywood, plant crops from the book, and conduct research on the plot and characters!

Developing an outdoor lab

At Isaac Dickson Elementary School in Asheville, NC (US), a partnership with the USDA Forest Service, an environmental NGO called Quality Forward, the state Department of Natural Resources, and the garden program MAGIC, resulted in the reestablishment of a nature trail by the school. This previously

Fig. 4.2 A workshop on 'literacy beds' helps teachers use school gardening to teach literacy, nutrition, and conservation education. This literacy bed used the book "Growing Vegetable Soup" to display props from the book in the school garden (photo by S. Wagner Booth).

abandoned trail on the outskirts of the urban school grounds became the focus of a partnership that involved 500 donated hours from adult volunteers and parents to bushwhack, mulch, and dig to restore the trail (NEETF 2000). The project included the development of curriculum resources and trail guides. Classes regularly use the trail to study wildlife habitats. Since the reestablishment of the trail, the project has expanded to include an archaeological study by students of a former African-American community that previously resided on the site (Figure 4.3).

Helping schools develop an outdoor lab, a schoolyard ecosystem, or an outdoor site for testing air or water quality can be one of the most direct ways to increase the positive outdoor experiences of youth. Involving a wide range of partners expands the level of publicity and community involvement and thus increases the sustainability of the project. Outdoor learning labs also provide a perfect context for infusing conservation education into the curriculum, as students can use the natural site for journaling, observing wildlife, restoring habitat, and graphing seasonal changes. A partnership between the Los Angeles County Fire Department, local businesses, concerned citizens, and the Chaparral Middle School in California (US) created a plant nursery for drought-tolerant species that could be grown locally. Thousands of trees were

Fig. 4.3 Students at Isaac Dickson Elementary School use a previously abandoned trail that volunteers, teachers, students, and parents turned into a nature trail for activities such as nature study and archaeological projects (photo by K. Daven).

planted as a result of this project, and the entire school is now involved in the nursery (Project WILD 2000).

4.2.3 Evaluation

The needs of school systems change over time, so regularly evaluate the impacts and effectiveness of new resources in the schools as a result of conservation education. Evaluation will ensure that you replicate success, not failure. Most donors or foundations want to see evidence of the effectiveness and impacts of funded programs. Identifying your evaluation questions and tools before starting a project will help you plan and will avoid leaving you trying to rationalize your program after the fact. Planning your evaluation in advance can help you focus your efforts and clarify your project to others (Chapter 1).

Evaluation tools will depend on the specific type of conservation education resources provided to the schools. If the resource is as simple as providing scholarship money for teachers to attend a conservation education conference, compare the number of teachers attending the conference before and after funding became available. A few months after the conference, ask the participating teachers how they have used the information obtained at the conference in their classroom. If their testimonies are positive, use them to attract more funding and applicants. For serving as a science fair judge, informal observation and discussions will indicate whether your participation has been a positive addition to the event.

Evaluation of more involved resources will demand multiple evaluation tools. If a project involved building an outdoor lab, for example, you can measure the knowledge of students about their schoolyard ecosystem before and after incorporating the lab into classroom activities. Students also can take photographs of a natural area before and after a conservation project to document environmental changes over time.

If you involved a steering committee, gather systematic feedback from them through interviews and feedback sheets about the effectiveness of the projects. This same feedback from students, through interviews, observation, and even short surveys can prove valuable not only in improving programs, but also in gathering anecdotal evidence of impacts to include in funding reports. Student portfolios provide additional concrete evidence of the effects of conservation education. Lastly, if the resources significantly affect delivery of content in the classroom, consider pre- and post-tests of knowledge and skills.

One strategy to planning your evaluation is to meet with your steering committee during the planning stages of your program and brainstorm evaluation questions. The Wildlife Clubs of Kenya, for example, conducted workshops on evaluation throughout the country and asked staff and teachers to brainstorm questions they most wanted an evaluation to address (McDuff 2001). The teachers and staff members also identified sources of evidence for the questions and indicators of success (Box 4.5).

Box 4.5 Example of evaluation questions developed by teachers with the Wildlife Clubs of Kenya

Evaluation question	Sources of evidence	Indicators (signs of success)
How effective is the implementation of your activities?	Photographs Artwork Surveys Interviews Letters of appreciation	Variety of activities conducted Number of members/renewal of membership Active involvement of students Gender balance among members Innovative conservation projects Accurate record keeping and minutes Submission of articles to Komba WCK magazine

Some other questions that can drive the evaluation include:

- What learning gains, if any, occurred as a result of the program?
- What instructional techniques contributed to the results of the program?
- What aspects of the learning environment contributed to the results of the program?
- How did the program impact the environment?
- How did the program impact other people in the school and community?
- Are the goals and objectives of the program appropriate?
- Were there unanticipated outcomes of the program? (Bennett 1988–89)

4.3 Connecting to academic standards

Student assessment and testing are linked to academic standards to provide a measure of accountability. Thus, a teacher is expected to cover the material in the academic standards, evaluated by measuring student achievement on tests based on these standards. In many school systems, the academic standards include measurable objectives or benchmarks that students must achieve, such as: "The students will be able to explain the process of decomposition." The teacher often has the flexibility to choose the activities and materials for teaching the concept of decomposition to achieve the objectives. For the purposes of accountability, academic standards are also linked to professional development for teachers, student advancement and graduation, and school funding. Conservation education activities can provide the context for building skills that lead to achievement of academic standards.

4.3.1 Planning

Reviewing academic standards

The structure of academic standards varies depending on the country or region. All academic standards, however, include what students should know or be able to do by the end of a specific grade. The organization of academic standards is typically by grade level and subject. Each subject or content area has broad goals with more specific concepts or benchmarks as objectives.

Knowledge and skills build upon each other, so students in primary grades may be expected to describe what plants need to grow (soil, water, sun, and air), while older students should be able to explain the process of photosynthesis. In some regions or states, the academic standards include specific objectives for the environment or environmental education, such as in Pennsylvania (US) and New South Wales, Australia.

The state of Pennsylvania (US) includes "The Environment and Ecology" as content area in its academic standards for grades four, seven, ten, and twelve. To access the standards, visit the Web site for the Pennsylvania State Board of Education (2005), where you also can retrieve the standards for subjects such as arts and humanities, civics and government, and mathematics. The academic standards for these four grade levels for environment and ecology include the following content areas with numbers for reference:

- 4.1 Watersheds and Wetlands
- 4.2 Renewable and Nonrewable Resources
- 4.3 Environmental Health
- 4.4 Agriculture and Society
- 4.5 Integrated Pest Management
- 4.6 Ecosystems and their Interactions
- 4.7 Threatened, Endangered, and Extinct Species
- 4.8 Humans and the Environment
- 4.9 Environmental Laws and Regulations

Each content area includes objectives for the specific grade level, so students build on their knowledge and skills in subsequent grades. Academic standard 4.1.7, for example, corresponds to "Watersheds and Wetlands" for grade 7 with goals and specific objectives. (Standard 4.1.4 would correspond to a fourth-grade benchmark on watersheds and wetlands.) By the end of grade 7, students should be able to:

1. Explain the role of the water cycle within a watershed
   - Explain the water cycle.
   - Explain the water cycle as it relates to a watershed.

2. Understand the role of the watershed.
   - Identify and explain what determines the boundaries of a watershed.
   - Explain how water enters a watershed.
   - Explain factors that affect water quality and flow through a watershed.

Note that the standards above focus on science content related to watersheds, yet the last objective opens the door for exploring the positive and negative impacts of human behavior on watersheds. Conservation education programs, such as the Watershed Learning Center, developed by the Brandywine Valley Association in Pennsylvania, explore this content by providing outdoor lessons to schools within walking distance to the school. As the oldest watershed association in the United States, the Brandywine Valley Association developed its outdoor lessons to correlate to the state standards (Kenny *et al.* 2003). This program also trains teachers through peer-teaching to conduct the outdoor activities, making it easier for them to cover the standards themselves on their own school grounds.

Implemented in 2002, the environmental education policy for New South Wales, Australia includes objectives for three areas: curriculum, management of resources, and management of school grounds (New South Wale Department of Education and Training 2001). The curriculum goals and objectives for kindergarten–12th grade focus on development of knowledge, skills, and values specific to environmental education (Box 4.6). The policy asks individual schools to conduct a curriculum audit to identify which EE objectives are currently covered in the mandatory syllabi and content. The policy requires schools to integrate environmental education issues to support other content such as English, math, and creative art and use special events and school community projects to enhance

Box 4.6 Environmental education goal and objectives for K-12 grades in New South Wales. (Adapted from New South Wales Department of Education and Training (2001). *Environmental Education Policy for Schools*. Retrieved June 1, 2005, from http://www.curriculumsupport.nsw.edu.au/enviroed/files/Env_ee_policy.pdf).

Aim of environmental education

This Environmental Education policy aims to foster students' understanding of the environment as an integrated system, and to develop attitudes and skills conducive to the achievement of ecologically sustainable development.

Environmental education objectives: knowledge, skills, and values

Students will develop knowledge and understandings about:
   The nature and function of ecosystems and how they are interrelated
   The impact of people on the environment
   The role of the community, politics, and market forces in environmental decision-making
   The principles of ecologically sustainable development
   Career opportunities associated with the environment

Students will develop skills in:

   Applying technical expertise within an environmental context
   Identifying and assessing environmental problems
   Communicating environmental problems to others
   Resolving environmental problems
   Adopting behaviors and practices that protect the environment
   Evaluating the success of their actions

Students will develop values and attitudes relating to:

   A respect for life
   An appreciation of their cultural heritage
   A commitment to act for the environment by supporting long-term solutions to environmental problems

the student learning outcomes (NSW Department of Education and Training 2001). Note that these objectives do not specify a grade level.

Unfortunately, environmental education is not a priority content area for the academic standards in most states or countries. Many dedicated groups, however, are working to integrate EE into legislation, educational policy, and capacity-building for teachers and nonformal educators who deliver programs to schools. Because the environment relates to both natural and social systems, connecting conservation education programs to traditional standards is not difficult. In the implementation section, we present examples of how to correlate lessons when environmental knowledge and skills are embedded in other standards, such as language arts, science, and healthy living.

Two strategies for addressing academic standards: infusion and insertion

The Department of Education in Queensland, Australia divides its curriculum into eight areas and recognizes environmental education as a "cross-curriculum" area of learning (Department of Education, Queensland 1993). The *P-12 Environmental Education Curriculum Guide* recommends infusing environmental education into four of these learning areas: Science, Studies of Society and Environment, Health, and Technology, while recognizing that subjects such as math, English, and arts also can use environmental contexts for skills development. The curriculum guide presents "characteristics of learning" and sample activities that could include an environmental focus. For example, students in years 4–7 should be able to "relate to people, places, and other creatures through literature" (Department of Education, Queensland 1993). A recommended activity is to read books, such as "Where the Forest meets the Sea" by Jeannie Baker and discuss the need for protected areas, as well as the impact of specific cultures on the environment.

The Outdoor and Environmental Education Centers in Queensland have correlated all activities in their programs with the academic curriculum, team building, and leadership objectives. Their programs highlight biodiversity and education for sustainability, two themes in the school curriculum. By matching their outdoor education programs with academic standards, the centers increase the probability that teachers can take the time to integrate experiential learning in unique environments such as forests, estuaries, and freshwater habitats with their teaching.

This strategy of *infusion* is one used by many conservation educators to integrate conservation education into an already crowded curriculum. Infusion incorporates conservation education content or process into established courses (Monroe and Cappaert 1994). You can use subject areas from science, social studies, art, math, and music for infusion. Some advantages of infusion are that conservation education does not have to compete with other subjects, and teachers do not need a new textbook as they often can use supplementary resources (Box 4.7).

As opposed to infusion, *insertion* involves creating a separate conservation education course or specific theme focused on conservation topics. For example, students in high school may be required to take an Environmental Science course, or teachers might explore a 2-week theme in social studies on environmental issues. As an individual course, teachers can build on specific concepts throughout the course. But this strategy also requires more in-depth knowledge and training of teachers. As both infusion and insertion have advantages and disadvantages, the ideal situation involves using both strategies within a school system when possible. With both infusion and insertion, however, conservation education must connect to academic standards.

Box 4.7 Benefits and barriers of infusion and insertion (Adapted from Braus, J.A., and Wood, D. (1993). *Environmental Education in the Schools: Creating a Program that Works*. Peace Corps Information Collection and Exchange, Washington, DC).

Infusion – Conservation education throughout the curriculum

Benefits

- Needs fewer resources, such as a separate textbook or teacher
- Does not compete with other subjects or times in the curriculum
- Encourages learning and problem solving across the curriculum
- Reinforces and build upon key environmental concepts

Barriers

- Requires extensive teacher training and effort
- Relies on motivated teachers for efforts to succeed
- Can dilute the environmental message to fit course objectives
- Can be challenging to evaluate success

Insertion – Creating a separate conservation education course

Benefits

- Easier to implement and evaluate as a single subject
- Allows teachers to present concepts in greater depth that build throughout course
- Places a priority on the subject

Barriers

- Requires trained teachers with in-depth environmental knowledge
- Takes time from standard topics in curriculum
- May imply that environment is not interdisciplinary
- Can limit the number of students exposed unless required

When you infuse conservation education into the curriculum, you must address the current academic standards and the conservation material that you want to infuse. Some conservation education programs work with an existing program to determine the overlap between their conservation objectives and the objectives in the academic standards. Other programs use the standards as a starting point for designing new conservation education programming for specific grade levels.

One planning strategy to link a conservation education program to academic standards is to sponsor a workshop that involves teachers, administrators, and subject matter experts to develop resources that match the curriculum framework or to correlate existing materials. Consider which teachers and other stakeholders would be most interested in such an endeavor, plan a time that is convenient for them, and find support for incentives such as a stipend for participation. Draft lessons developed in the workshop can be reviewed by other teachers and experts, pilot tested, and revised (Chapter 1).

4.3.2 Implementation

An on-line learning center in New Zealand, Te Kete Ipurangi, presents an 8-step process for using the New Zealand Curriculum Framework to develop environmental education programs in the schools (Box 4.8). Four options for connecting environmental education to the academic standards are available: (1) develop

---

Box 4.8 The following steps outline a process for planning environmental education programs within the New Zealand Curriculum Framework. (Adapted from Te Kete Ipurangi (2003). *Planning Environmental Education Programs within the New Zealand Curriculum Framework.* Retrieved October 9, 2004 from http:www.tki.org.nz/r/environ_ed/guidelines/plan_enviro_e.php.)

Step 1: Identify the needs of students.

Ask the following questions to identify student needs:

• What are the natural and built environments in which students live, work, and play?
• What environmental issues do students encounter in their daily lives?
• What attitudes, knowledge, and skills will students need in order to become involved with other conservation issues?

Step 2: Review current programs.

Which of the following aims of EE are found in current teaching and learning programs?

• Awareness and sensitivity to the environment
• Knowledge and understanding of the environment and the impact of people
• Attitudes and values that reflect feelings of concern for the environment
• Skills in identifying, investigating, and problem solving with environmental issues
• A sense of responsibility through participation in addressing environmental issues

Step 3: Identify new opportunities for the inclusion of environmental education within the New Zealand Curriculum.

Step 4: Identify possible links between school programs and initiatives undertaken by regional and local councils and community agencies.

Step 5: Decide how environmental education will be managed within the framework of the New Zealand Curriculum.

There are four basic ways to structure the environmental education program:

• Develop a program in one curriculum area.
• Develop a program based on cross-curricular themes, such as languages and math.

- Use an action-oriented approach to help students develop attitudes, knowledge, skills, and values to take environmental action.
- Develop an environmental studies course to enable students an extended study of environmental issues.

Step 6: Develop programs based on effective teaching and learning approaches.

Effective learning requires teachers who can achieve the aims of environmental education by:

- Planning programs based on the aims of the national curriculum statements.
- Using a variety of teaching and learning approaches such as: inquiry-based learning, games and simulations, case studies, community-based learning, experiential learning, education outside the classroom, investigation of local environmental issues, and evaluation and action in environmental problem solving.
- Providing opportunities for students to use their new knowledge, attitudes, values, and skills
- Creating and maintaining a supportive learning environment for students
- Assessing both cognitive and affective learning in environmental education

Step 7: Select appropriate resources to support teaching and learning.

Consider the following criteria for resources: appropriate content, skill development, language level, student interest, date, bicultural perspectives (Maori and non-Maori), multicultural perspectives, gender perspectives, and balance.

Step 8: Plan how to conduct evaluation.

programs in the context of achievement aims and objectives in one curriculum area, (2) develop programs with cross-curricular themes, from two or more curriculum areas, (3) use an action-oriented approach developing knowledge, attitudes, skills for taking action on an environmental issue, or (4) develop an environmental studies course (Ministry of Education, Wellington, New Zealand 1999).

Connecting to content or process

Remember, if you have an existing conservation education program, look for benchmarks or objectives in the curriculum that your program already addresses. When developing a new program, consider which objectives best fit your resources and skills. The following content objectives adapted from The EE Toolbox (Monroe and Cappaert 1994) provide examples of how to use environmental activities to meet existing objectives.

- *Physics objective*: Explain the laws of thermodynamics and give examples.
  Conservation education activity: Analyze the efficiency of energy production from renewable resources.

- *Math objective*: Be able to plot a curve.
  Conservation education activity: Plot the falling curve of a local endangered species population.
- *Art objective*: Express emotion through art.
  Conservation education activity: Express anger or happiness in a sculpture created from litter found at school.
- *Language arts objective*: Create a logical argument.
  Conservation education activity: Write a persuasive letter about the environmental health of a local stream.

One group of conservation educators that uses gardens to teach ecology and science wanted to develop a 1-day field experience with kindergarten classes from a local elementary school in North Carolina (US). Through prior communications with the teachers, they learned of the need for a spring field trip that corresponded to the study of plants in kindergarten. They first analyzed the kindergarten standard course of study and brainstormed potential activities that fit their site and resources. The standard course of study for science included the following goal and objectives:

Competency Goal 1: The learner will build an understanding of similarities and differences in plants and animals.

- Identify the similarities and differences in plants by: appearance, growth, change, and uses.
- Identify the similarities and differences in animals by appearance, growth, change, and purpose.
- Observe the different ways animals move from place to place, and how plants move in different ways.
- Observe the similarities of humans to other animals and their basic needs.
- Observe how humans grow and change.

They then met with the lead kindergarten teacher to discuss her priorities and ideas and share the results of their brainstorming. After this meeting, the educators developed a classroom lesson entitled "How does your garden grow?" which correlated to several objectives in science, language arts, and healthy living. The kindergarten students then traveled on a 10-min bus ride to a large organic garden to rotate through three stations of hands-on lessons where they dug in the dirt, transplanted their own tomato plants, read stories about what plants need to grow, and talked about the garden as a growing system, just like their bodies (Figure 4.4).

This 1-day project did not require a workshop with teachers or meetings with a steering committee. The teachers left with a packet of four lesson plans complete with all correlations, as well as a post-field trip lesson and instructions for transplanting the tomato plants which each student and teacher took home with them.

Fig.4.4 Checklists in hands, kindergarten students explore the dirt during a field experience designed to correlate with the kindergarten academic standards (photo by R. Griffin).

Conservation education activities also can address process objectives, such as critical thinking, creative thinking, problem solving, cooperative learning, and values clarification. Process skills are becoming increasingly important in educational policy today, with a focus in the schools on leadership, citizenship, and community development. An emphasis on the role of conservation education in promoting process objectives also helps you connect your program to overall educational goals. For example, students can explore a local environmental issue from the perspective of different stakeholders or attend a local hearing, analyze editorials, and make a choice to participate in a conservation campaign (Monroe and Cappaert 1994).

In some schools, administrators struggle with strategies to mitigate tensions between different socioeconomic or ethnic groups of students. By their nature, many conservation education activities emphasize group process and communication skills. Indeed, some residential conservation education programs schedule schools from predominantly different racial and ethnic backgrounds during the same week to work on diversity issues. While conservation education certainly is no panacea for these societal issues, bringing diverse students together to analyze shared environmental issues uses the process skills necessary for building healthy human and natural communities.

Helping teachers address academic standards through
new or existing resources

Conservation educators typically have a toolbox of existing activities or lesson plans that have not been correlated to academic standards. In this scenario, figure out what concepts or skills students would learn if a teacher followed the exact instructions for the activity. Find that concept or skill in the academic standards, and include the correlation in your lesson plan. Do not consider modifications or enhancements of the activity. Just correlate the activity to what the students will understand as a result of the lesson. Your lesson plan should list these correlations for easy reference for teachers. Typically, this section of a lesson or activity plan follows the overall lesson objectives.

A second typical scenario is that conservation educators must develop new activities or resources to help teachers address standards. As mentioned earlier, become familiar with the standards, and use them to explore the concepts you wish to convey. Choose a few standards to focus on, and develop activities or lessons that use environmental content to introduce and reinforce the standards. Use one of the experiential learning cycles to develop opportunities to experience, reflect, and apply the new information (Chapter 2).

4.3.3 Evaluation

Upon completion, have other teachers review the correlations between your materials and academic standards and revise them as needed. Give teachers or administrators who review your materials a free copy of the resources as a thank-you. Pilot-testing materials will likely reveal necessary changes that may affect the correlations. During implementation, one direct measure of the effectiveness of your material is whether or not teachers use the program. If not, investigate the reasons why and revise to address the problems.

The advantage of connecting programs to standards that are tested is that the quantitative measurements provide data for evaluation of the effectiveness of the program. In Florida (US), a study showed that environmental education lessons designed to correlate to the state standards improved student achievement, based on performance on a post-survey modeled after the state achievement test (Wilson and Monroe 2005). Writing scores of students significantly improved after using a set of biodiversity lessons with 132 ninth and tenth graders. The lessons engaged students in practicing writing techniques while learning about biodiversity, exotic species, and endangered habitats (Wilson and Monroe 2005).

4.4 Integrating conservation education into legislation and educational policy

In the long term, comprehensive conservation education in the schools will require the systematic inclusion of environmental concepts in educational standards, capacity-building for teachers and nonformal educators, funding, materials development, and legislative support for increasing the environmental literacy of students. Survey research shows that adults in the United States support teaching environmental education in the schools (ASCD 2001). A nationwide survey in the United States found that 95% of adults supported environmental education as a standard course of study (Roper-Starch Worldwide 2000). A survey of 1,000 adults in Minnesota (US) revealed that 90% of respondents support environmental education in the schools through state funding, and 80% consider the environment when they vote (Murphy 2004).

Despite this public support, the consistency of environmental content taught in the schools varies by grade level, with 44% of US secondary school teachers reporting that they include environmental topics in their teaching, compared with 83% of primary school teachers (Survey Research Center 2000). Given the current pressures of time and accountability in the schools, institutionalizing environmental education and supporting these efforts through training and funding remains paramount.

Many countries and states have made great strides in using a legislative agenda and educational policy to promote conservation education in the schools. This section provides a case study from the state of Minnesota (US) that highlights successes, challenges, and hopes of this approach.

4.4.1 Planning

Minnesota Statute &115A.073 presents the following four goals from the state's environmental education plan: Pupils and citizens should be able to apply informed decision-making processes to maintain a sustainable lifestyle. In order to do so, they should

(1) understand ecological systems and effect relationship between human attitudes and behavior and the environment;

(2) understand the cause and effect relationship between human attitudes and behavior and the environment;

(3) be able to evaluate alternative responses to environmental issues before deciding on alternative courses of action;

(4) understand the effects of multiple uses of the environment (Minnesota Office of Environmental Assistance 2000).

Background

The Minnesota Environmental Education Advisory Board (EEAB) consists of 9 state representatives and 11 citizens who provide guidance on the development and implementation of the state EE plan. According to educational specialist with the Minnesota Office of Environmental Assistance Denise Stromme, the EEAB plays an important role in advancing EE through legislation and educational policy. "This mix of 20 people have direct ears to commissioners of state agencies and departments like the Department of Education, and they advise the governor's office," Stromme said. "This board is a pipeline into policy."

With funding from the state legislature, the Minnesota EEAB developed the original state plan for EE. In 1993, the EEAB published *A GreenPrint for Minnesota: State Plan for Environmental Education*, with a second edition printed in 2000 with the assistance of the Minnesota Office of Environmental Assistance. The GreenPrint includes recommended outcomes, actions, and indicators of progress toward goals for EE.

Focus groups across the state informed the development of the first edition of "GreenPrint." Stromme notes that three specific priorities emerged for EE including: (1) a central clearinghouse for EE resources, (2) preparation for pre-service teachers, and (3) training for elected and appointed officials in understanding the field of EE. This input prompted the 1996 development of SEEK (Sharing Environmental Education Knowledge), an interactive, computer-based networking forum for EE. It also influenced 10 major universities in the state to develop coordinated EE courses for both pre-service and in-service teachers. The third item, however, was not implemented.

The state Office of Environmental Assistance includes three full-time and two part-time staff on the EE team. The state plan for EE aims for students and citizens to apply decision-making processes to create sustainable lifestyles, but no one had defined the core knowledge necessary to make those informed decisions. This deficit became the direction for the EE staff and the role they would play in promoting EE in the schools.

"We [leaders of EE in Minnesota] decided to develop an Environmental Literacy Scope and Sequence to identify (1) what students need to know and do to make such decisions—what is the scope or vision of what students should have achieved by the end of their schooling, and (2) what are age-appropriate steps from pre-K to adult to achieve that vision. In other words, what is the sequence of content?" Stromme noted. This terminology of a scope and sequence is familiar to teachers and makes EE more accessible and practical to both classroom teachers and nonformal educators.

4.4.2 Implementation

The 5-year process of creating the Environmental Literacy Scope and Sequence for Minnesota in the late 1990s involved a literal convergence of different groups wrestling to define benchmarks for environmental literacy. For too long, environmental educators talked about the interconnection of "everything," but had not articulated exactly what that meant in terms of content (Minnesota Office of Environmental Assistance 2002). "During the first year, we brought in all these experts, we struggled, but we kept coming back to the world as the interaction of natural and social systems," Stromme said. Finally, they defined EE as "the exploration of natural and social systems and their interactions." This definition stems from a report on improving student learning through environment-based education (Lieberman and Hoody 1997).

At the same time, the Minnesota Department of Children, Families, and Learning (DCFL) was writing new educational standards for students, with the inclusion of a standard on environmental systems, due to persuasion by the advisory council. Ironically, this standard was based on teaching about the interaction of natural and social systems. At the same time, a group of representatives from Departments of Education from 12 states were meeting to define the knowledge and skills needed for environmental literacy. This group, called the State Education and Environment Roundtable (SEER), came to this same conclusion about the importance of systems and later served as reviewers for the Minnesota Scope and Sequence project. These groups believed that concrete concepts, not environmental issues, were key to environmental literacy. "We don't teach students math skills that enable them to calculate a grocery bill, but won't help them balance a checkbook," said Stromme. "We teach the skills of addition and subtraction that apply to *any* situation. We need the same foundation of knowledge and skills for natural and social systems."

The former executive director of the GreenPrint council, Pam Landers spearheaded the scope and sequence project and noted that a breakthrough came in the process when Ed Hessler, from the DCFL realized that the American Association for the Advancement of Science (AAAS) had published their own "Benchmarks of Science Literacy." Those benchmarks included a sequence for teaching systems to students and adults (Minnesota Office of Environmental Assistance 2002). The group working on the scope and sequence project adapted the benchmarks to fit social and natural systems and their interactions.

"The results of the hours of work, communications and miscommunications, abraded egos, long phone calls, passionate e-mails, people entering and leaving the project, and moments of sheer gratitude, relief and satisfaction are in this publication," Landers writes in the preface to the Environmental Literacy Scope and Sequence. The document includes environmental literacy benchmarks by grade level with key systems concepts and supporting concepts. For example, the benchmarks for Grades 3–5 state that:

- In social and natural systems that consist of many parts, the parts usually influence each other.
- Social and natural systems may not function as well if parts are missing, damaged, mismatched, or misconnected.

For Grades 3–5, the key concepts include trophic level, cycles, migration, predation, structure, and function. The benchmarks and systems concepts build upon each other in a sequence of grade levels and can correlate directly to the state educational standards.

The scope and sequence was tested with hundreds of teachers and revised based on their feedback (Figure 4.5). The Minnesota Office of Environmental Assistance also has targeted nonformal educators in nature centers and parks by building their skills in using the scope and sequence when they work with students in the schools. "Until we get EE required in the state standards, we are integrating it into the schools through the back door," said Stromme. The office provided workshops for both formal and nonformal educators in using the scope and sequence as well.

Fig. 4.5 The involvement of teachers, environmental educators, administrators, and EE organizations resulted in the Environmental Literacy Scope and Sequence in Minnesota: a systems approach to integrating EE in the schools (photo by D. Stromme).

To build support from teachers, the board advised the state to create a permanent funding mechanism as an incentive for teachers who use the scope and sequence to implement real-world activities with their students. This Environmental Learning in Minnesota (ELM) fund was on hold for 2 years, but there are hopes to reinstate the proposal. "EE in legislation is tricky and political," Stromme said. "You have to have a long-term vision. Environmental education in the schools in this state is going to take time. Chicago math (a now accepted way of teaching math) took decades to get into the schools. We must maintain a long-term vision and continue to foster our early adopters."

4.4.3 Evaluation

How will Minnesota environmental educators know if the Environmental Literacy Scope and Sequence has influenced the integration of EE in the schools?

When asked that question, Stromme said she sees evidence of success from traditional EE programs, such as the Project WET curriculum in Minnesota, that have adapted their curriculum to fit the scope and sequence. In the future though, she gave the following indicators of the effectiveness of the initiative:

- Creating a permanent ELM fund as incentives for teachers using the scope and sequence.
- Presenting more sessions at the state conference dealing with content and the scope and sequence, rather than mere activities.
- Adopting the scope and sequence in other states. (SEER has approved the document and given anecdotal evidence that other states are using it.)
- Developing an EE certification program for educators in the state based on the scope and sequence
- Posting of other EE events on the interactive SEEK Web site, such as "fire-wise education" that references the environmental literacy benchmarks.

Evaluating the effectiveness of EE in legislation and policy involves the political process, as support for EE waxes and wanes based on the political times. National concerns about the environment in the late 1960s and early 1970s in the United States fueled both environmental quality legislation at a national level and subsequent state actions in EE. The last two decades have included what many describe as public "attacks" on EE to undermine state-level EE legislation. One positive aspect of these attacks, however, is that the field of EE in North America has developed guidelines for effective EE programs and materials, as well as organized state-level responses with legislative and policy agendas.

4.5 A sample of approaches to conservation education in the schools

The type of conservation education program you plan and implement will depend on a variety of factors, including the characteristics and needs of the school system. Historically, different labels, from place-based education to issue investigation, have characterized specific approaches to conservation education in the schools. The following case studies present three approaches for conservation education with schools: environment-based education, education for sustainability, and action projects. Chapter 7 explores additional techniques for connecting classrooms with conservation.

4.6 Environment-based education

Environment-based education describes instructional programs in schools that use local environments, either natural or social, as the context for a significant portion of students' educational experiences (NEETF 2000). The primary goal of environment-based education is student achievement in all subject areas. Driven by increased testing and accountability in the school systems, environment-based education uses the environment to promote measurable gains in student learning. Research has demonstrated that schools using environment-based education show not only improvements in student achievement, but also increases in student motivation, higher-order thinking skills, and a decrease in discipline problems (NEETF 2000).

Environment-based education serves as one model for teachers, administrators, parents, and nonformal educators interested in integrating environmental education into schools. In several states in the United States, conservation educators wishing to raise support for environment-based education have sponsored workshops for stakeholders in schools—from parents to department of education officials. Environment-based education can start with one study unit, one classroom, one team of teachers, or can involve an entire school.

In one classroom at Kruse Elementary School in Texas (US), teacher Libby Rhoden instructs a class of 19 first graders, mostly Latino children who live within 5 square miles of a refinery (NEEFT 2000). In contrast to other teachers at her school, Rhoden bases her lessons on experiences children have in their local environment. When the students go on a nature walk, they write about their observations. The students build math skills by observing and then tallying behaviors of birds near the classroom. While Rhoden's students reflect the same abilities as students in other classrooms, her students consistently perform higher than all other first graders in reading, language, and math (NEEFT 2000). We cannot determine the influence of the teacher on the achievement of the students, but other outcomes from environment-based education corroborate the findings from her classroom.

Teachers and administrators at Helen M. King Middle School in Oregon (US) restructured its curriculum into environmental themes to address poor academic performance and discipline problems. The school eliminated tracking students by test scores and adopted learning expeditions with themes such as "Rock the House" to explore the local environment. Results showed improvement in standardized tests in all academic areas, as well as a significant increase in parental involvement (NEETF 2002).

These findings have spurred the creation of additional resources to help teachers use the environment to improve environmental literacy while boosting achievement. One such example is the CD-ROM "Advancing Education through Environmental Literacy" (Archie 2003). Indeed, proponents view environment-based education as an avenue to achieve needed educational reforms. In practice, the differences between environmental education and environment-based education become blurred, but a focus on student achievement remains the defining characteristic of environment-based education.

4.6.1 Planning

A group of education agencies from 12 states in the United States coined the term EIC® (Environment as an integrating context for learning) to denote schools that use the local natural or social environment as an integrating context. This group, SEER, conducted an assessment to survey the impacts on student learning of using the environment in the K-12 curricula (Lieberman and Hoody 1998). All schools participating in the study had been designated by SEER as a school that integrates learning through the local environment.

EIC® schools use a form of environment-based education with the following characteristics: (1) interdisciplinary integration of subject matter, (2) collaborative instruction, (3) emphasis on problem-solving and project-based learning, (4) combinations of independent and cooperative learning, and (5) learner-centered and constructivist approaches (Lieberman and Hoody 1998).

The planning steps for EIC® schools apply to any environment-based education program and include the following:

- Build a team with like-minded teachers and designate one or two people in charge of communications.
- Allow ample time for planning before implementation.
- Seek guidance and support from the administration from the beginning.
- Create a network of support, including resources such as: businesses, parents, nature centers, zoos, museums, resource management agencies, university faculty, and city officials.
- Invest significant time as integrating the curriculum takes time.
- Start small, such as one teaching team and a one-month study unit.
- Build gradually by adding team members and study units.
- Plan to evaluate your progress, review your programs, and seek suggestions for improvement.
- Stay patient as experience shows that 3–4 years are needed to build a stable program (Lieberman and Hoody 1998).

4.6.2 Implementation

The "environments" used in environment-based learning can range from a small green space by an asphalt parking lot to 80 acres of woodlands surrounding the school. Students at Taylor County High School in Florida (US) study the nearby Ecofina River by researching its biology, ecology, chemistry, economy, and the influence of humans over time. In the Riverwatch program at Glenwood Springs High School in Colorado (US), students' attendance and participation at a community meeting sparked their interest in applying to the Greater Outdoor Colorado Grant to develop a riverside park. This group became the first high school application to receive funding, and students worked with city officials on the development of the park (Lieberman and Hoody 1998).

Some schools begin implementation of environment-based education with classroom experiences and later expand to the outdoors, such as the Endangered Species Program at Waldo Middle School in Oregon (US). Teams of teachers responsible for science, geography, math, language arts, and computer literacy coordinate a curriculum that uses endangered species as a focal point. Students gather information on endangered species through library and Internet research, interviews with biologists, and correspondence with conservation organizations. They then write a species recovery plan and conduct a public presentation with wildlife experts in the audience. Several years after implementation, the school expanded the program to include outdoor field visits (Lieberman and Hoody 1998).

Implementation of environment-based education typically involves teams of teachers working together to facilitate learning across different subjects. Teams from primary schools often include several teachers from one grade level, while secondary schools draw teachers from multiple disciplines. Usually these teams incorporate other stakeholders such as nonformal educators, conservation experts, students, and parents. One obvious constraint of implementing this approach is time. At some schools, teachers use their planning period for team meetings, constricting the time available for grading and paperwork. The importance of finding a core of dedicated members on the teaching team is critical to the sustainability of environment-based education.

4.6.3 Evaluation

The first qualitative assessment reported on 40 schools in the United States using EIC® through interviews with more than 400 students and 250 teachers and principals, as well as comparisons of standardized test scores (Lieberman and Hoody 1998). The findings reported in "Closing the Achievement Gap: Using the Environment as an Integrating Context for Learning" revealed benefits to EIC® programs such as reduced discipline problems, improved achievement on standardized tests, and increased motivation for learning. A more rigorous study compared eight EIC® schools with a control group of eight traditional schools and found that students in the EIC® schools scored higher on achievement tests than traditional schools (SEER 2000).

Evaluation research involving 11 environment-based programs in Florida (US) high schools showed a positive relationship between environment-based education and students' critical thinking skills, disposition toward critical thinking, and achievement motivation (Athman and Monroe 2004; Ernst and Monroe 2004). This study involved 586 high school students in a Pre-test–Post-test

Nonequivalent Comparison Group Design (9th grade) and a Posttest Only Nonequivalent Comparison Group Design (12th grade). These results corroborate findings from the EIC® schools and provide evidence for the role of environment-based education in educational reform efforts.

4.7 Education for sustainability

Many educators view education for sustainability as the future of environmental education (Elder 2003). The goal of sustainable development lies at the heart of education for sustainability, which brings together diverse fields such as systems thinking (Chapter 3), nature study, economics, and civics. The integration of social and economic systems with environmental systems is central to this approach.

Education for sustainability challenges students to find and evaluate solutions to the human influence on social and natural systems. In contrast to traditional environmental education, education for sustainability examines the connections between conservation, human rights, poverty, development, and the political process (Henderson and Tilbury 2004). In secondary and higher education, education for sustainability demands the integration across disciplines or departments.

Many countries and states with school policies for environmental education have used concepts from education for sustainability as a framework. The organization, Global Learning, correlated curricula on sustainability with the New Jersey state standards (Elder 2003). The Mekong River Education for Sustainability project promotes increased awareness and sustainable choices for students living on the banks of the Mekong River in Thailand, Cambodia, and Vietnam. In this initiative, students conduct projects in their local communities and share outcomes with students living in other countries who share the river as a resource (Projects International 2002).

There are unlimited ways that schools can address education for sustainability. One strategy in higher education focuses on campus greening initiatives, which contend that colleges and universities must set a positive example for sustainability to society. The University of British Columbia was the first university in Canada to adopt a sustainable development policy and a campus sustainability office. Their Web site shows the measurements in real-time of consumption of sheets of paper, kilowatts of electricity, and liters of water since September 2004. The site also measures the amount of resources conserved on campus by the sustainability initiatives (UBC Campus Sustainability Office 2005).

At the University of North Carolina, Chapel Hill (US), students approved a "Green Energy Referendum" that added a $4 per semester student fee increase to fund renewable energy strategies for the campus infrastructure (Elder 2003). Construction of new buildings and renovations follows the guidelines of Leadership in Energy and Environmental Design (LEED). A new fare-free transit system has increased ridership on the bus by 42%. Like many programs, a sustainability coordinator oversees education and outreach for these initiatives as well.

Especially for administrators, sustainability is attractive due to economic incentives. The National Wildlife Federation's Campus Ecology program calculated the annual revenues and savings of campus greening and conservation initiatives among 23 US campuses. The revenues and savings included $9,068,000 from energy conservation at SUNY-Buffalo, NY and $235,000 from new toilet and water fixtures at Columbia University, NY, US (Eagan and Keniery 1998).

4.7.1 Planning

One model for education for sustainability is the whole-school approach, which integrates the theme of sustainability into all facets of the school environment, including curriculum, pedagogy, facilities management, and school governance (Henderson and Tilbury 2004). This approach is found in primary and secondary schools throughout Europe, South Africa, Australia, China, and New Zealand. Three examples of such programs are discussed below: Enviroschools in New Zealand, the Green School program in China, and Eco-schools in South Africa. Box 4.9 compares the planning, implementation, and evaluation of these programs.

Initiated in the 1990s with three pilot schools, the Enviroschools program is managed by the Enviroschools Foundation, an independent trust with regional councils. This program offers two choices for schools: an award program or a 3-year program with extensive support materials.

The Green School project, initiated by China's Ministry of Education, centers on the international concept of ISO 14000, which establishes international standards for environmental management systems. Currently managed by the Center for Environmental Education and Communications, this program builds knowledge and skills in the management of school grounds and the larger environment. Schools follow a series of steps in order to apply for Green School awards, received by 15,000 schools (Henderson and Tilbury 2004).

The Eco-schools program, managed by the Foundation for Environmental Education (FEE), includes schools in 28 countries, including South Africa. Based on the principles of Agenda 21, this program is coordinated by the international nonprofit organization FEE in collaboration with national partners. Schools work through seven steps to achieve Green Flag certification. The initial steps in this program begin with the curriculum, rather than environmental projects in the schools.

The planning for all three programs involves developing action plans for the school. A committee of diverse stakeholders oversees the steps planned toward integrating sustainability into the school. For any whole-school program, partners such as the department of conservation, ministry of education, community groups, universities, and business, are key to the planning process. Partnerships in education for sustainability provide technical and financial support, connect the program to government policies, relate the program to community needs, and avoid duplication of efforts (Henderson and Tilbury 2004).

Box 4.9 A comparison of three approaches to education for sustainability (Adapted from Henderson, K. and Tilbury, D. (2004). *Whole-school Approaches to Sustainability: an International Review of Sustainable School Programs*. Report prepared by the Australian Research Institute in Education for Sustainability (ARIES) for the Department of the Environment and Heritage, Australian Government.).

	Enviroschools, New Zealand	Green School Project, China	FEE Eco-schools, South Africa
School focus	Kindergarten, primary and secondary schools.	Kindergarten, primary, middle, vocational, and special needs schools.	Kindergarten, primary, and secondary schools.
Funding and management	Managed by Enviroschools Foundations (funded by government and charitable contributions). Implementation supported by regional councils	Coordinated by the Center for Educational Education and Communications and local networks. Funded by State Environmental Protection Administration of China (SEPA)	Managed by Wildlife and Environment Society of South Africa in collaboration with South Africa Department of Education. Funded largely by corporate sponsor.
Framework	Connected to New Zealand Curriculum Framework and linked to activities of the Ministry of Education's EE Guidelines	A Government Framework based on the concept of ISO 14000 and the European Eco-schools. Connected to the National Action Program for Environmental Publicity and Education	Links to action detailed in the Millennium Declaration, signed at the 2002 World Summit on Sustainable Development. Correlated to the Revised National Curriculum Statements in South Africa
Focus and principles for planning	Partnerships through existing EE initiatives that integrate 5 principles: 1. Sustainability 2. Environmental education 3. Respect for diversity of people and cultures 4. Student participation and Maori perspectives	Based on Chinese social political foundation and encourages schools to use its educational resources to support the environment and integrate EE into the curriculum.	Focuses on curriculum-based action for a healthy environment. Schools choose three focus areas from the following themes: school calendar, environmental information and community knowledge, schools grounds and fieldwork, resource management, health safety, action projects and contests, adventures

	Enviroschools, New Zealand	Green School Project, China	FEE Eco-schools, South Africa
	5. Knowledge to enrich the learning process		and cultural activities.
Stages of implemen- tation	1. Develop an envirogroup, school policy, and partnerships 2. Create a whole-school vision map 3. Focus 5 themes in classroom learning and action Reflect through documenting and sharing lessons learned	1. Establish a Green School Committee 2. Focus on environmental management of resources 3. Integrate EE into curriculum 4. Provide in-service training for teachers 5. Ensure participation from entire school in environmental protection 6. Consider lifestyle choices 7. Participate in campus greening 8. Establish an eco-club for students	1. Create a working group 2. Audit the school environment 3. Choose 3 focus areas 4. Draft an environmental policy 5 Develop and implement lesson plans linked to sustainability 6. Take action involving the school and community 7 Develop a portfolio and review progress
Evaluation	Through questionnaires and reflection workshop	Not yet conducted	Not yet conducted
Achieve- ments	Documented environmental achievements in greening campus and reducing waste, water, and energy usage Integrated EE into science, language, social studies, technology, PE, and well being. Students involved in planning and evaluation.	Gave 15,000 Green School awards at different levels by the end of 2003.	Focused initial round of awards to Green flag schools for special environmental days, recycling programs, and a native plants garden.

4.7.2 Implementation

A review of the whole-school approach across different countries revealed factors that define sustainable schools. These include:

- School leadership and administration that value participatory decision-making with sustainability at the core of school planning.
- Integration of education for sustainability across the curriculum.
- Professional development for teachers, program partners, and facilities management staff.
- Greening of the school grounds.
- A decrease in resource consumption and an increase in environmental quality at the school.
- Participatory learning approaches that promote critical thinking and inter-cultural understandings.
- A focus on reflection and evaluation to guide future decisions, as well as encouragement of research to improve teaching practices (Henderson and Tilbury 2004).

The steps for implementation depend on the specific program (Box 4.9). For the Enviroschools program, schools form partnerships with the community, develop a vision map for the school, integrate sustainability into student learning and action, and reflect by documenting and sharing lessons learned and outcomes (Enviroschools 2004).

4.7.3 Evaluation

Some whole-school programs have conducted evaluations such as the Enviroschools in New Zealand, which used a questionnaire for schools and a 2-day reflection meeting with teachers, funders, and community groups to gather qualitative feedback. In terms of environmental impacts, schools participating in the Eco-schools program in Ireland reported a 50% reduction in waste during a 5-week waste monitoring survey in the schools (FEE 2001). Documented environmental achievements in the New Zealand program include greening campuses and reducing waste, water, and energy usage. Educational outcomes include the integration of education for sustainability into the subjects of science, language, social studies, technology, physical education, and well being (Henderson and Tilbury 2004).

Most whole-school sustainability programs, however, are characterized by a lack of evaluation. Despite the lack of systematic evaluation, success factors exist that appear to correlate with effective programs, including: connection with national government priorities; access to experts in education for sustainability; secure funding; professional development for teachers and partners; linking with established sustainability efforts; and partnerships that include a diversity of stakeholders (Henderson and Tilbury 2004).

4.8 Action projects

Many teachers and leaders in environmental education have used action projects to integrate conservation education in the schools (Stapp and Wals 1994; Robottom and Hart 1995, Stapp *et al.* 1996a,b). Action projects engage students in selecting, researching, planning, implementing, and evaluating a real-world environmental project. Created by Bill Hammond, the Monday Action groups in Lee County Schools, Florida (US) are one model for many teachers and conservation educators who use action projects in a school setting (Hammond 1995).

The program grew from a small group of students who met on Mondays with their teacher to a funded program with EE staff. The projects students have developed include the Sea Turtle Research and Conservation Program, the Bald Eagle Habitat Protection Ordinance, the Six-Mile Cypress Slough Acquisition, and the Florida Manatee Protection Act (Hammond 1995).

The GREEN program (Global Rivers Environmental Education program), an international water quality monitoring program, is based on action projects as well. Action projects have their roots in action research and the work of John Dewey and Paulo Freire. The interests and concerns of the learners in pursuing a topic for action dictate the direction of the project. The assumption behind action projects is that learners need skills and practice in addressing real environmental problems. Proponents of this technique believe that if schools neglect to integrate informed action skills, the school curriculum will never build the skills needed for students to address the real environmental challenges we face (Hammond 1996/97).

Nonformal educators, biologists, and ecologists have partnered with schools to plan, implement, and evaluate action projects, particularly when the interests of students coincide with pressing environmental problems. Likewise, teachers and students have instigated action projects in their schools and communities to help students build skills through concrete, well-informed efforts.

4.8.1 Planning

Planning an action project involves a sequence of steps initiated by students but facilitated by teachers, and often outside experts. Teachers should never use students to advocate for their personal causes. Rather, they can involve students in researching a variety of perspectives and possibilities and making their own informed choices for action. If you are developing action projects as a focus for a school system, planning also may involve building a constituency of supporters, such as through an advisory council. The following steps for planning are geared toward teachers and students working together (Easton *et al.* 1996; Project WILD 1995):

1. *Investigate potential project topics.* Students should explore topics in their community, talk to experts, read about issues in the local newspaper, walk around the school and neighborhood, and record observations.

2. *Assess interests, skills, and resources.* Students involved should write down a list of their own interests, skills, and resources for the action project, which will help guide the planning process.

3. *Develop a list of possible projects.* Next, groups of students should create a list of project topics. If they notice that air quality is a problem in their community, potential projects to address the issue could involve air quality monitoring and an education campaign to local families.

4. *Narrow the project choices.* Each group should discuss the feasibility of the project and narrow down the choices to the top three to five projects. Choosing a small piece of a complex problem is often a feasible choice. Avoid choosing an action project that takes too much time, since projects always take more time than anticipated!

5. *Become experts and pick a project.* Teams research their top choice and present their reasons for supporting the proposed project to the class. Teachers can help by bringing in experts such as air quality specialists to talk about the potential projects. Finally have the class decide on the final choice either by a vote or by discussion and consensus. Students should look at projects that suit their skills, meet a community need, and generate enthusiasm among the group.

6. *Write an action plan.* Next, instruct the class to develop a goal for the project and specific, measurable objectives. Use a task and timeline to keep track of responsibilities. The action plan should identify the environmental project, the goal, starting and ending dates, concrete tasks and deadlines, team members responsible for each task, resource people, ideas for marketing, and a plan for measuring your success.

4.8.2 Implementation

Implementation is the time when students put their plan into action. With the help of a teacher or facilitator, students must keep careful records of their tasks and accomplishments for efficiency. Be sure to publicize their efforts by having them contact the media to cover their work. For action projects involving water quality monitoring, resources such as GREEN's "Investigating Streams and Rivers" provides specific guidelines and technical advice (Stapp *et al.* 1996b).

Action projects can range in time, effort, and skills from a clean-up project to lobbying for legislation to protect habitat. Consider the following three levels of environmental action as students implement their action projects and make decisions about the level of action to take in the community (adapted from Hammond 1996/97):

Level 1: Action projects that result in a direct product or outcome in a short period of time. Examples of such projects include river clean-ups, building a nature trail or butterfly garden, or creating a school-wide recycling campaign.

Level 2: Action projects that require long-term work and require additional skills and efforts for sustaining commitment. These projects persist after the group of students who initiated the action project leaves the school. As such, the action project demands skills in communication, organization, and leadership. Projects such as creating and maintaining a wildlife conservation area or developing environmentally sustainable purchasing patterns for a school fall into this category.

Level 3: Action projects that implement changes in regulations, policies, or laws. Such projects require more skills, persistence, and support from outside groups than the previous levels, although the impacts are long-lasting. Often groups of students in subsequent years adopt the project to build on earlier efforts. Another option is for students to continue their project given the collaboration of teachers at different grade levels. Examples of such projects include lobbying to pass laws for habitat protection or promoting changes in school board policy such as the formal adoption of green purchasing patterns.

Whatever the level of action implemented by students, your job as facilitator is to provide ethical guidance as they proceed through the project. The Lee County program developed this framework to guide students in any grade level (adapted from Hammond 1996/97):

1. *Be for something.* Avoid being "against" something, but rather reframe your interests in terms of a solution.
2. *Do your homework.* Always research a problem or issue. Read several papers about the topic and interview experts to become a knowledgeable expert (Figure. 4.6). (You should know more about your topic of interest than a local county commissioner.)

Fig. 4.6 Interviewing community members is an excellent way for students to gain knowledge and perspective about an action project (photo by M. McDuff).

3. *Learn to understand those who support or oppose your views.* Try to understand the interests of your opposition by interviewing them to discover any common interests.
4. *Treat everyone you meet as a person of integrity and high moral value.* Treat people as you would like them to act.
5. *Avoid stereotyping others.* Stereotyping results in poor communication and limits possibilities of reaching solutions and learning from others.
6. *Avoid blaming others.* Accept responsibility for the success or failure of your actions and move on.

4.8.3 Evaluation

Students, teachers, and other stakeholders involved in an action project should evaluate the success of the project and suggest improvements for the future. Reflect on whether or not the project accomplished its goals and objectives. What were the most and least successful parts of the project? How did the project affect student learning?

Evaluation is a critical part of the learning process, not only for students involved in the project, but also for teachers and others who will undertake action projects in the future. During the planning stages, ask students to identify indicators of success for them. How will they know their project is successful? Possible indicators could include changes in environmental quality, effective group process, adequate support from the school and community, achievement of objectives, and efficient time management. Results from an action project of Fort Myers Beach Elementary students in Florida (US), for example, resulted in a state and local law limiting the collection of live echinoderms and mollusks to two specimens per day without a special permit (Hammond 1995).

At the end of the project, engage students in a discussion to evaluate the project, identify successes and challenges, and suggest improvements. You can modify the evaluation activity in Box 4.10 to conduct a similar discussion at the mid-point of the action project and make changes during the project (Stapp *et al.* 1996b).

4.9 Summary

Conservation education in the schools will only succeed when benefits accrue to all partners. When conservation education helps teachers and administrators do their jobs more effectively, this approach becomes more practical. Taking the time to understand the realities of school systems, use effective communications, serve as a resource to schools, and connect your programs to academic standards will provide the building blocks to effective conservation education in the schools. Documented impacts of these efforts provide evidence to use when lobbying for a place for conservation education in legislation and educational policy.

Some of the most successful conservation education programs in the schools have found strategies to meet educational needs through involving students in local conservation needs. Education in the schools should include learning about

Box 4.10 Lesson plan for an evaluation activity for action projects to gather feedback from students. (Adapted from Stapp, W.B., Cromwell, M.M., Schmidt, D.C., and Alm, A.W. (1996b). *Investigating Streams and Rivers* Kendall/Hunt, Dubuque, IA.)

Objectives:
By the end of the lesson, students will be able to evaluate successes, challenges, and areas for improvement for an action project

Participants:
Students involved in the action project

Materials:
Flipchart paper
Markers

Procedure:

1. Evaluation of successes, challenges, and improvements.
   a. Post a piece of flipchart paper on the board for a large-group discussion or give smaller groups a piece of flipchart paper each. Divide the paper into three columns and label them "plus," "minus," and "change."
   b. Ask the students to think about this question: How well did our action project address the problem or issue we identified? Have students list the things that worked well with their action project under the "plus" column, the things that didn't work under "minus," and use the "change" column to list specific improvements they suggest.
   c. As facilitator, you can offer your feedback as well.
2. Individual reflection of the problem-solving process.
   Ask each student to write down:
   a. Summary of the problem or issue they decided to address
   b. Outline of the steps followed to address the issue.
   c. Description of the action plan.
   d. Summary of the results of their action.
3. Group discussion and reflection.
   Use the following questions to guide the discussion:
   a. How are perceptions of the action project within the class similar or different?
   b. How effective were the communications between class members, teacher, and others?
   c. What part of the action project was most challenging or difficult? What would have made it easier?

d. What part was most enjoyable?
e. How did you feel when you completed the action project?
f. Did you meet your indicators for success? Would you describe the project as successful? Why or why not? Were there any changes in environmental quality as a result of the project?

Evaluation:
Ask students to reflect on the effectiveness of the assessment. Did the individual reflection and large-group discussion capture their concerns and successes? Observe class to ensure that everyone played some role in the activity.

the local environment, but too often, academic demands preclude the chance to study the natural world outside the classroom. When conservation education links academic demands to the study of the natural and social systems and their interactions, everyone benefits.

To implement conservation education within schools, you can choose from a variety of approaches, from environment-based education to action projects. The evaluation of any conservation education initiative ensures that you gather feedback from all stakeholders to continue elements of success and make improvements for the benefit of the students and local environment.

Further reading

Environmental Education in the Schools: Creating a Program that Works, by Judy Braus and David Wood (1993). Peace Corps Information Collection and Exchange, Washington, DC.

Whole-school Approaches to Sustainability: an International Review of Sustainable School Programs, by Kate Henderson and Daniella Tilbury (2004). Report prepared by the Australian Research Institute in Education for Sustainability (ARIES) for the Department of the Environment and Heritage, Australian Government.

Closing the Achievement Gap: Using the Environment as an Integrating Context for Learning, by Gerald Lieberman and Linda Hoody (1998). State Environmental Education Roundtable, Science Wizards, Poway, CA.

Environmental Literacy Scope and Sequence: Providing a Systems Approach to Environmental Education in Minnesota, by Minnesota Office of Environmental Assistance (2002), Minnesota Office of Environmental Assistance, St. Paul, MN.

Investigating Streams and Rivers, by William Stapp, Mare Cromwell, David Schmidt, and Andy Alm (1996). Kendall/Hunt, Dubuque, IA.

Taking Action: an Educator's Guide to Involving Students in Environmental Action Projects, by Project WILD (1995). Council for Environmental Education/Project WILD, Gaithersburg, MD.

5

Making conservation come alive

Remember how a powerful story told by parents or grandparents makes family history come alive? By immersing listeners in the rich details of their past, storytellers create a sense of ownership and excitement about local history. We all know stories that we continue to tell to others because of the inspirational pull of the tale.

Like a good story, the techniques in this chapter bring conservation education to life, infusing energy and excitement into environmental conservation (Box 5.1). These techniques celebrate the wonder of nature, immerse learners in different perspectives, and use the outdoors as a context for discovery. These educational strategies do not dwell on the gloom and doom messages that we often hear, but rather, reveal the joy and complexity of the environment around us. Each technique presented in this chapter engages the audience in learning, whether through a simple environmental game or an extensive field studies course. The educational techniques also reflect the experiential learning cycle (Chapter 2), as the techniques need reflection with participants to make learning more effective.

Techniques, such as hands-on activities foster discovery of the sheer awe of the natural world. Storytelling, games, case studies, and role-plays promote understanding of the interests and positions of multiple players in conservation issues. Conservation educators can use these techniques to build on each other. For example, the characters in a case study might inspire you to create a role-play for your participants. Some games, for example, also have a compelling story as their introduction.

This chapter features incentives such as the River of Words contest that brings conservation alive by recognizing quality art and poetry. We also include the nuts and bolts for organizing field trips—an exploration of a nearby urban park or a semester-long field studies course. Lastly, the chapter provides tips for using wilderness skills to promote awareness and responsible use of the backcountry. Such field experiences give real meaning to conservation, discovered not in a magazine or TV documentary, but in the real environment.

5.1 Hands-on activities

Hands-on activities in conservation give learners first-hand experience exploring a new phenomenon or concept in the natural world. Students dissect the parts of

Box 5.1 Many techniques bring conservation to life by engaging learners in direct experience.

Technique	Purpose
Hands-on activities	Involve learners working with and manipulating materials to explore a concept or solve a problem
Storytelling	Transmits knowledge through an oral tradition and allows audience to experience the emotions and environment of characters in the story
Games	Demonstrate and teach environmental concepts through fun but structured play, involving elements such as challenge, collaboration, movement, and quiet concentration
Case study	Uses narratives about individuals and communities facing environmental dilemmas to engage participants in studying the characters, identifying lessons learned, and proposing potential solutions
Role-playing	Assigns learners to specific roles in a scenario to "act out" the situation and gain new perspectives, self-awareness, and often communication and problem-solving skills
Contests	Provide incentives for environmental action or involvement to raise public awareness of an environmental issue, resource, or organization
Field trips	Provide first-hand experience for participants with a physical site and resources in a community
Wilderness awareness and skills	Promote a connection to the earth by increasing awareness and skills for self-reliance and low impact in the outdoors

a flower before learning about the concept of pollination. Adults explore the shoreline of the ocean, collecting shells to document species of aquatic life. Hands-on activities involve learners working with and manipulating materials to solve a problem or explore a concept (Haury and Rillero 1994).

Children develop and grow through hands-on discovery of real people, places, and objects (Moore and Wong 1997). Children learn about their environment through exploration and play. Not surprisingly, research shows that young children learn most effectively through these direct interactions with their environment, rather than passive learning activities (Katz 1987). Hands-on activities build on our innate, natural processes for learning.

Schools in the low-income East Feliciana Parish in Louisiana (US) recently adopted a hands-on math and science program called Project Connect. The program uses hands-on activities for place-based education in this rural area where 85% of students qualify for free or reduced-lunches due to the high levels of poverty. The hands-on activities, which include identifying aquatic organisms and exploring the pine forest ecosystem, are correlated to the state education standards. Two years after the expansion of Project Connect, the number of students passing the science portion of the state standardized tests increased by 13% in one year (Null 2002). These findings provide evidence that some students who have not thrived in the traditional classroom environment are more effective learners with hands-on activities.

Hands-on activities typically use inquiry and constructivist theories of learning (Chapter 2). Learners construct new understandings by combining previous knowledge with new discoveries (Moore and Wong 1997). When you use hands-on activities in conservation education, you become a facilitator in this learner-centered technique. For this technique, we will explore strategies for conducting three types of hands-on activities: outdoor hands-on activities, experiments, and manipulative exercises and collections.

5.1.1 Planning

Planning will connect the hands-on activity to the attitudes, knowledge, and skills you are exploring with your learners. Ask yourself a few questions when you consider hands-on activities in conservation education: What information, concepts, or skills do you want to teach? How long will the activity take? What is the context for this new concept?

Hands-on activities reflect the experiential learning cycle described in Chapter 2, as learners typically first explore a new concept, materials, or the environment with minimal guidance from the facilitator. For example, the Minnesota Department of Natural Resources Nongame Wildlife Program developed hands-on activities that facilitate learning about the state bird, the Common Loon (*Gavia immer*). The initial stages of one activity allow students to compare the bones of a loon to a goose and record their observations. After the hands-on exploration, the facilitator takes a more active role, introducing conservation concepts, and then learners apply their new knowledge to a new situation or problem.

One Project WILD (2000) activity, Owl Pellets, gives students in a classroom the chance to explore the contents of an owl pellet, the regurgitated, indigestible parts of an owl's prey. The facilitator gives an owl pellet to students in pairs or teams and asks them to use a toothpick and their hands to examine and dissect the owl pellet. Students quickly find bones and bone fragments from the owl's prey. Their findings lead to rich discussions of predator–prey relationships, and students then use the bones to identify the species of prey, such as voles, mice, and even skunks (Figure. 5.1).

Hands-on activities usually require more organization, creativity, and logistics than traditional instruction, but they often generate more enthusiasm in learners. Follow these simple steps before implementing your activity:

• Try each activity yourself, so you know what to expect with your learners.

Fig. 5.1 Students explore the contents of an owl pellet in this hands-on activity that explores predator–prey relationships (photo by S. Cross).

- Collect all materials, and have extra materials on hand in case participants need to try the activity a second time.
- Prior to the activity, divide all materials into correct amounts for each group. Cooperative groups provide more opportunities for participants to interact with materials, and groups are easier for the facilitator to manage. Consider the number of participants you want in each group. Many conservation education programs keep their materials for hands-on activities in boxes, so they can grab the "lesson box" with all the materials before the activity.
- Think about local organizations or businesses that might donate materials for your program or school.
- If you work with your participants on a long-term basis, build some excitement about the upcoming activity by displaying a picture or object related to the hands-on activity.
- Practice giving clear instructions for the activity in multiple ways, such as verbally and in writing. Before participants begin exploring, remind them to make a prediction about what they think will happen during the activity.
- Consider the location for the activity. Is this an indoor lab experiment or a nature walk with outdoor exploration? (WGBH 2000).

5.1.2 Implementation

Outdoor hands-on activities

Many hands-on activities involve learning in the outdoors—from identifying wildlife from their tracks to journaling while walking a nature trail. If you are an educator used to working indoors, implementing hands-on activities outdoors without boundaries can seem daunting. If you are a conservation outreach specialist used to working with small groups of adults, the prospect of taking 30 youngsters to explore a marsh may feel overwhelming. But a few guidelines for implementing hands-on activities outdoors can address most concerns (adapted from Grant and Littlejohn 2001).

- *Have clear expectations.* With both youth and adults, discuss behavioral expectations before you conduct hands-on activities in the outdoors. Such an atmosphere, similar to a group contract, ensures greater willingness to follow the guidelines. Make a list of behaviors and keep them positive, such as "Use quiet voices" instead of "Don't make noise."
- *Plan the logistics*
  o Have a clear signal for getting everyone's attention, such as a birdcall or three handclaps.
  o Discuss where you will gather once you are outside. When using trails, establish clear meeting places at trail heads.
  o Explain that if participants get separated, they should sit in one place and wait for someone to find them.
  o Decide who will lead the group.
  o Use a variety of group sizes, from pairs to large groups, throughout the hands-on activity.
  o Give specific assignments to focus attention.
  o Discuss safety and any hazards. Bring a first aid kit.
  o Be prepared for any kind of weather by bringing rain gear, coats, and water bottles.
  o Evaluate at the end of the hands-on activity outdoors. Discuss what went well, what did not, and changes for future activities.
- *Practice and model activities.* Clear instructions for the hands-on activity will help focus your learners when outdoors. Practice gathering together with your signal before venturing outdoors. When possible, model your assignments by becoming an active participant yourself.
- *Be flexible.* When teaching outdoors, natural distractions will divert participants' attention. Be flexible when these compelling distractions occur and try to create natural transitions. Sometimes the bird flying overhead will give the perfect stimulus for the next discussion. Use these teachable moments!
- *Communicate strategically.* Be prepared to face noise, wind, sun, and other distractions you cannot control. Keep the following in mind:
  o Keep the sun in your eyes, rather than your learners' eyes.

○ Stand so the wind is to your back, which will push the sound of your voice toward the participants.

○ As you talk, try to reduce the distance between you and your participants. This may mean kneeling down when talking to younger participants.

○ If you are on a narrow trail, have the participants form a double-file line. When you stop, have those in the front row kneel down, so everyone can see or hear without going off the trail.

- *Carry props.* Carry a prop bag with natural artifacts to redirect learners' attention. You also can use tools, such as hand lenses, to get a closer look at the natural world.

Experiments

Hands-on activities also include lab or field experiments that allow learners to handle, manipulate, or observe a scientific process related to conservation. For example, members of the Wildlife Clubs in Central African Republic worked with ecologists to compare the germination and growth rates of seeds collected from elephant dung with seeds collected on the ground. This simple experiment exposed a critical role of the elephant dung in promoting the growth of plants in the forest ecosystem.

As a warm-up activity, ask your participants to write the safety rules for the experiment on the board or a poster. Safety is a critical factor during any experiment. You are responsible for ensuring that all participants follow the safety rules. For both lab and field experiments, consider the following ideas to ensure both a fun and educational experiment:

- *Participant-designed experiments.* Split the group into small teams and ask each team to design an experiment to test one variable. For example, if you are testing for factors that affect the rate of erosion, have the participants identify all the factors that could influence erosion, such as slope, water flow rates, and soil type. Each group can then isolate one factor and test it one at a time.

- *Lab stations.* If you want to explore several different concepts, set up numerous stations around the room. Give participants written directions and have them circulate to each station.

- *Silent/interactive lab.* Have participants work silently in the lab or in pairs and then give them time to interact and share results (Center for Talented Youth 2002).

Manipulative exercises and collections

Science classes are known for the use of manipulative materials and collections, such as models of chemical structures or leaf collections. Similarly, conservation educators can use such hands-on activities to explore ecological concepts with youth and adults. For example, bring in an unfamiliar object, such as an aquatic sponge, and have participants explore the object and guess its origin. Have learners create models of ecological concepts, such as a model of their watershed. Another option is to ask participants to collect different seashells and identify the marine species for each shell. The possibilities for using manipulatives and collections are only limited by your imagination!

5.1.3 Evaluation

Your own observation will be a powerful evaluation tool as you assess the effectiveness of the hands-on activities. An observation checklist is a tool for documenting student engagement and learning from the hands-on activity. Your checklist might include a rating of factors, such as level of participation, engagement, on-task behaviors, clarifying questions, contribution to discussion, consistency with behavioral expectations, connections made between hands-on observation and ecological concepts.

You can collect your own evaluation data comparing traditional lecture teaching with hands-on activities. Try teaching one of your classes or groups of participants with more traditional teaching while another learns the same concepts through hands-on activities. Give each group the same pre-tests and post-tests and compare results. You can enlist the help of evaluators in the education departments of your local college or university to review your methodology and evaluation instruments.

One zoo education program in Florida used evaluation as a tool for assessing the efficacy of hands-on activities for preschoolers. The Tots program challenges the traditional approach to zoo education that simply exposing children to wild animals promotes learning. This educational program involves hands-on activities, such as direct interaction with the animals, as well as the involvement of parents or guardians in all activities. An evaluation that used observation and interviews showed that the Tots program was effective at promoting knowledge about animals and environmental awareness among the children (Pringle *et al.* 2003).

5.2 Storytelling

"Stories form a link between our imagination and our surroundings" (Caduto and Bruchac 1997). Indeed for most cultures, oral traditions and stories transmit local knowledge and wisdom about the environment, our relationship to the earth, and our relationship with each other. Stories allow us to become the characters and experience their emotions and their environments.

Why is storytelling so effective at transmitting information—from the latest gossip at work to the medicinal uses of local plant species? Storytelling is effective because it is:

- Simple. No one needs training to tell a story.
- Timeless. Storytelling has been and will be around forever.
- Empowering. Stories give people information, so they can better live their lives or perform their jobs.
- Appealing to different audiences. Everyone, from the very old to very young, likes hearing a good story.
- An excellent strategy to pass along traditions.

- A form of recognition. Including someone in a story gives that person a presence in the community or audience.
- Fun. Who has not heard the request, "Tell me again the story about . . ." (Armstrong 1992).

For centuries, families have used stories to promote cultural and behavioral norms. Likewise, storytelling can teach vital conservation lessons and inspire environmental action. To preserve environmental stories, organizations such as the Wildlife Clubs of Kenya, (WCK) have collected traditional stories in print. Ibrahim Ali, director of the WCK, remembers sitting by the fire as a child listening to stories at the end of each day with his grandmother and grandfather, sisters, brothers, and cousins. WCK staff coordinated the creation of "Trees, Myths, and Medicines" (WCK 1997) a book of stories, myths, and artwork related to trees in Kenya as told by Kenyan youth (Box 5.2, Figure. 5.2). Such compilations of stories are a rich resource for environmental educators who want to use storytelling as a technique for conservation education.

5.2.1 Planning

Think about the specific objectives of your storytelling and the interests of your audience as you choose stories to fit your program's theme. Do you want to inspire

---

Box 5.2 Wildlife Club of Kenya collected stories and artwork from Kenyan youth about the trees and forests in their community. This story, entitled, "The People of Coco," was written by Zeenat Parpia, when she was a 16-year-old student at Star of the Sea High School on the coast in Mombasa (WCK. (1997). *Trees, Myths, and Medicines*. Jacaranda Designs, Nairobi, Kenya. Used with copyright permission granted from the Wildlife Clubs of Kenya and their publisher, Jacaranda Designs Ltd. PO Box 76691–00508 Nairobi, Kenya. info@jacaranda-africa.com)

The People of Coco by Zeenat Parpia

"Oh God of hunger please help us, the people of Coco. It's over a month now since famine struck our village. Our herds have died and our land has become barren. We are now on the verge of losing our people," prayed the chief of Coco. The effects of the famine were evident. He was so skinny and fragile that one could see the bones protruding out of his skin. Suddenly a monkey-faced like structure fell out the sky. Terrified, the chief ran home to take refuge.

Next day, the cock doodled very early and the chief went to pray before the sun rose. But on reaching the place where he had been worshipping the day before, a surprise awaited him. There before him stood a miraculous and giant tree. Since then the coconut has grown naturally and has solved the problem of the people of Coco. The coconut in fact received its botanical name—*Cocos nucifera*—from the people to whom it was given.

Nazmina Panju

Shil Shanghavi

Jatinder Panesar

Preeti S. Shah

Fig. 5.2 Artwork by Kenyan students from the coastal region of Kenya illustrates the importance of the coconut tree, as told in the story "The People of Coco." (WCK (1997). Trees, Myths, and Medicines. Jacaranda Designs, Nairobi, Kenya. Used with copyright permission granted from the Wildlife Clubs of Kenya and their publisher, Jacaranda Designs Ltd. PO Box 76691–00508 Nairobi, Kenya. info@jacaranda-africa.com).

awe at the power and beauty of nature? Is your objective to instill an environmental ethic? Do you want your listeners to laugh and have fun? Are you looking for a calm evening story at the end of a day of activities?

Rich resources exist for storytelling about the environment, including a large body of both fiction and nonfiction books with natural resource themes. A diverse range of books can inspire storytelling from tales such as "Kenju's Forest" (Monmoto 1991) to "The Giving Tree" (Silverstein 1964). Books such as the "Keepers of the Earth," "Keepers of the Night," and "Keepers of the Animals" (Caduto and Bruchac 1989, 1994, 1997) focus on Native American stories about the environment with corresponding activities. The book, "Nature and the Environment for Children and their Teachers" (Bang 1997) provides a reference guide for environmental books for children.

You can draw on the cultural traditions and local stories in your area. Folklore of native plants, for example, can provide stories and inspire learning of local species. Two examples of folklore books include "Wildflower Folklore" and "The Folklore of Trees and Shrubs" (Martin 1984, 1992).

Another source for storytelling is the life stories of naturalists or environmental heroes. Joseph Cornell (1989) recommends studying the biographies and stories of your own heroes and telling their story in your own words. You even can develop stories that describe the actions and lives of local environmental heroes. The EcoTeam, an environmental education curriculum for third-grade students in the United States, includes one lesson about local citizens who have made a difference in the environment. These stories show students that every person in the community has a role to play as an environmental citizen. Newspapers and environmental magazines often feature environmental heroes too. The magazine "Yes! A Journal of Positive Futures" highlights modern environmental success stories across the globe (e.g., Anthony 2005).

When telling a story about an environmental role model, select props to portray the historical times and unique traits of your character. Depict the dress of the character with simple props, such as a hat or tie. Memorize passages from the written or spoken words of your character and learn the plot lines of key events in his or her story (Cornell 1989). Avoid memorizing the entire text and instead memorize the key points while you "become" the character for the audience. Listen to recordings of other storytellers to hone your skills.

If you are telling a story from a written text, read the story aloud to yourself several times before you tell it from memory or even read it to an audience (Caduto and Bruchac 1997). One technique you can use to help plan your stories is a bag of items that represent mnemonic devices—with each item representing a key plot line in a story (Caduto and Bruchac 1997).

Another storytelling strategy for both children and adults is guided visualization, which can help your listeners become more aware of their environment.

Guided visualization is the process of describing a scene and helping the listeners relax and imagine that environment (Murdoch 1994). Used in meditation, guided visualization begins with the listeners sitting or lying down with their eyes closed as you begin to paint a scene. Ask the audience to describe, write, or draw the images in their head during or after the story.

As you practice your stories, think about the traits of good storytellers that you have heard in the past. What made them special and why do you remember their faces, voices, and their stories? One key to planning a good story about conservation is to choose stories and characters that give you real joy (Cornell 1989). If you find joy in the story you are telling, your audience will also experience the energy in your voice and the movements of your body.

5.2.2 Implementation

A few tips will help you involve your listeners in a successful story and have fun while you are telling your tale. Remember that part of becoming a good storyteller is finding a style that suits you! Enjoy experimenting with the power of stories to evoke attitudes and actions about the environment.

Tips for Successful Storytelling

- Have your listeners sit in a circle, if possible, so everyone can make eye contact.
- Incorporate responses from the audience to involve your listeners and help with pacing of the story, (e.g. when you say "Ho" they respond with "Hey!").
- Teach any chanting or hand movements before the story begins.
- Memorize the beginning and ending points of your stories to increase your confidence.
- Try to feel the quality and mood of each story before you begin.
- Describe the setting with specific details so listeners can "see" it in their minds.
- Maintain eye contact with your audience.
- Use your hands to express emotions and actions.
- Establish and preserve the timeline of your story by using key phrases, such as "then," "next," "finally," and "but at the same time . . .".
- Alter your rhythm and speech to maintain interest in your stories, and vary your voice to suit different characters.
- Use humor to relax the audience and keep them open to your message.
- Adapt your stories to different age groups. Adults focus on verbal messages while children are more visual and physical.
- Remember to share the joy you feel in your stories with the audience (Cornell 1989; Caduto and Bruchac 1997).

As you model effective storytelling, engage your students or participants in telling their own stories. Ask learners to write and tell their own fictitious environmental stories or ecological autobiographies that describe their connection to the natural world. Telling these tales can empower learners in the role of storyteller to teach others. Older students or even senior citizens can practice their storytelling with younger audiences.

5.2.3 Evaluation

Begin assessing the effectiveness of this technique before telling stories to your actual audience. Start by practicing your story in front of a few people to get feedback. Ask them specific questions: When were they most involved in the story? What points were most and least interesting? What could you do to improve your stories?

After you adjust your storytelling based on the feedback, you may want to record your storytelling session and listen to yourself. Listen for places when you speak too quickly or slowly. Do you enunciate clearly throughout the tale? Put yourself in the place of a listener and evaluate your storytelling.

When you tell your stories with a "real" audience, try to observe the points where the audience seems most engaged. Are there points where your listeners seem to drift? Did any external factors help or hurt your delivery, such as weather, lighting, or outside noises? Try telling your stories to different age groups and audiences to adjust your delivery and increase your confidence. Lastly, ask yourself if you truly became your characters and lived the lessons they shared!

5.3 Games

Games are a powerful technique to demonstrate and teach ecological concepts to both children and adults (Project WILD 2000). Games can energize a teaching environment, immerse learners in the material, promote collaboration with other players, and encourage students to learn from their mistakes (Teed 2004). Just as important, games are fun! Through games, a generation that can memorize over 100 Pokemon characters with all their characteristics and history also can learn the names and interactions of 100 wildlife species in their habitats. Games provide a useful strategy to introduce or reinforce concepts taught through other conservation education techniques. Because games often use movement and action, these techniques appeal to kinesthetic learners of all ages (Figure. 5.3).

Many conservation education games build on those played by children of all cultures and times, such as freeze tag or hide and seek. The fast-paced game, "Quick, Frozen Critters," (Project WILD 2000) for example, allows participants to simulate the actions of predators and prey as they find food or avoid capture. Others, like environmental board games, use logic and memory to teach environmental concepts.

Environmental games for adults are often adaptations of childhood games, books, and game shows. Charades, bingo, Monopoly, Trivial Pursuit, and even "Jeopardy" have been adapted for environmental topics. Bingo can be adapted to any topic by making bingo cards that have environmental categories. Renditions of Monopoly include "National Parks Monopoly," "Astronomy Monopoly," "National Geographic Mountaineering Monopoly," "Monopo-tree," and "Wild

Fig. 5.3 Games that involve movement and fun can engage learners of all ages (photo by M. McDuff).

Animalopoly" where players are animal caretakers that learn animal facts and trade for land, food, water, clean air, and habitat.

An engaging activity for adults, "The Council of All Beings" was created by Australian rainforest activist and deep ecologist John Seed and scholar Joanna Macy. The Council of All Beings is a series of rituals aimed at establishing a deep bond with nature. The rituals include mourning, remembering, and giving voice, and include guided imagery, movement, dance, and role-playing (Seed 2001). On-line games are also available, such as The Great Green Web Game from the Union of Concerned Scientists (2000). Suitable for adults or high school students, this environmental trivia game calculates the impact of a player's consumer choices on the environment.

Conservation educators working with adults may encounter administrators who have difficulty appreciating the value of "games" as an instructional tool. Referring to games as "activities" in your agenda for a workshop or training can overcome this perception. Whether you work as a teacher in a primary school or as a conservation educator with a forestry department, games inject a dose of fun into learning.

5.3.1 Planning

Full value contract

Before playing active games, you should establish rules with your participants to ensure a safe, fun experience. One strategy is to establish agreements or a group contract, sometimes called a "full value contract." If you are facilitating a group for a short period of time, you can introduce three simple agreements for a full value contract: Play safe, play fair, and play hard.

Talk with your participants and have them give examples of what it means to "play safe." Even children as young as 4 years old know that "playing safe" means no hitting, pushing, or name calling. If time permits, represent the agreements on a flag or stickers and have each person sign their name to show their commitment. Ask the group if they want to add any other agreements. You can accomplish a full value contract in 15 min or less.

If you have more time, the group can develop their own contract with examples, such as "be respectful to others," "be safe with words and actions," "listen to others," and "give 100% effort" (Malone 1999). Since everyone agrees to follow the full value contract, the facilitator can call a time-out in the game if players violate the contract.

If you are playing quiet, sedate games, such as board games, it is still a good idea to establish rules before playing. With these expectations, participants young and old assume the responsibility for a fair and respectful game.

Content and location

The content of your game will depend on the ecological concepts you want to explore with your participants. Follow these steps when planning your games (Teed 2004):

- Define objectives. What do you want the participants to learn?
- Decide the type of game you want and the storyline. Do you want an active game or an indoor activity? Do you want to use an established game, such as a trivia game or relay race, as a foundation? Will participants play in teams or alone?
- Divide your objectives into challenges. You may choose to have multiple levels of challenges, so participants progress to more difficult tasks during the game.
- Design rewards, if appropriate. Link incentives to learning or applying the material.
- Create the game, with needed props or cards.
- Test the game. Ask players to try the game and give you feedback on the duration of play, clarity of instructions, and engagement of players.
- Play the game!

Conservation agencies and educational organizations publish numerous resources, such as "Project WILD" (2000), "Project WILD Aquatic" (1992), and "Wild Adventures" (Malone 1999), which contain environmental games. There are also books with games you can adapt for environmental themes, such as "Everyone Wins!: Cooperative Games and Activities" (Luvmour 1990).

One game, "Hooks and Ladders," simulates a salmon run, while another "Migration Headache" demonstrates limiting factors affecting populations of migrating birds (Project WILD Aquatic 1992). A challenge in facilitating games is to ensure that players identify with their role, while maintaining a link to the concepts. You can accomplish this connection by planning a focused discussion of the activity that reinforces the concepts. The discussion you have at the end of a game should establish that the game is a "model" that does not represent all the real-world complexities (Chapter 2) (Project WILD 2000).

Games are perfect for outside locations, such as a field. However, you can play most active games indoors as well. Before you begin a game in an outside area, walk the field and look for any items that could cause injury to players, such as broken glass. You may need to create boundaries and can use sticks, ropes, or cones to mark the space.

5.3.2 Implementation

Looking at examples of games will give you ideas for the breadth of concepts possible to demonstrate with these activities. From carrying capacity to population growth, games model and reinforce concepts important to conservation education. Challenge your participants to create their own games as teaching and learning tools with unique adaptations to the ecology of your area.

Environmental board game: monopo-tree

Modeled on the board game Monopoly, Monopo-tree players toss a dice and move their game piece around the board to collect "property." Instead of collecting houses and apartments, however, players collect tree species representing different layers of the forest. Players also must answer questions related to forest ecology to advance in the game. The level of detail and strategy make this game perfect for adult learners. Monopo-tree was developed by college students for a Forest Biology class as a course assignment. You could model the game for your local forest ecosystems or enlist adult learners in creating the board game themselves as a learning experience.

Camera

Divide the participants into pairs, and instruct one player to take the role of photographer, while the other plays the role of a camera. The photographer leads the camera, with her eyes shut, to find a striking picture. When the photographer wants to take a picture, he points the lens of the camera (his partner's eyes) and presses the shutter button (by tapping the shoulder) to open the lens. The camera's "eyes" stay open only for three to five seconds, until the photographer taps the shoulder again to close the lens. This simple activity often has a powerful effect, with people remembering their pictures for years. Partners should switch roles so each can experience being the camera. When adapted for different ages, both adults and youth enjoy this game. Participants should sketch their "photograph" after taking the "picture" (Cornell 1989).

Stalking the shoe

Choose one person in the group to be "It" and instruct her to sit in the middle of the circle with a blindfold over her eyes. Have everyone else take off their shoes and place one of their shoes around her. The rest of the group are the "stalkers," who start about 30 paces from the blindfolded person. The stalkers must advance as quietly as possible and try to reclaim their shoe without alerting the person in the middle. Instruct the stalkers to pretend they are walking through a wooded path and must walk with complete calm and quiet. You can place branches or leaves on the ground to increase the difficulty.

When the blindfolded person hears a stalker, she points in that direction. If she points to someone, that person must freeze (with the help of a referee to make the call). The stalker who is frozen must return to the outside of the circle to begin again. Those who succeed in getting their shoe have a chance to be in the middle of the circle. Talk with participants about the value of walking quietly while in nature.

5.3.3 Evaluation

Before you try a new game for the first time, try a "dry run" by practicing the directions for the game with a colleague or friends. Ask for feedback on the clarity and flow of your instructions. When you are playing a game with participants, observe the activity with the following questions in mind: Are all players involved? Are players making the connection between the game and the concepts? Are they listening and understanding the instructions?

Each game should include some form of debriefing to evaluate the immediate impact of the activity, just as described in nature awareness activities (Chapter 12). Debriefing uses questions to allow reflection by participants after the activity. During a workshop, for example, one teacher playing the "Camera" game said the activity allowed her to slow down and observe the natural world for the first time in months. Such comments during an effective debrief reveal rich qualitative data for evaluation.

Structure the debriefing around three types of questions: "What?" "So what?" and "Now what?" Start by asking questions that allow participants to talk about what happened during the experience. "What did you just experience?" Then ask questions that allow the group to reflect on "What does that mean?" Lastly, engage the group in identifying applications of the activity to their lives. "Now what can you do with this knowledge?"

Debriefing can last just a few minutes or expand into a longer discussion. This method of debriefing follows the experiential learning cycle (Chapter 2). Debriefing also provides the opportunity of checking the participants' understanding of the ecological concepts explored in the game. With practice, games become a practical tool to teach and energize!

5.4 Case studies

Learners in an introductory environmental studies course at Oberlin College, Ohio, US, read about ecological, legislative, and economic issues associated with land development and subsequent wetland loss. The class presents the perspectives of four different stakeholders' views on expanding the airport and then writes a report giving recommendations to a fictitious mayor. At Framington State College, students review published data on the effect of coyote removal in Texas, US. Teams analyze the data and predict how changes at one trophic level affect populations and communities at other trophic levels (Case Studies in Science 2003).

These examples of case studies immerse the learners in real-life problems to develop decision-making skills about conservation issues. Disciplines such as medicine and business have traditionally used case studies to teach the applied skills of decision-making and problem solving. Case studies present dilemmas with the background of people, institutions, and problems to engage learners with the characters and circumstances of the story (Herreid 1994; Stanley and Waterman 2002). Although case studies are a type of story, they are usually interactive, open-ended, and encourage participants to learn from the experiences and ideas of others. Case studies sometimes involve role-playing, a technique presented in the next section.

With a case study, learners typically collaborate to identify issues, locate resources, pose pertinent questions, and support their conclusions. The goal of case studies is to teach students *how* to think about conservation, rather than memorizing the content of conservation issues. As such, critical thinking skills (Chapter 2) are at the heart of case studies as a teaching technique. Using Bloom's taxonomy (Chapter 2), case studies deal more with comprehension, application, analysis, synthesis, and evaluation, than with knowledge (Herreid 1994).

5.4.1 Planning

When planning, decide whether to write your own case study, use an existing one, or use published documents, such as newspaper articles, magazines, or journals. Choose what teaching format to use with your case study. Your choices will depend on what skills you want learners to gain and the time needed to prepare your case study.

Using Existing Case Studies

Resources such as "Case Studies in Science," provide environmental case studies with teaching notes that you can adapt for your own teaching objectives and local environmental conflicts (Case Studies in Science 2003). Box 5.3 presents an example of a case study entitled "Goodbye Honeybuckets" that analyzes environmental health issues of native Alaskans.

Examples of other cases from this on-line resource include "The River Dammed" and "Exotics." "The River Dammed" explores the removal of the lower Snake River dams in the United States. Students must decide the vote of Congresswoman Madeline Gibson as she weighs the perspectives of stakeholders who represent farmers, local tribes, sports fishermen, and government agencies. The "Exotics" presents students with the ecological, political, social, and economic factors surrounding exotic species, and the role of resource management in natural resource policy.

Box 5.3 The case study "Honeybuckets" follows specific steps as participants analyze environmental health issues among native Alaskan residents (Adapted from McNeil, L. (2001). Goodbye honeybuckets: Retrieved October 13, 2005 from http://serc.carleton.edu/introgeo/icbl/strategy1.html)

---

1. Introduce the case

Give the case to participants and ask a volunteer to read the case aloud while others read along silently.

More than 20,000 rural Native residents live in Alaska in communities without running water and where homes and offices use plastic buckets for toilets—called "honeybuckets." The waste from these toilets is often spilled in the process of hauling it to disposal sites and these spillages have led to the outbreak of epidemic diseases such as Hepatitis A.
 An Alaskan challenge: Native Village Sanitation, US Congress, 1994

 Even today, there are still <u>villages without a municipal sewage system.</u> John Kepaaq is a member of the <u>Tribal Council</u> in Icy Valley, and he is concerned about the type of sewer system being considered, since outside developers often do not realize the <u>problems of construction in the Arctic.</u> Icy Valley is a village of 200 people who live with <u>permafrost,</u> darkness, and long cold winters. John wants to ensure that the sewage system is appropriate for cold temperatures and <u>safe for the tundra environment.</u>

2. Recognize the issues

Ask participants to read the case again by themselves and note key words or phrases that are important to understanding the case. (See <u>underlined</u> phrases above).

3. Identify major themes

Ask this general question: What is the case about? You will probably get a diversity of answers. Post 5–7 responses on flipchart paper on the board, which might include: tundra ecology, health issues in rural Alaska, arctic climate, multicultural perspectives, sewage treatment, groundwater and permafrost, and Alaska geology.

4. Pose specific questions

Ask participants to use a chart listing *what they know* and *what they need to know*. This strategy is a productive way to generate questions. For this case, allow 10–15 min in small or large groups. Ask each group to identify three to five key questions they feel are essential for understanding the case.

Sample responses to Know/need to know analysis

 Know: Alaska is cold and sparsely populated. There is oil in northern Alaska. Sewage treatment systems are common in the "lower 48." Different sewage treatment systems exist, such as city sewers, outhouses, septic tanks.
 Need to know: How do honeybuckets work? What is the environmental impact of the current system? Why is there no municipal sewage treatment system in the

village? How does the tribal council work? What is a tundra? What are the seasons like in Alaska? What is the soil chemistry and composition? What is the effect of permafrost on sewage treatment? What are the major limiting factors due to tundra climate and soils? What are the feasible sewage treatment plants?

5. *Obtain additional resources*

Help learners obtain additional resources to help them answer their questions. Resources could include: textbooks, library materials, computer simulations, journal articles, interviews with Tribal elders, relevant maps, climate data on the tundra, interviews with local experts, and websites.

6. *Define problems*

As participants learn more about a case, they can better define problems and frame questions for investigation. Examples of defined problems for this case include: comparisons of climatic conditions in different biomes; temperature effects on decomposition; role of microbes in decomposition.

7. *Design and conduct scientific investigations*

Students might locate or generate datasets, conduct additional interviews, or perform lab and field experiments.

8. *Produce materials to support conclusions*

Ask participants to identify different ways to present their conclusions, such as a public presentation on the problems and solutions to arctic waste treatment; an evaluation of existing sewage treatment facilities in the Arctic; a marketing report on the effectiveness of composting toilets in Icy Valley.

The basis of a case study can be as simple as an article in the local newspaper, which you use to probe students with questions as they analyze the issues, interests, and consequences for action. Other existing material includes novels, television shows, magazine articles, cartoons, and advertisements. The topics illustrated in Gary Larson's *Far Side* cartoons provide simple, but rich material for an ecological case study. Ask students in groups to explore an environmental issue like the loss of biodiversity depicted in such a cartoon by asking, "What is the issue presented?" "What do you currently know about this issue?" "Who are the stakeholders involved?" and "What are potential approaches or solutions to this issue?" Some educators collect a series of articles on one topic, such as global warming, and ask students to analyze the issue through a series of questions about the actors, their values, the conflict, and the options (Herreid 1994).

When you consider an existing case study, ask yourself the following questions:

- What are the learning objectives I want to address with this case?
- How easily can I modify the case study?
- How difficult or complex are the issues in this case study?
- Will my participants care about these issues?

- What product will I ask my participants to produce?
- Is the case too short or too long for the time I have?
- What other resources will be available for this case (Stanley and Waterman 2002)?

Writing your own case study

While more time-consuming than using an existing case study, the advantage of writing your own case study is that you can tailor the case to meet specific objectives. You can choose from three types of case studies (Reynolds 1980):

- *Decision or dilemma cases* present a central character who must make a decision or solve a problem. The case study begins with an introduction to the problem and the character, as well as a background section. An additional section includes recent developments in the situation and appendices of documents, tables, and letters. An example might be the head of the Environment Ministry in Colombia deciding whether to allow logging in a national forest reserve.
- *Issue cases* teach students to analyze issues. There is no central character or pivotal decision to make. Rather, the case study asks the students to figure out what is happening in the case. Students might analyze several descriptions of the Valdez oil spill in Alaska, for example.
- *Case histories* are completed stories that illustrate conservation issues in action. This type of case study is often less exciting than decision or issue cases, but case histories allow students to analyze historical events in perspective, such as the recovery of Osprey after the ban on DDT.

Planning assignments or outcomes from case studies

Outcomes for the case study will flow from your objectives. Before you decide how to teach the case study, think about the desired outcomes:

- For learning content, ask questions that require learners to pull from back-ground documents.
- To explore multiple perspectives, stage debates, on-line discussions, trials, or discussions.
- For critical thinking and analysis, ask for opinion papers, recommendations, or statements of the problem.

To synthesize concepts, require participants to write a proposal or create a multi-media presentation (Teaching and Learning with Technology 2003).

5.4.2 Implementation

The following formats are commonly used to present case studies (adapted from Herreid 1994):

- *Discussion* involves presenting learners with a decision or issue case. Typically, law and business schools use this format, with the facilitator asking questions and students identifying issues, problems, solutions, and consequences.

- *Debate* is useful when an issue involves two opposing views, such as "globalization is a threat or a benefit to the environment." Both teams prepare written briefs of the issue and come prepared to argue either side. One side presents for five minutes, and the other side presents for the next five minutes. Then a five-minute rebuttal follows on each position. Lastly, each side presents a three-minute summary.
- A *public hearing* is used by many regulatory and public agencies. The format allows the expression of a variety of opinions. Structure the public hearing format with a panel that listens to presentations by different groups role-playing specific positions. For example, participants could be involved in deciding the protected status of an endangered species in their country.
- A *trial* uses two opposing sides, their attorneys, witnesses, and opportunities for cross-examination, in a case such as opponents versus supporters of damming the Three Rivers Gorge in China. To ensure involvement from all students, ask students to prepare position papers in support of each side, as well as reaction papers to the trial.
- *Problem-based learning* is a method used by many medical schools. A teacher works with a small team of four to five students and presents a written account of a patient. The students identify what they understand about the symptoms and additional information they need before formulating a diagnosis. Modify this format to involve an ecological scenario, such as the health of an ecosystem.
- *Scientific research team* mimics the scientific method where learners pose questions, make hypotheses and predictions, test those predictions through experimentation, and draw conclusions. Student research projects are often based on actual studies, such as collecting samples in several bodies of water to assess water quality. Learners write a research paper of their results and peer-review the papers of other teams.
- *Team learning* involves organizing a group into teams of four to seven students and dividing the content for a course or workshop into learning units. For each learning unit, participants read individual assignments that cover the basic facts of the unit. Each student takes a short test covering the essential points of the reading. The teams of students immediately take the same test together. The teams discuss their answers with each other and the teacher, who then clarifies points from the test and reading. Then, the student teams apply the facts to a problem or case.

Note that several of these formats, such as the public hearing and trial, involve role- play of characters in the case study. Another format used with adults working to resolve a local conservation issue involves developing a case study by finding examples of communities facing similar problems, studying their situation, researching relevant documents, and inviting speakers from that community. Adults then write up a case study to help them make decisions about environmental issues in their community.

5.4.3 Evaluation

When evaluating your use of case studies, assess the skills you want participants to learn from their engagement with the case. Some skills that learners may acquire during case studies include:

- Participating in group work
- Identifying issues
- Developing questions
- Locating resources
- Conducting research and investigations
- Producing materials, such as brochures and Web sites
- Making presentations, negotiating, and debating (Stanley and Waterman 2002)

Research also suggests that case studies, examples, and stories are an effective way to teach environmental problem-solving skills (Monroe and Kaplan 1988). Case studies present many opportunities to assess student products, such as requiring peer evaluations of a debate and a group self-evaluations. You also may use assessment forms that allow students to rate their own work experience with peers and your teaching (Stanley and Waterman 2002).

Use the following questions as a guide for evaluating the effectiveness of your case study:

- How effective was the case study as a learning tool with students?
- What were the stumbling blocks in the case study for students?
- Did student discussion address the objectives of the case?
- Were students able to locate additional resource materials?
- How useful were supporting materials and background information?
- Was adequate time given to the case study?
- How well did the case study fit with other components of the course or workshop? (Stanley and Waterman 2002)

5.5 Role-playing

As children, we first experience the perspectives of others through pretending. Through play, children become dogs, mothers, fathers, babies, or even monsters! Children negotiate resources like toys, space, and the plot lines in their imaginary games. Adults often use role-plays to test systems, such as a mock emergency in a community. Role-playing as a technique in conservation education builds real-world skills and understanding of different perspectives by assigning roles to learners in a scenario. Role-playing can range from young children imitating the movements of a deer to adults conducting a mock environmental summit on global warming.

5.5.1 Planning

Role-playing is a useful technique to choose if your learning objectives involve self-awareness, teamwork, initiative, communication, empathy, and problem solving

(Blatner 2002). In addition, students are often more motivated to seek solutions to complex issues when role-playing than with traditional assignments, such as writing a research paper. Two other techniques in this chapter, "games" and "case studies," may involve role-playing by participants as well.

The objectives of the role-play

As with other techniques, first define the objectives of the role-playing exercise:

- What topics do you want the role-play to include?
- What awareness or skills do you want the participants to gain?
- Do you want to use or adapt an existing role-play or write your own?
- How much time is there for the role-play?
- What outcomes (such as presentations, reports, research) do you expect from participants?
- Do you want students to role-play separately or as a group (Teed 2003)?

If your primary aim is to increase sensory awareness, create simple role-plays that allow participants to imagine the sensations of other life forms. Ask participants to choose a plant, tree, or animal and pretend to be that creature. As you plan these sensory role plays, begin with simple images, such as the life cycle of a tree, starting with the seed in the ground, growing into an adult tree, rotting and falling, and then becoming part of the soil (Cornell 1979). You can use role-plays to visualize concepts, such as the food chain, by asking learners to role-play different organisms in the food chain. Draw from the natural settings around you—a nearby creek or an urban park—as a source of inspiration for choosing the "characters" in your role-plays.

If you want participants to explore how to solve an environmental problem, planning the role-play will be more involved. As a facilitator, provide the setting and characters for the participants, but allow them to decide the dialogue and direction taken by the role-play. Try asking the learners to conduct research in order to make informed decisions in their role.

Types of role-plays

You can plan either individual or group role-plays. In an individual role-play, learners take on a character assigned to them. Instruct the characters to present their issue in a format such as a letter to the editor. Provide learners with some background information, as well as sources for additional research.

Interactive or group role-plays ask participants to prepare for specific roles and engage with other characters on the environmental issue, such as through a debate, negotiation, town meeting, or summit. Participants usually role-play different stakeholders, such as landowners, politicians, developers, and scientists who are affected by the same conservation issue. "Wheels versus Wildlife" is a role-play that gives participants the viewpoints of different stakeholders in a neighborhood where the city council plans to build a skateboard center in the local park (Murdoch 1994). The facilitator distributes the following roles:

- A skateboard enthusiast who supports the skateboard center and will represent her town in the national skateboarding championship.
- A conservationist who knows that birds and animals rely on the park for habitat.
- A local resident who uses the park daily for walks and relaxation.
- A senior citizen who has been knocked down by skateboarders using the sidewalk.
- A youth club leader who supports the development and thinks young people need a place for recreation.

Allow participants to plan their arguments and come up with some recommendations as a part of the exercise. The amount of preparation and research depends on the time available and the complexity of the issue.

Conservation education facilities, such as Pullenvale Nature Center near Brisbane, Australia, use role-plays to bring environmental history to life for its visitors. For example, fifth-graders recite the following poem:

> *Travel the Path of time unknown*
> *A destiny awaits you but not on your own*
> *Your strength is the group a bonding of trust*
> *Seize now this challenge before time turns to dust.*

And then with a snap of their fingers, the students become Bush kids in the 1890s who solve an environmental mystery. Facilitated by staff with dramatic skills, the students use props and their imagination to interact with characters who have devised an evil plan to capture birds and sell them to a museum. The students identify evidence, make inferences, draw conclusions, and develop an action plan to save the birds. Students wear costumes and use quill pens and ink, while creating the strategic steps in their plan.

Before immersing themselves in the role-play, the students read a fictional story about how Bush kids stopped a plan to dam a creek, and thus saved a habitat for wildlife. The story reflects the habitat surrounding the nature center. When the fifth-graders arrive at the center, the staff assumes character roles to encourage students to work together to address issues involving native wildlife.

Sources for role-plays

Reading existing role-plays can give you ideas for writing your own, or you may find published role-plays that fit your learning objectives. One excellent source for environmental role-plays for adults is the Program on Negotiation at Harvard Law School (PON 2004). These negotiations build from single-issue, two-party negotiations to multi-issue, multiparty negotiations with a focus on conflict resolution (see Chapter 8). For example, "Bradford Development" involves a negotiation between a mayor and a land developer on an open space fund for developers. More complex role-plays, such as Lake Wasota Fishing Rights, involve multiple interests, from native peoples to scientists, and ask participants to interpret complex environmental documents and data.

Other sources for role-plays include the Web sites of departments of conservation or environment in different countries. For example, the Web site for the New Zealand Department of Conservation (2005) includes a detailed description of the well-known role-play "Build-a-Tree." In this role-play, participants assume the parts of the tree, including the trunk, taproot, lateral roots, xylem, phloem, and bark. Even adults enjoy simulating the different parts of a tree and their functions! The Web site for the Australian Government Department of the Environment and Heritage (2005) includes instructions for a role-play called "Coastal creatures in crisis" that helps participants understand the impact of litter on sea birds and marine animals.

If writing your own role-play, decide on a setting for the characters affected by an environmental problem. Provide background information about the problem and the characters, and define the goals of each character. Data and maps can be provided to aid understanding. Decide what background information will be provided to all characters and what will be given only to individual players. Differing access to information in the role-play illustrates that peoples' perspective often depends on the information they have about an issue. This also adds the element of surprise and simulates the real world of nuance and power.

5.5.2 Implementation

When beginning any role-play, promote a safe and supportive environment so participants feel comfortable getting into their character. Introduce the role-play by providing characters with all background information and goals. Articulate your expectations for the participants, including any research or reading they must accomplish before the role-play.

If you are conducting an interactive role-play, assign the roles and put participants in groups. There may be different groups role-playing the same scenario at the same time. This situation provides rich material for discussion, as each group may arrive at different conclusions. All the students playing the same character can meet for 10 min before the role-play begins to discuss any questions they have about their character.

Discuss the challenges of playing a character with very different views from the participants' before the role-play begins. Encourage everyone to assume the role of actors, getting into the feel and emotions of their character. Additional research you require before the role-play also can enhance implementation, as you can ask participants to interview people similar to their character (Box 5.4). For example, the role-play "Model United Nations" asks participants to talk to the embassy of the country they represent for advice on the role of the UN (Teed 2003).

Set any ground rules before the role-play begins, such as time limitations. As facilitator for an interactive role-play, observe each group and make notes about the process and interactions between characters to use during the discussion.

Box 5.4 A mock environmental summit is an example of a role-play for adults

After studying global warming, learners role-play representatives from different countries and organizations at an international summit focused on global warming. Preparation for the role-play includes studying the IPCC report on Global Climate Change, the Kyoto Protocol, and data on human impacts on the environment. The teaching materials for this role play include information about the characters, lecture material, lab activities, writing assignments, links to reading assignments, and the proposed amendment to the Kyoto Protocol, which is the focus of the role-play (materials available at http:// serc.carleton.edu/introgeo/roleplaying/examples/envsumit.html).

Remember that one challenge of implementation is to let the role-plays unfold as an "unstructured drama" (Dallman-Jones 1994). The job of facilitator is to provide enough structure to achieve the learning objectives but allow the participants leeway to dialogue as they learn.

If you have any adult participants who may resist the idea of role-playing, describe the activity as a scenario or simulation, and maintain a professional attitude yourself. Role-plays like the negotiation simulations from the Program on Negotiation at Harvard Law School stem from a vast body of research. Sharing such information with adult learners can increase their investment in the activity.

5.5.3 Evaluation

Before attempting a role-play you have written, ask a colleague to read it and give you feedback on clarity. If possible, test the role-play before implementing it with a larger group. Practice giving instructions, even if you are conducting a simple role-play of a fish swimming down a creek.

One of the most important elements of a role-play is the discussion at the end. During this discussion, learners should describe what happened in their groups, the interactions and motivations of characters, their conclusions, and what they learned from the activity. Often, participants will be amazed to learn the motivation of another character from pertinent background information revealed during the discussion. They can write a reflective essay about the process and products of the role-play. To assess the learners' involvement, evaluate how they worked to promote the goals and perspective of their characters.

As you evaluate the effectiveness of the role-play, ask yourself if the role-play helped to achieve your stated objectives. Consider indicators of an effective role-play. Were there any stumbling blocks in the role-play? Did the learners understand the background information? Did they conduct the required background research? Did participants get "into" their characters? Did all characters interact with each other? Did learners gain a new appreciation for different perspectives? Were there any unintended outcomes from the role-play? What would you do to improve it in the future?

5.6 Contests

Avoid driving your car during the *Strive not to Drive* campaign and enter your name in a drawing for a free bike! Race your adventure team down the Green River and win cash and prizes. Write a winning essay about the value of wildlife for the African Wildlife Foundation and get a free trip to Nairobi National Park.

Contests and competitions can raise public awareness and knowledge about your environmental issue and the mission of your organization. Contests are a fun strategy to involve people in an environmental cause with concrete incentives for participation (see Chapter 9). You can use contests in collaboration with other conservation education techniques to increase public awareness and involvement. For example:

- Each year, Kinabalu National Park in Malaysia, hosts a race up a mountain, the Mount Kinabalu International Climbathon. Local media promote and cover the event, which increases awareness of the park and conservation activities.
- The World Population Film and Video festival conducts an international video competition for college and high school students. This competition awards critical thought and expression through video of resource consumption, the environment, and population growth.
- Another contest, the G-8 Summit Sea Turtle Naming Contest, gives students in grades K-5 the chance to name sea turtles fitted with satellite transmitters and follow the movements of the sea turtles on a Web site.

As you research other environmental contests, pay close attention to the sponsorship of the contests. Many large corporations, for example, sponsor environmental competitions to promote a green image. International Paper Company, for example, gives a $10,000 award for environmental educators who teach their students an understanding of the relationship between economic growth and environmental protection. Discuss with your partners or a steering committee the types of sponsorship you want for the contest.

5.6.1 Planning

Think about the target audience and outcome for your contest. Do you want to target schoolchildren or adults? Do you intend to sponsor an art, video, essay, photography, speech, or music contest? Or do you want to organize an athletic event, such as a bike race, a triathlon, or an adventure race? Would your organization prefer to sponsor a contest aimed at increasing environmentally responsible behaviors, such as reducing energy use in the home or reliance on automobiles?

Decide on the scale of your contest—will you advertise on a local, regional, national, or international scale? Look for successful models of other organizations that have sponsored similar events. With the scope and scale of your contest decided, begin identifying partners and promoting the competition.

Identify partners and create a timeline

Your partnerships with other organizations, from schools to the media, are key to a successful contest. Meet with local organizations that will help attract participants and promote the contest (such as schools, corporations, youth clubs, newspapers, radio, and athletic teams). Ask groups connected with your target audience if the contest appeals to them. What could you change to maximize participation? If working with students, enlist them in planning the logistics of the contest.

Meet with partners who could provide external funding for prizes and logistics. Coordinate recognition for donors, such as names on T-shirts or Web sites. Once you get one donor, others are often more willing to contribute. Send letters to partners, such as schools, explaining the contest and offering to visit schools and facilitate integration of the contest into the curriculum. Work with partners to identify the site, equipment, and any needed permits for your competition.

If your competition requires volunteers, use your partners to recruit and train your volunteers, such as judges for a photography contest or timekeepers for a race. As you select judges, ensure they reflect the diversity in your community. With the help of partners, create a realistic timeline. Consider how much time people need to prepare for the contest, publicize, recruit participants, and receive sponsorships. Cosponsoring an event, such as a contest, can be the start to a long-term partnership (Chapter 8).

Promote the contest

Submit news releases and Public Service Announcements to media contacts (Chapters 9 and 10). Ask other environmental organizations for the names of helpful reporters at local newspaper, radio, and TV stations and invite the media to cover the event itself. Work with volunteers or students with graphic experience to design flyers for the contest, and ensure that your partners post links to the contest on their Web sites. Circulate announcements of the contest well in advance of the deadline for submission or application. For art and essay competitions, begin advertising at least 4 months in advance. Remember to use the media to engage people in the competition and announce the winners.

5.6.2 Implementation

Contact your volunteers or judges to remind them prior to the contest. If your contest does not involve on-site judging and you are collecting entries, distribute copies of the entries to each judge. Ensure that participants know what to do in case of inclement weather, if your competition involves the outdoors, such as a kayak race or run. Many competitions hold their races, rain or shine.

Distribute the awards for a contest to at least the top three winners, and give certificates to all participants. If possible, send thank-you notes to all participants to promote participation in future contests. Write thank-you notes to all partners, and send them a copy of any media coverage, such as photos of the winners to celebrate a successful contest!

The range of contests and competitions is as broad as the different objectives of conservation groups. The following examples reflect that diversity and can provide ideas for your own contests:

- The nonprofit group Population Connection held a contest challenging secondary school participants to write a television script about the six billionth baby born in the world.
- The Stockholm International Water Institute awards the Stockholm Junior Water Prize recognizing a young person involved in water-environment issues on a regional, national, or international level.
- The United Nations Environment Programme hosts an International Children's Painting Competition on the Environment each year for children aged 6–15. The contest asks participants to paint a picture of a healthier, cleaner, environmentally friendly world.
- The river conservation group, Riverlink, hosts a triathlon along the French Broad River, US.
- The Dublin Nature Center in Ireland coordinates an art contest, culminating in the production of a calendar used in a fund-raising campaign.
- The River of Words contest, awards prizes and recognition to youth who create visual art and poetry depicting rivers in their communities.

5.6.3 Evaluation

The day after your contest, meet with your planning committee to discuss the successes and limitations of your contest. What worked well? What would you do differently in the future? Did you receive an adequate number of entries? Was the media coverage sufficient? Review your objectives for the contest and implement strategies for measuring your success.

Ask partners to identify the strengths and weaknesses of the contest. What recommendations for improvement would they give? For competitions, a three-question survey can be given to all competitors at the end: (1) What did you like most about the event? (2) What did you like least about the event? (3) What recommendations would you make to improve the event? Participants can complete the short survey while standing in line for donated food and drinks!

If one of your objectives is to increase membership in your organization, see how many people joined after the competition. Membership applications can include a question asking how the new member heard about the organization. For larger organizations, the information in your advertisements can include a specific phone number or e-mail address, so you can measure the number of inquiries about your organization from the advertisements for the contest.

5.7 Field trips

Field trips can involve a variety of locations and participants: senior citizens taking a tour of their local nature center, preschool children riding the bus to hear an environmental storyteller at the local library, or politicians touring a toxic site cleaned up with pressure from community members.

Field trips provide first-hand experience with physical sites and resources in a local community. Whether you arrange a field trip for your own learners or design a field trip to attract participants, field trips take advantage of the physical environment to enhance learning about conservation.

Field trips can range from a 10-min visit to investigate the biodiversity of a backyard or schoolyard (Russell 2001) to a semester-long environmental field studies course exploring the ecosystems of another country. There are even virtual fieldtrips (see Chapter 11), such as those offered by LEARNZ, an on-line education program in New Zealand that offers 16 virtual field trips for schools. Field trips provide opportunities for giving information, receiving public input, and providing first-hand experience with on-site sensory activities (Chapter 12). Field trips often occur in unique settings, such as zoos, museums, and parks.

With adults, field trips are often used in environmental planning processes to allow people to see a site for a proposed development or park.

Field trips are a particularly useful tool when:

- multiple stakeholders are involved in a process that has a geographic focus;
- citizens require information best explained on-site;
- educators or planners desire interaction between key stakeholders; and
- the need exists to increase public knowledge of a local issue (Citizen Science Toolbox 2004)

No matter what the duration or scale, effective field trips include trip selection, logistics planning, pre-visit and post-visit activities, and evaluation. The Ridgefield National Wildlife Refuge in New Jersey, for example, previews three field trips on its Web site, which includes pre-visit activities, field maps to study before the visit, and on-site activities. The field trips include "Refuge Habitats," "Birds of the Refuge," and "Refuge Cultural History." The pre-site planning page also includes a description of basic ecological concepts that visitors should know before their field trip.

5.7.1 Planning

One certainty of learning in the field is the unexpected. Changes in the weather, the interactions of participants, budget, transportation, access to the field site, and administrative support can affect your field experience in both positive and negative directions (Crimmel 2003). Detailed planning with contingency plans can help reduce the possibility that the unexpected will ruin the trip. You cannot anticipate everything, but you can identify the most critical elements of the field trip or those most likely to change (like the weather) and develop back-up plans. Thorough planning using the following step-by-step process can help you anticipate the challenges of taking learners into novel environments (adapted from CampSilos 2002):

Trip selection
- Identify the objectives, rationale, and evaluation plan for the field trip.

- Select the site to visit. Always contact the site to arrange the date and time. Obtain any pre-visit materials.
- Visit the field site to become familiar with the area. Take photos if possible to share with learners or use for publicity shots. Get ideas for pre-visit activities that connect with your curriculum or your organization's goals. If working with students, the field trip and its pre-visit activities should correlate with academic standards (Chapter 4).

Logistics

- Get approval for the trip from the appropriate supervisor of both your institution and the location (such as director of organization or departmental chairperson at school).
- Arrange transportation and any needed funding to cover costs.
- Make arrangements for lunch and other meals as needed.
- Develop a schedule for the day. Select times to suit the largest number of participants (after-hours for full-time workers, school times for student groups, daytime for retirees or parents with small children).
- Publicize the field trip and agenda with local media. Place posters at stores and libraries.
- Arrange for special equipment, such as film, video camera, digital camera, and journals.
- Prepare nametags for participants, chaperones, and visitors. If participants will be divided into small groups, make that designation on the nametag through a color label or sticker.
- Collect money for admission fees and pay the field trip site.
- Write a letter of permission for parents, if children are your participants, including the following information: date and location of field trip and transportation arrangements; educational purpose of field trip; provision for special needs students; cost and scholarships if available; lunch or other meal arrangements; money needed; schedule; materials or gear, such as rainy weather gear; information if child needs prescribed medicine administered; emergency contact information; and parent/guardian signature. Usually school systems will have a standard permission form for field trips.
- Request needed chaperones by sending a letter to parents or including a request in a school or organizational newsletter.
- Take copies of permission slips, extra food and water, and a first aid kit.

Leave a plan with a supervisor that includes the transportation route, contact names with phone numbers, and location of the nearest hospital.

Preparing participants for the field trip

Discuss the connection of the trip to the current unit of study or the goals of your organization. Introduce any pre-visit activities to familiarize participants with vocabulary, concepts, or issues covered in the field trip. You can provide a visual introduction to the site by showing photographs or exploring the Web site of the field site. The virtual field trips (Chapter 11) provided by many environmental centers can facilitate research and discussion while planning the actual field trip.

Think of strategies to introduce observation skills to participants, such as asking them to describe an ordinary object, such as a pencil or paintbrush to their peers. Brainstorm open-ended questions to gather information during field trips, and ask participants to come prepared to record responses in a journal. Consider assigning groups of participants in the role of "researchers" to explore a specific aspect of the trip and report to the others. Remember to ask the group to brainstorm expectations of behavior on the field trip as well. Finally, review the schedule for the field trip with participants (Camp Silos 2002).

Monterey Bay Aquarium in California includes a planning section on its Web site for youth leaders, Elderhostel leaders, and teachers preparing a field trip. More than 80,000 schoolchildren participate in programs at the aquarium each year, and many teachers start their planning at the aquarium Web site. Monterey Bay Aquarium field trips include facilitated and self-guided programs, with choices including "Splash Zone" for preschoolers to second grade, "A World of Sharks" for grades 3–8, and "Discovery Lab Investigations" for grades 6–8. Each field trip topic includes correlations to the California science standards. The field trips are available in both English and Spanish. The on-line registration form allows leaders to list their top 10 choices for dates and their top choices for field trip programs.

5.7.2 Implementation

The field trip itself is the time to watch participants directly engage in the landscape around them—from watching sea turtles return to their nesting sites to viewing a proposed greenspace in an urban community. The field trip immerses participants in the experience, a critical part of the experiential learning cycle (Chapter 2). One such field trip, a "Toxic Tour" in the Roxbury neighborhood in Massachusetts shows tourists, environmental educators, students, politicians, and citizens reclaimed and cleaned hazardous waste sites in this low-income community outside the city of Boston. Residents in these communities deal with environmental injustices ranging from lead contamination in buildings, dumping in vacant lots, and toxics in the air and groundwater.

More than a thousand diesel vehicles are housed in the fourteen bus and truck depots within a 1-mile radius of Roxbury. The low indoor and outdoor air quality in Roxbury has resulted in the state's highest rate of hospitalization from asthma, more than five times the Massachusetts average. A non-profit organization called ACE, Alternatives for Community and Environment, developed this "Toxic Tour" that features a walk that highlights polluting businesses, some of which have been closed due to efforts by ACE.

Youth interns with ACE lead this field trip with detailed maps showing the toxic sites in their neighborhood and stories about successful community efforts to mitigate these sites (Figure 5.4). The tour also stops by the EPA air monitoring

Fig. 5.4 Youth interns with the organization ACE (Alternatives in Community and Environment) lead politicians, tourists, and residents on a "toxic tour," a field trip of toxic sites in their community that conservationists and community members have worked to mitigate (photo by M. McDuff).

station now established in the area, which shows that particulate matter levels are 20% higher there than at the Harvard School for Public Health, only 1 mile away. Several of the interns who lead tours have planned careers in environmental law, prompted by their experiences with the field trips.

Helpful hints

During field trips, remember to take the following steps: Ensure that all particip-ants have nametags and are aware of any safety and behavior considerations on the trip. Divide the group into small groups or partners. Use a speaking style that engages the entire group. For example, if you are leading a group along a trail, stop in a place wide enough for the group to gather, and move toward the middle before you start talking (Chapter 12).

Give specific activities to the participants, such as sketching pages in a journal based on their observations, following a mystery with clues, writing postcards to themselves at the end of the field trip, recording answers to prepared questions in field notebooks, and recording resources seen with a video or digital camera (Camp Silos 2002).

Ask participants open-ended questions as they make observations, such as:

- How are these objects or resources different from each other?
- Describe observations including the setting and the object.
- Describe your favorite scene from the field trip.

- Pretend you are an archaeologist in the future observing this scene. What would you be able to conclude about this culture and the ecology of the area?

Many successful field trips draw on issues happening in the local community. Students in Wildlife Clubs in Bayanga, Central African Republic, for example, had questions about the researchers working with World Wildlife Fund and Wildlife Conservation Society. Who were these foreigners living in the rainforest with the elephants? Were they mining gold in the forest? Conservation educators developed partnerships with the researchers through weekly field trips for the wildlife clubs to the research camps. The elephant ecologist taught the youth in the wildlife clubs to compile research data by observing elephants and identifying individual elephants by the shape and markings on their ears. In post-trip evaluations, the wildlife club members reported that they understood much more about the elephants, often regarded as pest species. The assessments also revealed more positive student perceptions of wildlife research and conservation.

Field studies courses

The field studies course is an extensive field trip, in which facilitator and students live in the field for an extended period of time. Some field courses are of a short duration, such as the 2-week international field experiences of the WorldWide courses at Warren Wilson College in North Carolina. Students spend a semester in academic study preparing for the field and then travel to experience diverse cultural and natural resources in destinations such as Ireland, the Bahamas, and China.

The United Kingdom and several European countries have a long history in field studies courses. Scotland has the Aigas Experience offering field studies at the Aigas House in Inverness in the Highlands of Scotland and abroad. The Field Studies Council has a network of 17 field centers in England and courses overseas. The Barcelona Field Studies Center in Spain offers course in geography, biology, and environmental studies covering topics such as coastal processes and management, ecosystems, hazard management, river processes and management, rural depopulation, sustainable development, tourist impact and management, urban management, and volcanic landforms and landscapes. These courses provide hands-on study of academic content in the field, rather than only in a traditional classroom.

5.7.3 Evaluation

Post-visit activities and evaluation are critical to the success of any field experience. This aspect of the field trip allows for reflection and application of learning in the experiential learning cycle.

Post-trip activities

Facilitating or providing post-trip activities makes the field experience more than just a fun trip, as these activities provide a context for and extend learning. The following guidelines provide examples of post-trip activities:

- Share observations and reactions from participants about the field trip.
- Create a Web page, joint journal, or bulletin board about the field trip.

- Share assignments or journal entries completed on the field trip.
- Link field trip observations and activities to the curriculum or the goals of the organization through reflections and discussions.
- Send thank-you notes to facilitators, chaperones, and donors and include special information gathered during the field trip.
- Create a news report about the field trip and submit it to the local newspaper or organizational newsletter.
- Present a public multimedia presentation about the field trip.
- Document lessons learned by participants about the field trip (CampSilos 2002).

Assessment activities

You can use multiple methods to evaluate the effectiveness of your field trip. Pre-and post-tests are one method to assess gains in knowledge and attitudes as a result of the field trip. Staff with the Wildlife Clubs of Kenya asked students to draw and list the species and habitats they thought they would see in Nairobi National Park before a field trip to the park. The students completed a similar drawing after the field trip. The drawings revealed that a significant number of students increased their knowledge about the park as a result of the field trip.

Interestingly, field trips to the national parks were an impetus for the creation of the Wildlife Clubs in Kenya more than 30 years ago. Students in local Kenyan schools told a Peace Corps volunteer that Kenyan students never got to see the wildlife in their national parks, while tourists were the primary visitors to the parks. The youth negotiated with the Kenyan government to allow all members of Wildlife Clubs free access to the national parks, prompting many clubs to raise money for field trips to the parks.

As the facilitator or leader for a field trip, use the following questions as a guide for documenting your own observations about the field trip. Learners could even edit their own video footage to address specific evaluation questions. Involve your participants, volunteers, and administrators in these questions during a reflection about the trip:

- What was of unique educational value of the field trip?
- Did the participants meet the objectives and expectations?
- What benefit, if any, did participants derive from learning in the field?
- What aspects of the field trip provided the most educational value to participants?
- Was there adequate time?
- Was there adequate staff supervision?
- What would improve a visit to this site in the future?
- What points could be emphasized next time?
- What problems arose that should be addressed in the future (Camp Silos 2002)?

Use the photos or video you took of the field trip as a prompt for feedback during the debriefing session. Collecting information for evaluation from a variety of stakeholders involved in the field trip will give you valuable data for improving the quality of the experience.

5.8 Wilderness awareness and skills

The increase in urban living worldwide has resulted in a decrease in our direct connection to the earth for survival. Food for many people comes from grocery stores and fast food restaurants, rather than gardens and farms. At the same time, outdoor recreation is expanding, but recreationists to natural areas often lack adequate backcountry skills and ethics, which can result in human injury and harm to natural areas.

Wilderness awareness and skills—often called primitive skills or backcountry ethics—teaches a connection to the earth by increasing awareness and skills for self-reliance and low impact in the outdoors. You do not need a large tract of wilderness to teach skills such as nature observation, wild edibles, shelter building, and animal tracking.

Some skills, such as identifying edible plants, can be introduced in an urban area. Watch as participants change their perspective of dandelions from pesky weeds to tasty snacks. Including wilderness skills in your toolkit of techniques also connects us to ancestors and indigenous people today who use these same skills and keen awareness for daily survival.

5.8.1 Planning

Planning to teach wilderness skills involves assessing the needs and knowledge of your participants and identifying local resources, such as people and places. Even if you are not an expert in animal tracking or plant identification, field guides and local experts provide an ideal starting point for your lessons. Remember that the focus of wilderness skills is discovering the natural world and strategies for reducing human impacts on it. As an educator, you can build your skills along with your participants.

Content

During planning, you can choose to focus on sensory awareness activities or specific backcountry skills from fire-making to shelter-building. Many workshops and extended teaching sessions begin with sensory awareness and progress to concrete skills. Or you may choose to teach a wilderness skill to meet a group process goal, such as enhancing teamwork among participants. Creating a shelter like a debris hut can enhance group skills among participants along with wilderness skills.

Books written by Tom Brown (Brown 1983a, b) provide rich resources for planning lessons or workshops on wilderness skills. Known as "the tracker," Brown was raised in the United States under the guidance of Stalking Wolf, a displaced Apache Indian. Brown's wilderness school, books, and educational philosophy draw heavily on Native American tales and knowledge. His books include field guides to wilderness survival, nature observation and tracking, city and suburban survival, wild edible and medicinal plants, and nature and survival for children (Brown 1984, 1985). While the guides include North American species, the techniques and educational focus apply to any geographic area.

Another source for content is the Leave No Trace Center for Outdoor Ethics, an organization that conducts workshops across the globe in the seven principles of Leave No Trace (Box 5.5). With an emphasis on backcountry skills, this non-profit organization has trained 1700 master educators worldwide who in turn educate others in responsible outdoor recreation (Figures. 5.5, 5.6). Alice Cohen, a former Wilderness Ranger and current educator with the USDA Forest Service, facilitates these Leave No Trace workshops for adults. "Too often recreationists have a desire to go to 'wilderness' areas when they don't have the skills or knowledge to rough it," she explains. "Often they really just want recommendations and skills to camp outdoors." Trainings, such as the Leave No Trace workshops, help minimize the impacts on our public lands.

Box 5.5 The Leave No Trace principles reflect the importance of wilderness skills and outdoor ethics (used with permission from the Leave No Trace Center for Outdoor Ethics (2005). For more information about Leave No Trace, visit www.LNT.org).

Leave No Trace principles

1. Plan ahead and prepare
   - Know the regulations and special concerns for the area you will visit.
   - Prepare for extreme weather, hazards, and emergencies.
   - Schedule your trip to avoid times of high use.
   - Visit in small groups. Split larger parties into groups of 4–6.
   - Repackage food to minimize waste.
   - Use a map and compass to eliminate the use of marking paint, rock cairns, or flagging.

2. Travel and camp on durable surfaces
   - Durable surfaces include established trails and campsites, rock, gravel, dry grasses, or snow.
   - Protect riparian areas by camping at least 200 ft from lakes and streams.
   - Good campsites are found, not made. Altering a site is not necessary.
   - In popular areas:
     ○ Concentrate use on existing trails and campsites.
     ○ Walk single file in the middle of the trail, even when wet or muddy.
     ○ Keep campsites small. Focus activity in areas where vegetation is absent.
   - In pristine areas:
     ○ Disperse use to prevent the creation of campsites and trails.
     ○ Avoid places where impacts are just beginning.

3. Dispose of waste properly
   - Pack it in, pack it out. Inspect your campsite and rest areas for trash or spilled foods. Pack out all trash, leftover food, and litter.
   - Deposit solid human waste in catholes dug 6–8 in. deep at least 200 ft from water, camp, and trails. Cover and disguise the cathole when finished.
   - Pack out toilet paper and hygiene products.
   - To wash yourself or your dishes, carry water 200 ft away from streams or lakes and use small amounts of biodegradable soap. Scatter strained dishwater.

4. Leave what you find
   - Preserve the past: examine, but do not touch, cultural or historic structures and artifacts.
   - Leave rocks, plants, and other natural objects as you find them.
   - Avoid introducing or transporting nonnative species.
   - Do not build structures, furniture, or dig trenches.

5. Minimize campfire impacts
   - Campfires can cause lasting impacts to the backcountry. Use a lightweight stove for cooking and enjoy a candle lantern for light.
   - Where fires are permitted, use established fire rings, fire pans, or mound fires.
   - Keep fires small. Only use sticks from the ground that can be broken by hand.
   - Burn all wood and coals to ash, put out campfires completely, then scatter cool ashes.

6. Respect wildlife
   - Observe wildlife from a distance. Do not follow or approach them.
   - Never feed animals. Feeding wildlife damages their health, alters natural behaviors, and exposes them to predators and other dangers.
   - Protect wildlife and your food by storing rations and trash securely.
   - Control pets at all times, or leave them at home.
   - Avoid wildlife during sensitive times: mating, nesting, raising young, or winter.

7. Be considerate of other visitors
   - Respect other visitors and protect the quality of their experience.
   - Be courteous. Yield to other users on the trail.
   - Step to the downhill side of the trail when encountering pack stock.
   - Take breaks and camp away from trails and other visitors.
   - Let nature's sounds prevail. Avoid loud voices and noises.

Fig. 5.5 Participants in a Leave No Trace workshop prepare to naturalize a fire ring on public lands (photo by A. Cohen).

Fig. 5.6 Mission accomplished! (photo by A. Cohen).

Location

The location for teaching wilderness skills depends on local resources and objectives. Even the most urban localities have potential as teaching sites. You can teach a skill like orienteering using a map and compass in any location, from a city block to a rural farm. In urban areas, such as Ann Arbor, Michigan, (US) adults participate in orienteering clubs on the weekends as a recreational activity that promotes a backcountry skill.

If you have time and access to wildlands, such settings give a real-life context for teaching. Instructors from Outward Bound, an international outdoor education organization, take participants on wilderness courses ranging from 4 days to 3 months. During these courses in locations, such as Scotland and Chile, participants learn skills, such as rock climbing, kayaking, and orienteering, along with reliance on their group for navigation. At the end of the course, they participate in a "solo" experience, where they spend at least one night alone in the backcountry. Another international organization with an emphasis on wilderness skills is the National Outdoor Leadership School (NOLS), with schools in countries as diverse as Kenya and the United States. These organizations tailor programs to specific audiences, including at-risk youth, high school and college students, corporate leaders, and adults.

5.8.2 Implementation

You may decide to focus your content and location simply on building an awareness of the natural world. One resource for starting such activities is a tool called the "Tourist Test," published by Kamana Naturalist Training Program (Young 2001). Naturalist Jon Young showed 125 public school students slides and sound tracks of plants and animals in their neighborhood, and the majority of students could not identify the flora and fauna. To counter this lack of knowledge, he created a curriculum guide appropriate for rural, urban, suburban, or backcountry settings. The objective is for participants to go from "tourists" in their own neighborhoods to "natives" with an intimate connection to the natural world around them.

One activity involves instructing learners to find a "secret spot," a place in their backyard, their community, or a local park that they can visit everyday. Ideally, they must visit the secret spot frequently through different seasons. Participants record observations regarding specific questions about their secret spot, such as "Is there water in the area? How big is the area? Do you feel safe in this place?" (Young 2001). Students then create maps of their secret spot and sketch the topography, hydrology, soils and rocks, hazards, inspirations, trees, mammals, and plants they can see from the spot. Each subsequent map increases the students' awareness of their place in the natural world.

Another program that focuses on wilderness skills is 3-D Life Adventures, which brings together culturally diverse high school students in extensive expeditions of outdoor adventure, environmental education, and cultural exploration.

The 21-day Appalachia to Atlanta Expedition includes backcountry skills, backpacking, whitewater canoeing, the culture of the Cherokee Indians, pioneer culture, an urban immersion experience, and a service project.

5.8.3 Evaluation

When you design a lesson, allow someone with local knowledge of your area to give you feedback before implementation. After teaching the wilderness skill, include some follow-up activities so learners can practice using their skills in their daily lives. You may have them keep a journal about edible plants they see in their neighborhood or schoolyard. Also, have your participants give you feedback on which aspects of the lessons were most informative. What improvements could you make? Facilitators of the Leave No Trace workshops use peer evaluations of teaching presentations given by participants on specific backcountry skills. Such evaluation tools involve the learners in evaluating specific aspects of the instruction.

If you are working with students, you also can involve teachers in observing how often school-age participants use the knowledge and skills from your lessons. During a wilderness skill workshop at Warren Wilson College (US), environmental education students designed lessons on edible plants and animal tracking for local fourth-grade students. On the college campus, the fourth-graders collected dandelions for a salad and pine needles to make a tea containing more vitamin C in a cup than fresh-squeezed orange juice (Brown 1983a). One month later, teachers observed that the fourth-graders still snacked on dandelions and wild onions collected from their own playground! (The teachers also enforced a rule that students had to wash the wild edibles before eating.)

Another important evaluation tool is pre- and post-lesson assessments to measure achievement gains in knowledge and skills. You can use both qualitative and quantitative measures, ranging from journal entries, maps, and written tests. The program 3-D Life Adventures uses pre-expedition and post-expedition surveys to measure the effectiveness of its programs that explore both ecological and cultural diversity. The staff worked with local college professors to design the surveys that measure both changes in knowledge and attitudes. The survey results revealed a significant increase in environmental and cultural knowledge among participants, as well as more positive environmental and cultural values at the end of the program (3-D Life Adventures 2003).

5.9 Summary

Making conservation come alive can mean discovering the natural world around us through a neighborhood scavenger hunt or understanding the interests of an industry group by researching their perspective for a role-play. Many of the techniques in this chapter emphasize the experiential approach to conservation education, such as hands-on activities, field trips, and wilderness skills. The aim of these techniques is to immerse the participants in exploring the outdoors or an environmental concept. Other techniques bring conservation alive through a minds-on approach, such as storytelling, games, case studies, role-playing, and contests. Most of the techniques in this chapter involve an element of fun.

Planning these techniques involves both research and logistics, such as selecting the site of an outdoor hands-on activity or researching appropriate case studies for a group of learners. This chapter contains many helpful hints for implementing the techniques, including tips for engaging an audience in your storytelling to encouraging participants to identify with a character in a role-play. Lastly, evaluation allows you to measure impacts and collect feedback for improvement. And every technique—from storytelling to contests—engages the audience in learning through direct experience.

Further reading

Tom Brown's Field Guide to Wilderness Survival, by Tom Brown (1983a). The Berkley Publishing Group, New York, NY.

Keepers of the Earth: Native American Stories and Environmental Activities for Children, by Michael Caduto and Joseph Bruchac (1989). Fulcrum, Inc., Golden, CO.

Sharing Nature with Children II, by Joseph Cornell (1989). Dawn Publications, Nevada City, CA.

Project WILD K-12 Curriculum and Activity Guide, (2000). Council for Environmental Education, Houston, TX.

Ten-Minute Field Trips: A Teacher's Guide To Using The Schoolgrounds For Environmental Studies, by Helen Ross Russell (1998). NSTA Press, Washington, DC.

6

Using the arts for conservation

Failure of people to engage in sustainable land-use or consumption patterns may be partially due to our focus on technology and technical dissemination of information. Conservation education and outreach ideally promotes interdisciplinary understanding of the natural and built environment through the sciences, arts, and humanities. Yet, teaching materials often emphasize a science-based understanding of topics, while other ways of knowing are overlooked (Turner and Freedman 2004). Some believe this technocentric approach does not engage people in reflecting upon their values or personal behaviors (Job 1996). While science may be viewed as a creative process of discovery, the arts serve as another mode to acquire and interpret knowledge of the world. Different perspectives help people realize that we are all engaged in the search for understanding the world around us. Interdisciplinary programs help learners appreciate the wonders science reveals to the artist, and the philosophical and historical context that the arts and humanities provide for science (Rous 2000).

Environmental arts programs are usually experiential as well as interdisciplinary. Many educators believe that people learn best by doing. Exploration, observation, reading, writing, environmental monitoring, and problem-solving activities embedded in environmental art projects can make topics relevant to learners. Art can help students examine people's impacts on the environment. Incorporating art into the classroom can engage students who would not otherwise excel, and reveal students' hidden talents (Ford Foundation 2005).

Researchers have found that using multiple ways to teach, such as drawing or drama, in addition to written activities, enhances long-term memory. Brain-based learning theory suggests educators use multiple modalities—such as the arts—to enhance learning (Weiss 2000). Within formal education in the United Kingdom, environmental education is a cross-curricular theme in the National Curriculum for England and Wales. Activities, such as art, music, poetry, and creative writing, offer opportunities to address students' attitudes, beliefs, and emotions (Gurevitz 2000).

Most conservation issues are relevant to personal values and ethics, as well as science and economics. Conservation educators have used the arts sporadically to inform public opinion. In Tasmania, the photographs of Peter Dombrovskis brought the Tasmanian wilderness to the public's attention. The Tasmanian Wilderness Society printed a poster of his photograph of the Franklin River. It became the symbol for a campaign to prevent flooding the river. The photograph was published in full page, full color newspaper advertisements just before a federal election. The caption, aimed at the standing government party, read: 'Would you vote for a party that would destroy this?' The image and the campaign helped oust the government and halt the damming of the Franklin River (Grant 2001; Curtis 2003a).

The arts, such as literature, painting, photography, theatre, music, and dance, offer a way to make an emotional connection to people. Art can transcend beliefs and cultures. Using the visual and performing arts can help conservation educators reach new audiences. Art can provoke reactions that typical education and outreach methods do not. Art has the potential to inform audiences or participants in a new way about conservation topics. It can stimulate new dialogues and catalyze new perspectives.

Artist Paulus Berensohn, a ceramicist, focuses on touching people's emotions.

We're constantly being touched . . . by nature, and it's only our imaginations that can help us know that. That's how I would save the planet. I would start with the senses and imagination. Now that may sound naïve and impossible, but it can start in kindergarten . . . I've been interested in environmental problems for a long time, and I just don't see economics solving it or politics solving it, because they're all bandages, and it seems to me that if we are going to have a new connection to the environment it will have to happen in individual hearts and souls . . . the artist can help us fall in love with the earth again.

(Berensohn 2002)

6.1 Emotion, art, and learning

When emotional input is added to learning experiences, it makes them more memorable and exciting. The brain deems the information more important and enhances memory of the event. Presenting facts alone is less likely to result in long-term changes in feelings and behaviors. Good teaching engages feelings (Weiss 2000; Cable and Ernst 2003). Because we do not completely understand our emotional system, however, we do not fully incorporate emotion into environmental curricula in schools or into environmental outreach programs (Sylwester 1994).

Greater understanding of the psychobiology of emotion should eventually enable educators to apply this knowledge in and out of classrooms. Emotion is often a more powerful determinant of our behavior than our brain's logical/rational processes. Many more neural fibers project from our brain's emotional centers up into the logical/rational centers than the reverse (Sylwester 1994). Actions such as purchasing a lottery ticket (the odds of winning are ridiculous) or wearing an unflattering style because it is fashionable, provide daily evidence that rational thought does not rule our behavior. Emotions allow us to respond quickly to incoming information by avoiding conscious deliberation. This leads to both irrational fears and foolish behaviors, and to immediate and appropriate action.

Our brain stem and limbic system along with the cerebral cortex regulate our emotions. The limbic system is the brain's main regulator of emotion and helps process our memories. The limbic system is powerful enough to override both rational thought and innate brain stem response patterns. Memories formed during a specific emotional state are easily recalled during a similar emotional state later on (Thayer 1989). For example, during an argument people can remember similar previous arguments. Educational techniques, such as simulations and role-playing activities (see Chapters 5 and 11) enhance learning because they provide emotional prompts that tie memories to the kinds of emotional contexts in which they will later be used in the real world.

Our emotional system is modulated by molecules that travel throughout our body and brain (Sylwester 1994). Some of these molecules respond to the arts! For example, music as well as positive social contact can elevate endorphin levels, creating a feeling of pleasure (Levinthal 1988). By engaging multiple senses or by emphasizing social interaction, the arts can provide emotional pleasure and support to participants. The arts also seem to be important to the development and maintenance of the systems that initiate and conclude cognitive activity (Sylwester 2000).

Religious and political leaders throughout history have used the arts—from frescoed church ceilings to political theme songs—to reinforce beliefs and promote certain behaviors. Research into the emotional and health benefits of the arts is receiving increased interest. Using the arts in medical practices has taken many forms. Art therapy engages the patient in dance and movement, music and sound, and painting or other visual arts to promote physical or psychological healing (Lewis 1993). Artists create art for patients, patients become artists themselves, or patients cocreate with artists. Therapists have found that incorporating movement, sound, art, and journal writing into their therapeutic relationships helped patients identify and be in touch with feelings, explore unconscious material, gain insight, and solve problems (Rogers 1993). The arts have enhanced the practice of medicine as well. Physicians at medical centers have introduced daily doses of poetry as part of medical students' training. Writing and reading poems has helped medical students better understand a sick patient's feelings and their own relationship to disease and healing (Grace 2004).

Research carried out in behavioral sciences has shown that images of nature and some types of music are among a small number of things that can reduce stress and hold attention in positive ways across cultures and different personalities (Friedrich 1999). Researchers have found that paintings and photographs of nature can decrease stress in patients in health care settings. In Sweden, postoperative heart patients randomly assigned to rooms with a nature scene of still water surrounded by trees had a more rapid recovery than patients who viewed abstract rectilinear forms or no art at all (Freidrich 1999).

Conservation problems require creative solutions, so it makes sense to access multiple fields of knowledge and many ways of knowing the world in order to take care of it.

6.2 Visual arts for the protection of natural areas

The visual arts—painting, sculpture, photography, and other media—can inspire environmental protection, help with fund-raising, influence political activities, stimulate a new perspective, or improve classroom instruction. Although participating in an art event may not directly stimulate changes in environmental behaviors, it can make participants more open to information or engage their positive feeling in support of an organization or a cause, such as the protection of a natural area.

When geologist Ferdinand Hayden made plans to explore the Yellowstone area in the western United States with his US Geological and Geographical Survey team in 1871, he recognized the value of having artists accompany him. Hayden always tried to employ artists to enhance his scientific observations and reports. He understood the popular appeal and scientific value of photographs and paintings. For him, photographs and pictures of the American West were as important as scientific specimens in studying the region and presenting it to the public for their understanding and comprehension (Hassrick 2002).

Hayden used Thomas Moran's wood engravings to illustrate his first published accounts of the exploration of Yellowstone that were printed in a popular magazine, *Scribner's Monthly*, and to illustrate his government report (Hassrick 2002). Moran's images stirred great interest and helped advocate that the Yellowstone area be set aside as a national park (Figure 6.1). Moran's watercolor paintings as well as photographs taken by his traveling companion, William Henry Jackson, proved

Fig. 6.1 Wood engraving, *Rock Pinnacles Above Tower Falls*, by Thomas Moran; printed in Scribner's Monthly in May 1871 (courtesy of the Huntington Library, San Marino, CA).

remarkably effective props for Hayden and others who used them to persuade Congress to take the historic step and establish the world's first national park.

Moran's paintings and Jackson's photographs "did a work which no other agency could do and doubtless convinced everyone who saw them that the regions where such wonders existed should be preserved to the people forever," wrote Corps of Engineers Captain Chittenden. Jackson wrote that the watercolors and photographs made during the survey "were the most important exhibits brought before the [Congressional] Committee." The "wonderful coloring" of Moran's sketches, he wrote, made all the difference (National Park Service n.d.). Moran created a monumental painting, over 3 m long, of the Grand Canyon of the Yellowstone. His intent was to "satisfy the myth of a bigger, newer America," and to enlighten the nation about the magnificence of the region, as well as entertain them (Hassrick 2002). Congress purchased the canvas for the US Capitol building. It was the first landscape painting to hang in the Capitol.

Artists continue to be influential in promoting conservation and exposing the unique beauty of wildlands. Conservation organizations can often solicit paintings or photographs from professional artists to use in conservation campaigns, to accompany news releases, or to illustrate publications or interpretive kiosks. Regardless of the source, obtaining high quality images is one key to the success of using the arts to help achieve conservation objectives. Box 6.1 provides some tips for taking or selecting effective photographs.

Sometimes amateur pictures capture the uniqueness of a place and can stir the imagination. The Nature Conservancy launched a project, *Photovoice*, in the Tibetan hamlet of China's northwest Yunnan province. Although most residents could not

---

Box 6.1 Tips for taking good photographs to promote conservation.

1. Get close and fill the picture frame with your subject.
2. Balance the composition; follow the rule of thirds by dividing your frame into three parts horizontally and vertically, like a tic-tac-toe board. Place the subject or landscape elements where these lines intersect.
3. Be selective; crop out extra elements that are distracting.
4. Focus on your subject.
5. Capture interesting light, such as during dawn, dusk, or storms; use side and back lighting to create special effects.
6. Use a tripod for long distance shots to ensure your camera is still.
7. Put people in your pictures. People like looking at other people.
8. Photograph animals doing something—eating, yawning, running—not just sitting there.
9. Have your image tell a story.

read or write, with a camera in their hands, 223 people from 64 villages took more than 50,000 shots. The effort gave local communities a voice in such issues as unsustainable tourism and rapid development. They spoke volumes about their rich cultural heritage, the landscape's stunning diversity and the hardships they endure.

The story of how powerful photographs contributed to the debate about protection of a remote Alaskan refuge illustrates the capacity visual images have for stirring people's emotions. The photographic exhibit and book, "The Arctic National Wildlife Refuge: Seasons of Life and Land," by Subhankar Banerjee helped stir concern for Alaska's Arctic National Wildlife Refuge, a place few people ever venture. Controversy over drilling for oil and gas has surrounded the 8 million ha refuge for decades. Proponents of oil development often describe the refuge as a wasteland. Defenders of the pristine ecosystem call it "America's Serengeti," for its vast caribou herds and other wildlife. Policy-makers and environmental groups used Banerjee's stunning pictures to counteract some oil developers' claim that the refuge is a "flat, white nothingness."

6.2.1 Planning

Banerjee became interested in the Arctic Refuge after visiting other parts of Alaska and reading a US Fish and Wildlife Service report about wildlife in the refuge. "I was stunned by the biodiversity of the refuge and how little had been documented photographically. I was inspired by what I read and dreamed of doing a year-round exploration and documentation of the refuge," Banerjee explains in his book.

"I thought, My God, it is the most debated public land in the US. Every magazine, every newspaper, every TV station has done multiple stories on the place, and yet, believe it or not, while there had been pockets of studies by biologists and botanists, it had not been visually documented in a way that was comprehensive and included all four seasons. I realized I had a tremendous opportunity," said Banerjee (Sischy 2003).

The refuge is the calving ground for a herd of 120,000 caribou that migrate there each year. They join 36 species of land mammals, nine marine mammals, and 180 bird species, some migrating from 6 continents to nest. Magnificent displays of wildflowers carpet the refuge each summer. Banerjee found the landscape filled with life, even in subzero winter when he photographed polar bears near their dens.

Banerjee spent several months carefully planning the development and publication of his photographs and book. His lofty goal and seriousness of purpose impressed the publisher of Mountaineers Books during their first meeting. Banerjee described his objective for the publication of his book: "I want to see official, permanent wilderness designation for the coastal plain of the Arctic National Wildlife Refuge." He brought with him an exquisite, though incomplete, collection of images. He also had commitments from a number of well-known environmental writers and biologists, such as Peter Matthiessen and George Schaller, to provide authoritative essays to accompany his photographs.

Among his supporters was the Blue Earth Alliance, a nonprofit organization dedicated to supporting photographic documentary projects that educate the public about endangered environments, threatened species, and current social issues. Banerjee credits the organization for providing him with advice about shooting the photographs, managing a project of this magnitude, writing grant proposals, and hosting a photo exhibit of his work.

As political interest in oil exploration and development increased in 2001, Banerjee raised funds to return to the Arctic to complete his photographic project. Dozens of organizations and individuals supported and contributed to his work. His association with the Blue Earth Alliance made fund-raising easier. Supporters could now give tax-deductible donations to his project. Ultimately the project cost over US$200,000.

6.2.2 Implementation

Banerjee's book and photographic exhibits provide a visual document of reasons to preserve the Arctic Refuge in its pristine state. Proposed oil exploration and development may threaten the delicate ecosystem and the subsistence livelihoods of indigenous people, adding to other threats, such as global warming. Banerjee spent 14 -months in the field to investigate the year-round story of the landscape and the indigenous cultures, the Gwich'in Athabascan Indians and Inupiat Eskimos, who depend upon it. During this time, Banerjee, with an Inupiat guide, traversed more than 4000 miles by kayak, raft, snowmobile, and on foot through this harsh, yet stunning refuge. He understood why the Gwich'in people call this coastal plain "the Sacred Place Where Life Begins."

Word of Banerjee's photography spread and the Alaska Wilderness League asked him to bring some of his images to Washington, DC, to support the campaign to prevent oil drilling in the refuge (Figure 6.2). His photographs were enlarged into mounted pictures, over a meter long, and were used by members of the House and Senate at press conferences and debates over the energy policy during the 2001–2002 legislative sessions. During Senate debate on the administration's plan for drilling in the Arctic refuge, a senator held up one of Banerjee's photos and referred to the area's beauty in her arguments to ban oil and gas exploration in the refuge. She urged her colleagues to visit Banerjee's photo exhibit on display at the nearby Smithsonian Institution to see the refuge's pristine wilderness for themselves.

6.2.3 Evaluation

The measure was defeated in the Senate. Victory, at least that particular battle, was won by Banerjee and other supporters of the Arctic Refuge. The foreword to Banerjee's book was written by Jimmy Carter, former US President and Nobel Peace Prize winner. Carter writes, "It will be a grand triumph for America if we can preserve the Arctic Refuge in its pure untrammeled state. To leave this extraordinary land alone would be the greatest gift we could pass on to future generations."

Banerjee's book and exhibits provide a good example of how the arts can help inspire concern for the natural world. "I don't look at myself as an activist," says Banerjee. "I'm an artist bringing information to the public. What I hope to bring

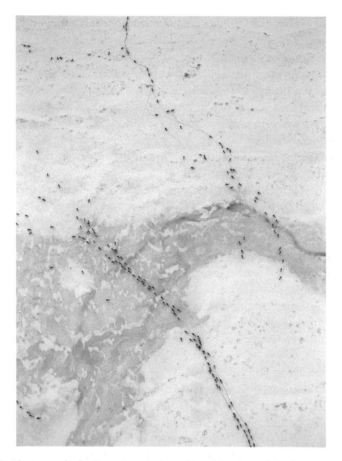

Fig. 6.2 Photographs by Banerjee, such as this aerial view of the Porcupine caribou herd in the Arctic National Wildlife Refuge, were used to sway congressional support for protection of the Refuge (photo by S. Banerjee).

back to people is the land itself, and to raise awareness of its importance. The debate about this issue is not an informed debate, and I want the public to know the truth (Griggs 2004)." In 2005, a new bill passed the Senate that allowed oil exploitation in the refuge. Opponents are still fighting for the protection of this pristine wilderness, with the arts as well as politics.

6.3 Art exhibits

Organizations like the San Isabel Land Protection Trust in Colorado (US) rely on an annual art sale to help provide funds for accomplishing their goal of protecting ranch and forest lands. With proceeds from the sale of regional oil paintings, watercolors, pastels, woodblock prints, and bronzes, the Trust was able to purchase over

$100,000 in conservation easements to protect scenic beauty and wildlife habitat. The art exhibit brought together artists, donors, landowners, and others at a fun social function, and involved many in the organization of the exhibit. Well worth the effort, the art event required careful organization. Planning for the art sale involved 50 local businesses and individuals as sponsors for the show. A number of volunteers, from marketing assistants to clean-up crews, helped make the event successful. Twenty-nine artists provided 149 pieces of art and over 200 people came to bid on the work.

Art exhibits have the capacity to provide more than fund-raising and social benefits. They also can engage the audience in contemplating new perspectives and ideas for viewing the environment. An art show provided a way for Forestry Extension faculty at Oregon State University to communicate with new audiences about forest management (Withrow-Robinson *et al.* 2002). Extension agents were accustomed to talking with traditional clientele, such as forest landowners, but had little experience with an urbanized public, people who were becoming more involved in natural resource policy debates in the northwest US. To reach this audience, extension foresters decided to use a traveling art exhibit to engage the public in a discourse about forestry. "The idea behind this exhibit is to stimulate visitors with provocative images to get them thinking and talking about values, forests, and forestry issues," said project team leader, Withrow-Robinson.

6.3.1 Planning

The main educational objectives of the art exhibit were to: (1) increase viewers' awareness of the complexity of forest issues and stimulate a dialogue about them, (2) reach new audiences, and (3) to learn about the public's understanding of forestry and promote the university extension program. The extension team developed a template for the content of the exhibit. They selected 10 issues to illustrate the scope and complexity of regional forestry concerns. The topics included wildlife habitat, aesthetic beauty, harvest methods, jobs, forest health, recreational use, water resources, fire management, conflict resolution, and urban sprawl.

In addition to topical content, Withrow-Robinson and his team wanted to include images that represented different regions of the state and different ecological zones. They also selected a diversity of artistic media, including painting, photography, ceramic sculpture, Native American carvings, quilts, and furniture. Six venues were selected for the traveling exhibit around the state. Locations were based in part on having an extension agent who could serve as a local coordinator. The local coordinators searched for locations with many public visitors. Each site needed to be secure so the art was safe from theft or vandalism. The sites physically needed well-lighted space, walls that could be nailed into, and room for 3D items.

Planning for the art show began 18 months ahead of time in order to book exhibit halls and contact artists to make arrangements. When the art show first was developed the team combed art galleries and perused regional art books to try to identify artists from whom they could solicit work. By the third year of organizing the art exhibit, the Oregon Council of Arts collaborated to send out mailings to local artists' guilds with solicitations for work that would address the forestry theme and content areas (Box 6.2). Artists submitted slides or photographs of their work. A selection committee made up of team members and art department faculty reviewed the materials. They rated and selected artists' work based on its congruence with the program's educational themes—the link between consumption and management of forests, and the illustration of a diversity of forestry issues. The

---

Box 6.2 Guidelines to artists submitting works to the "*Seeing the Forest*" traveling exhibit (provided by Withrow-Robinson, Oregon State University Extension Service, Corvallis, OR).

Exhibit Content: Artwork relating to all aspects of forests and forestry will be considered for *Seeing the Forest*. However, the central theme we wish to explore is how Oregonians, by their actions as consumers, are linked to and have an impact on forests and forestry (additional information provided in a prospectus). Work that relates to the central theme will be given preference in the selection process. We will also select artwork supporting a secondary theme and so will still consider artwork relating to a broad range of forest topics. The secondary topic is to illustrate the complexity of forestry issues and diversity of individual values and cultural perspectives. To accompany the artwork, the exhibit will include artist's statements, supporting text, brochures, a questionnaire, corkboards for visitors' comments, and a publicity poster.

Selection Process: *Seeing the Forest* is an educational art exhibit. We will select artwork of the highest artistic quality that allows us to fulfill our educational goal by illustrating the themes of the show. We hope to include artwork in a broad range of media, and seek representation from various regions around Oregon. Ultimately, we wish to produce a traveling show that engages and appeals to a diverse public audience. Artistic approaches may include abstract, conceptual, expressionistic, fantasy, representational, portraits, and still life, among others. A Jury made up of Extension Faculty and members of University Art Departments will select art.

Submission: There is no entry fee. Both two and 3D artwork in a range of media will be accepted. Please remember each art piece should be easily transportable, available for sale, and available for touring for the entire 7-month show. To submit work, send 2″ × 2″ slides or 4″ × 6″ prints to the *Seeing the Forest* Project.

Please put your name and title of the work on each slide or print. Please fill in and return the "Slide Entry Form" with your contact information and facts about artwork represented in each slide. Remember the slides and prints are all the committee will have to judge your work, so professional quality is vital. If you enclose a self-addressed stamped envelope, your slides and/or prints will be returned to you. Along with your slides, submit a brief written statement for each piece of art submitted (125 word limit). Explain how the piece relates to this year's central or secondary show themes. These statements will be available to the jury during the selection process to help in interpreting art under consideration, and some may also be used when selected work is displayed in the exhibit.

Deadline for slide and statement submission is 3 months prior to show.

Artists whose work is selected for the show will be notified by 2 months prior to show.

Additional Submission Restrictions:

- All work must be prepared for display. 2D work must be framed and under Plexiglas (not glass). Frame clips are not allowed. Flat pieces must be framed and wired for both wall hanging and free hanging from a rail and wire system. For wall hanging preparation, insert a ring on each side and string wire between them. Free hanging preparation must ensure pieces hang straight from a rail and wire system. Fabric art must arrive ready for display (i.e. with a metal, wood, or plastic dowel rod for hanging support).
- Artwork is limited to 20 kg or less.

Agreements: If your art is selected, you must . . .

- Send your artwork to *Seeing the Forest*, at your expense. Deadline for delivery in Corvallis is [2 weeks prior to show]. Specific details on location will accompany your letter of acceptance.
- Pack your art piece in a sturdy, reusable container. Containers are required even for hand-delivered artwork. (At the end of the tour, your art will be shipped back to you at your expense unless other arrangements are made.)
- State a valid price, reflecting current market value for your artwork.
- Handle any sales transactions directly with the purchaser, and be responsible for any gallery fees. (OSU Extension Forestry does not charge a commission, but some of the venues we will be exhibiting do. Commissions range from 0 to 35%.)
- Keep the art piece in the exhibit for the entire 6 months.

selection committee also judged the work based on its relevance to a general audience, effectiveness of the artistic image, technical quality, use or depiction of recycled materials, and diversity of subject matter and geographic origin. "We provided the selection committee with a scoring sheet to rank each art piece," said Withrow-Robinson. "We received several hundred entries. Of course, one challenge is that the artists didn't necessarily depict the entire range of issues we wanted to explore with the public. We worked with what we got."

6.3.2 Implementation

The art show was produced by the efforts of the central leadership team, local coordinators, and a host of volunteers from the arts and forestry communities. About a dozen forestry extension agents around the state helped to find sites for the show. These local coordinators communicated with local papers for promotion of the show and handled other logistical details of the local show. In order to reach a general audience with little awareness of forestry issues, the exhibit was displayed in public spaces, such as libraries, community centers, and government buildings, in addition to art galleries.

Scheduling was handled centrally. Moving an art exhibit is a major task. It involves carefully packing the artwork, loading, traveling, unloading, hanging, and then tearing down and hanging again at the next site. Transition time of a couple of days was allotted to complete these tasks between each 1-month show. The team had restricted the size and weight of the art in order to make a traveling exhibit feasible. The team also recognized the importance of involving artists in the set-up and hanging of each show. "We're foresters by training," explains Withrow-Robinson. "We always needed artists to help layout the artwork in each exhibit space, based on light, space, proximity of other pieces, and a host of artistic concerns."

Extension agents promoted each show using several media. An attractive poster was distributed to media outlets and posted on public buildings and meeting places around towns. News releases (see Chapter 10) were sent to local media ahead of the exhibit.

More than 65,000 people viewed "Seeing the Forest" in 6 Oregon communities (Figure 6.3). The show used several methods to communicate with the audience. The obvious communication was through the 53 pieces of art—the visual communication from the artist to the viewer. Most artists also supplied a short written statement to accompany their image. For example, artist Stev Ominski

Fig. 6.3 Visitors to the art exhibit, "Seeing the Forest," were exposed to images showing the complexities of forest management (photo by V. Simon-Brown).

contributed an acrylic painting of a tree stump and sawyer in a forest. He called it, "Predator." Beneath his painting appeared his artist's statement: "This piece was painted as a personal 'crying out in disbelief' at the loss of a grove of Douglas fir I used to enjoy visiting. The fact that the trees were privately owned and ready for harvest did little to dampen the fire of my feeling of outrage. One visit, a forest; the next, gone! Admittedly, I was fascinated by the process I had missed . . ."(Withrow-Robinson *et al* 2002).

The extension team produced a brochure that was available at the exhibit sites. It explained the goals of the show and key points for consideration. This information also was displayed on text panels scattered throughout the exhibit. "We realized some people wouldn't bother looking at the brochure," said Withrow-Robinson. "So we produced the text panels on poster board to make sure the purpose of the exhibit was clear and the viewers' experience more informed."

The team also provided ways for the audience to communicate with them, with each other, and with the artists. The public could meet and discuss the show with the team and the artists at a reception hosted at each site when the exhibit opened. Comment cards were available to viewers to post at the exhibit for everyone to read. These could be posted on corkboards distributed throughout the exhibit.

6.3.3 Evaluation

Comment cards allowed participants to share feedback with other viewers. Researchers also provided a 6-question survey inside each exhibit brochure asking viewers for their anonymous feedback (Box 6.3). A box was placed at the exit of the show for viewers to place their completed surveys.

The extension team received many informal comments during the course of the exhibits. Comments pasted on the corkboards for other viewers to read were often emotional, polarized statements. For example, one response to the painting called "Predator," complained that it was ". . . inflammatory and ill conceived . . ." while another thought it was "fabulous." Overall, the extension team was impressed with the thoughtful dialogue that was generated in this informal setting. The following series of comments provides an example (Withrow-Robinson *et al* 2002):

"I don't like the trees being cut down. Trees give us air."

"Where do you think the wood for your house came from?"

"I think people are too harsh. Our forests are really valuable, and it's important that we do care. They are a renewable resource, but it takes a long time to grow one tree. It's important that we don't get careless."

"Yes, renewable—but only when used wisely. Past practices should be learned from, not repeated. [The forest] was here thousands of years before us and should be here for thousands more!"

"Logging is necessary; we need the wood to build, but we also need to let the old-growth alone."

"Antilogging propaganda does no one a service other than spreading falsehoods and an unrealistic reality."

"I've read all the comments here, and I don't see any antilogging comments except for the first one. People are just saying, 'Be careful, use it well' "(Withrow-Robinson *et al.* 2002).

Box 6.3 Viewer questionnaire used in evaluation of the *Seeing the Forest* Exhibit (Withrow-Robinson *et al.* 2002).

---

1. Did Seeing the Forest succeed in illustrating the diversity of forest issues in Oregon?

    ☐Yes   ☐No

2. Which issues did you see illustrated?

    Harvest methods   Conflict resolution   Wildlife habitat   Jobs
    Aesthetic beauty   Recreation   Water resources
    Urban encroachment   Forest health   Fire control

3. I think *Seeing the Forest* increased my understanding of the complexity of forestry issues.

    Strongly agree   Agree   Disagree   Stongly disagree

4. Where do you live? (list of regions provided.)

5. Which art image did you like the best? Why?

6. Which art image did you find the most intriguing or thought-provoking? Why?

The written survey included in the exhibit brochure was completed by 365 viewers. The extension team compiled statistics on the responses to the multiple-choice questions. They found that 85% of respondents thought the art show successfully illustrated the diversity of forest issues in Oregon; 77% agreed that viewing the show increased their understanding of the complexity of forest issues. The team performed a content analysis on the open-ended questions to identify recurring themes in preferences and provocations of specific art pieces. Because the respondents were self-selected, bias is a problem in interpreting the results. Yet, there were a wide range of responses, from positive to negative, that displayed the gamut of attitudes about forest management and the art in the show. Viewers were aware of forest issues and able to identify them in the images. Both this quantitative data as well as the qualitative data indicated that forest issues, such as aesthetic beauty, harvest methods, wildlife habitat, forest health, and conflict resolution, were the easiest issues for people to recognize in the artwork.

The team was satisfied that they had done a good job of selecting artwork to represent a diversity of issues in the show. They had reached new, non-forestry audiences across the state by using art as an alternative educational approach. Project leader Withrow-Robinson felt the show had moved their Forestry Extension program forward to engage individuals and communities in a dialogue about natural resource issues. It cost several thousand US dollars to produce a traveling show, not including the salary costs of the teams. "It was a very rewarding experience," said Withrow-Robinson. "After three years of producing the show, the only way we could keep doing it would be to hire a coordinator dedicated to the program. We're experimenting with other art forms now, such as theatre. There's a lot that can be done!"

6.4 Hands-on arts programs

Some art groups have embraced environmental protection as a meaningful artistic subject. Japanese artist Tadashi Tonoshiki and 100 residents of Niinohama collected litter that had washed up on the beach, some from as far away as Taiwan and Siberia. The artist's "ecological action art" stimulated participation in a beach cleanup by people who may not have gotten involved otherwise. Director Floria of the International Friends of Transformative Art comments: "Certainly we still need to act politically—we can demonstrate and march and write letters to Congress. But art is unique—it has a certain power, an interaction with our consciousness in a way that we might not even understand yet" (Cembalist 1991).

The creation of art has been used in therapies to help patients identify feelings, gain insight, and solve problems (Rogers 1993). In a similar way, art can be used as an educational tool to help participants explore new landscapes and examine feelings and concerns they have for the environment. The opportunity to produce arts and crafts can attract nontraditional visitors to a site and can entertain traditional audiences and reveal new perspectives. An innovative example is a program developed by The Nature Conservancy that combined art with lessons about urban ecology.

6.4.1 Planning

The Nature Conservancy, in partnership with the New York Foundation for the Arts (NYFA), created a community art and urban environment project entitled *Wild New York: Creating a Field Guide for Urban Environments*. Offered free of charge to members of The Nature Conservancy, the project aimed to increase awareness of urban ecology through a hands-on artistic experience (The Nature Conservancy and NYFA 2004).

The idea stemmed from collaboration between The Nature Conservancy and the NYFA to develop a community environmental art project that could creatively illustrate people's interactions with their environment. Project Leaders, Molly Northrup and Chriss Slevin determined the objectives, audience, and budget for the project. The format was a series of four field trips to different parks around New York City. Naturalists and an artist guide would accompany participants, explain ecological phenomena and introduce a new art project at each site. A final meeting would allow participants to piece together their art creations into a "guidebook."

The audience initially was recruited from among the many dedicated bird-watchers/ naturalists in New York City. The Nature Conservancy works closely with various organizations that coordinate public lectures and bird/nature walks for their members throughout the year. Working with this group of regular birders and constituents involved in the NYFA, the project proposed to create a multidisciplinary activity that illustrates the importance of open spaces for wildlife and people in New York City. By creating art from their experience in nature, the participants could address urban environmental issues in a creative way that would capture the attention of people that might otherwise be unconcerned. For those people who were already engaged in the issues, they could see them illustrated in a new way. It also would be an opportunity for the birders to create a document that arises from their wildlife observations and passion for nature.

Some issues that would be addressed on the field trips:

• New York City as part of a major migration corridor for birds.
• Urban encroachment on many important natural areas, such as wetlands in Jamaica Bay.
• Historical and natural changes in parks and natural areas in the city, including invasive species.
• Biodiversity in New York City.
• The role of the human community in maintaining and supporting open space and natural areas.
• Use of art to educate and engage New Yorkers about their natural environment.

Participants would create a field guide of their observations while taking nature walks in areas around New York City. The field guide would be loosely designed and structured by the NYFA artist and then assembled and created by the participants. There would be three two-hour nature walks led by naturalists and the artist. A final afternoon would be devoted to assembling the books and discussing the results. Ten to fifteen participants would be accepted into the program.

Schedule

March: Staff interviewed and selected a NYFA artist to serve as instructor (five artists applied for the short-term position) and worked out logistics: dates, materials, and guidelines.

April: Staff sent program invitations to potential participants with details and a schedule. Due to a small budget, TNC limited advertisement of the program to their Web site, e-mail, and flyers mailed to people who had participated in bird-walks offered in the past. Northrop was surprised by the response: "Interestingly, none of our regular birders joined us," she said. "It was a new group of participants: younger, some with artistic interests, and a couple of teachers. I think this was partly due to offering the program on the weekend, as well as the content." The NYFA also advertised to their Trustees; none attended, although a number expressed interest. Fifteen participants were accepted on a first come, first served basis. Participants were required to reserve space in advance, and were encouraged to attend all four sessions if possible.

May–June: Program begins. The 2-h nature walks took place on three Saturdays. On the fourth Saturday participants assembled their field guides and discussed their experience in a discussion.

Budget

The staff calculated that the program would require a budget of several thousand dollars. Expenses included:

- Artist—$1200
- Materials—$1200
- Transportation—$400 van rental for distant field trip site
- Mailing/marketing—$400 for copies and postage

6.4.2 Implementation

Artist Maddalena Polletta and two New York City naturalists led the series of guided walks through some of New York City's parks and wildlife areas. At each walk, the naturalists discussed the general ecology of the area and Polletta introduced participants to simple art techniques for capturing images from the natural world. Each session involved a hands-on activity, designed for new artists and those with little or no art-making experience (Figure 6.4). Over the course of the project, participants created simple, hand-made field guides documenting their experiences observing bird life and diverse natural habitats, to create an artistic record of their field observations.

The project was free of charge, and all artistic materials were provided to participants. Each participant was asked to fill out a survey before and after the

Fig. 6.4 Participants in The Nature Conservancy Program in New York create their own artistic guides to urban environments (photo by TNC).

series to provide formal feedback on the impact of the project. Their typical schedule follows.

Session 1

Site: Central Park, Manhattan
Activity: Leaf and Flower Prints

Participants explored a pond and meadow in Central Park, looking for examples of native species used in landscaping. The naturalists discussed nonnative, invasive plants and the problems they cause. During the walk, participants collected samples of some nonnative, invasive plants. At the end of the walk, Polletta demonstrated a simple printmaking technique to create leaf or flower prints from the invasive plants.

Session 2

Site: Inwood Hill Park, Manhattan
Activity: Sunprints (Cyanotypes)

Inwood Hill Park offers the last natural forest in Manhattan. In this woodland site, participants searched for the typical oak and hickory trees that characterize this forest and provide habitat for birds. They collected leaves, grasses, branches, and other specimens to use in creating sunprints (cyanotypes), a photographic process that creates silhouettes of objects by using natural sunlight on treated paper. The cyanotype paper was obtained from an art supply store, but directions for making your own paper are available in photography books and Web sites.

Session 3

Site: Jamaica Bay Wildlife Refuge, Brooklyn
Activity: Watercolors

The diverse habitats of Jamaica Bay offered an inspiring landscape for creating watercolor paintings. The refuge, which includes a salt marsh, upland field and woods, fresh and brackish water ponds, and an open expanse of bay and islands, is renowned as a prime birding spot where thousands of water, land, and shorebirds stop during migration. Using reference materials and personal observations, Polletta instructed the group in techniques for creating drawings and watercolor paintings of the birds and the landscapes they inhabit.

Session 4

Site: New York Foundation for the Arts, Manhattan
Activity: Creating and Assembling the Field Guide Portfolio

At the last session, Polletta led an indoor workshop on bookbinding. Using techniques for creating handmade books, participants learned yet another art form. They each assembled a handmade portfolio, incorporating the various pieces of artwork created during the guided walks. The portfolio serves as a vivid record of each participant's relationship to and experiences in nature throughout the program.

6.4.3 Evaluation

Project leaders conducted a written before-and-after survey to assess changes in participants' awareness and their perceptions of their experiences. At the beginning of the first session and the end of the last session, they asked participants to complete a short survey to provide feedback on the impact of the program.

Most participants had a fair amount of experience visiting natural areas prior to registering for the program. In contrast, most of the participants' responses regarding their level of artistic experience were lower. A majority reported "not a lot" of experience.

When asked what they hoped to gain from the program, participants desired increasing their knowledge of both the natural and the art world. Some examples of participants' expectations were:

- "To be more familiar with plants and birds, and to learn new art techniques in a natural setting."
- "A memorable experience, a little more knowledge, closeness with nature, some satisfaction in executing the craft elements."

At the end of the project all participants rated their experience as a 5, on a scale from 1 to 5, with 5 being the most positive. All responded that they learned from both the guided walks and from the art activities. They reported that they will be more likely to do more art projects in the future and that they will be more likely to visit natural areas more often.

The enthusiasm for the course is apparent in the responses to the questions: Did the activities and the guided walks change your experience of visiting natural environments in any way? If so how?

- "The walks with Sean made what would otherwise seem like not very wildlife-filled walks seem to be teeming with a variety of species. He demonstrated the value of close attention. So did Maddalena's instruction to always be noticing and drawing details."
- "Absolutely. I pay more attention to trees, the leaves, the birds and have more of an interest in learning about them."

The program also had other impacts. It introduced TNC to new audiences that have not been attending the traditional conservation lecture programs and nature walks, said Northrop. The participants tended to be younger and have an interest in art first, with a secondary curiosity or appreciation for nature. It encouraged an artist and environmental educator to call Northrop and propose a new TNC art and nature program for children that will be conducted in partnership with another organization. TNC also helped promote a lecture series, *Human/Nature: Art and the Environment*. After further disseminating information about their art activities, Northrop received numerous requests, mostly from teachers, for copies of their program guide.

6.5 Environmental literature

Writers have long aroused concern for the environment. Science writers have contributed to public understanding of the connections between people and a healthy environment. In the United States, Rachel Carson's book, *Silent Spring*, awakened the citizenry to the dangers of chemical pollution from far-reaching effects of toxic pesticides in the early 1960s. She warned of impacts to the food chain, and the threat of springtime without the songs of birds. Legislation to protect the environment followed her publication. Internationally, scientists contributed to the publication of the *Millennium Assessment Report* (2005). It provides ample evidence to help instigate policy-makers to take action to protect the global environment.

While some environmental writing provides stark facts to increase awareness, other types arouse emotions and enhance people's appreciation for nature's diversity and beauty. Naturalist writers often combine these approaches and provide a rich source of literature to explore people's relationship with nature. Reading journals and stories of naturalists' experiences can provide new perspectives and build awareness of different ways of being in the world. Environmental writers provide unique learning experiences, from demonstrating the value of uninhibited inquiry to developing an appreciation for observation skills (Matthews and Bennett 2002). In classroom settings, courses on nature writers have helped participants examine their sense of place or provided training in history or language arts. One such class in the United States might read the works of authors mentioned in Box 6.4. Every region has writings and stories that stimulate reflection about the environment.

Writers also have inspired people to examine more closely the natural world and our relationship to it. Well known poems, such as William Blake's (1789) "Songs of Innocence," provide simple metaphors to help people view the world:

> *"To see a world in a grain of sand*
> *And a heaven in a wild flower*
> *Hold infinity in the palm of your hand,*
> *And eternity in an hour."*

Some poetry helps readers reflect on the relationship between the environment and their spiritual or emotional well-being. It reminds us of the solace nature offers. Wendell Berry's (1985) poem, "The Peace of Wild Things," captures this introspection.

> *"When despair for the world grows in me*
> *and I wake in the night at the least sound*
> *in fear of what my life and my children's lives may be,*
> *I go and lie down where the wood drake*
> *rests in his beauty on the water, and the great heron feeds.*
> *I come into the peace of wild things*
> *who do not tax their lives with forethought*
> *of grief. I come into the presence of still water.*
> *And I feel above me the day-blind stars*
> *waiting with their light. For a time*
> *I rest in the grace of the world, and am free."*

> **Box 6.4** Environmental literature and writing programs explore books, poetry, or journals to increase participants' environmental concern and knowledge.
>
> ---
>
> A list of US books might include the following:
>
> Abbey, Edward (1968). *Desert Solitaire: Season in the Wilderness*
>
> Berry, Wendell (2002). *The Art of the Commonplace: The Agrarian Essays of Wendell Berry.*
>
> Dillard, Annie (1988). *Pilgrim at Tinker Creek*
>
> Ehrlich, Gretel (1985). *The Solace of Open Spaces*
>
> Heinrich, B. (1994). *A Year in the Maine Woods.*
>
> Hubbell, S. (1987). *A Country Year: Living the Questions*
>
> Leopold, Aldo (1966). *A Sand County Almanac*
>
> Lindbergh, Anne (1978). *Gift from the Sea*
>
> Maclean, N. (1983). *A River Runs Through It*
>
> Muir, John (1911). *My First Summer in the Sierra*
>
> Snyder, Gary (1974). *Turtle Island*
>
> Thoreau, Henry (1854). *Walden*

An international nonprofit organization, River of Words, combines the use of poetry and art with observation-based nature exploration to excite youth about their watersheds. The River of Words conducts training workshops for teachers, park naturalists, community groups, and resource agencies to use this multidisciplinary, hands-on approach. The program targets children's literacy, critical thinking skills, and creativity. Participants engage in outdoor activities to explore their watersheds and enter an annual environmental poetry and art contest. In 2004, River of Words received over 20,000 entries from around the world (Pardee 2005). The poetry and art portions of the curriculum help make the science relevant and personal. After studying a spider's web, an 11-year-old transformed his observation into a poem: "Dawn's reflection / honeycomb of light / bound by diamonds / caught overnight" (Pardee 2005). The River of Words program has resulted in community partnerships in support of education and watersheds, creek clean-ups and restorations, and student environmental poetry clubs and art exhibitions (River of Words 2005).

Both reading and writing about the environment promotes reflection about our place in the world. Teachers may use environmental prose or poetry to stimulate students of any age to examine other perspectives about the environment or as examples for creating their own stories and exploring their personal experiences. To be effective, environmental writing assignments should be planned in advance and integrated with related activities. The following questions provide a useful outline to consider when developing a writing project (Lindemann and Anderson 2001, Brew 2003):

1. What is the purpose of the writing assignment—to prepare for a discussion, reveal attitudes, reinforce a skill, or encourage reflection?
2. When should learners complete the assignment—after a specific experience, at intervals during the experience, at the end of the experience?
3. Can you provide a relevant example to help guide the assignment?
4. How should learners complete the assignment—on their own, with a partner, in groups?
5. Who is the audience for what is written—the learners' themselves, the class, a publication, the instructor?
6. What will be the response to what the learners have written—a discussion, written comments, a grade?

Once these questions are answered, an instructor can provide a detailed description of the assignment for the learners to follow. Clear communication between teacher and students will help ensure that the writing is integrated with other environmental activities to promote reflection and learning. The following section on keeping a journal provides a model for planning, implementing, and evaluating a writing project.

6.6 Keeping an environmental journal

Environmental journaling usually involves a combination of careful description and personal thoughts and feelings. In describing an object, scene, or phenomenon, journaling can provoke the writer to delve deeper into the scientific basis of nature and stimulate a greater ecological understanding. At the same time, journaling can help participants reflect on their thoughts, feelings, and impressions of the environment. Whether the smell of a walk in the woods or the sight of a polluted wetland, positive and negative responses can be examined and shared. Through writing and reflecting, journaling can help people discover why something is important to them. Sometimes it can reveal deeper meanings in a childhood memory and reconnect them with a place.

Writing in a daily journal is often prescribed to keep writers' prose loose and fluent. Similar to a physical activity, practice keeps a writer in shape. Journaling is used in the classroom to help learners overcome any fear of writing they may have and to provide a chance to practice skills they are acquiring. Environmental writing projects are a wonderful opportunity to infuse environmental education across the curriculum.

6.6.1 Planning

Many guides to journaling suggest key steps in making the process successful. If keeping a journal is a school assignment, teachers often specify the type of writing instrument to use, from notebook paper to computer programs. In a class, time is often divided between the outdoors, for inspiration, and inside, where thoughts can be shared and work critiqued. Outside of a structured class, whatever materials feel comfortable may be used, whether it is jotting with a pencil in a decorative book or typing on a laptop computer. The physician–poet, Dr William Carlos Williams wrote some of his famous poems on prescription blanks (Barnet and Stubbs 1977).

As an educational technique, journal keeping involves more than telling learners to look at the world and write what they see and think. Few students can produce beautifully worded details and insights without facilitation and practice. To stimulate ideas, successful examples of writings from current or prior classes can be shared with participants. Another approach is to read and analyze examples from published environmental writers, nonfiction and fiction, prose and poetry. Teachers also may use structured writing assignments to help students observe and think in new ways about the world around them.

People keeping a journal outside of a structured class find it helpful to establish a set time to write in a journal—when awakening in the morning, during lunch break, or before bed. Ten to fifteen minutes a day is a minimum recommendation to maintain a writing habit. Entries can be just a few phrases or a few pages. The important thing is to write for the recommended amount of time. Many authors recommend writing freely, without correction or revision; without worry about spelling, vocabulary, or punctuation (Barnet and Stubbs 1977). They encourage journal keepers to write in whatever idiom or voice that feels comfortable.

As part of a structured class, it is useful to provide students with a suggested word count to set a minimum goal for their writing requirement. Rous (2000) suggests 700 words per week for upper level school students. Writing workshops or weeklong courses for adults encourage daily entries.

6.6.2 Implementation

The content of the journal may capture an arresting sight or a significant moment. Journals are different from diaries. A diary is a record of things that occurred, such as "had a picnic lunch with Blema and Perry." A journal reflects on the event and gives a sense of why it was meaningful. A journal can provide a record of the rich details of daily experience. By jotting down reactions and ideas about the environment and the people or wildlife sharing a place with them, writers may come to a deeper understanding of their relationship.

A variety of exercises can assist learners as they explore ideas for journal keeping. Journals offer a good opportunity to practice writing literal or impressionistic descriptions—short, medium, and long—of persons, places, and things. A popular exercise for nature writing is to describe a natural scene, whether a still pine forest or a stormy ocean. With enough detail, these descriptions can form word pictures, which are common in nature literature. Writers use words and paper to describe a scene in the same way artists use paint and canvas. Word pictures help the reader visualize the world you are creating with words.

Box 6.5 Henry David Thoreau's journal entry on November 1, 1855 paints a word picture through his description of a wading bird on the Musketaquid River (Thoreau 1906).

As I pushed up the river past Hildreth's, I saw the blue heron (probably of last Monday) arise from the shore and disappear with heavily-flapping wings around a bend ahead; the greatest of the bitterns (Ardeoe) with heavily-undulating wings, low over the water. Seen against the woods, just disappearing, with a great slate colored expanse of wing. Suited to the shadows of the stream, a tempered blue as of the sky and dark water commingled. This is the aspect under which the Musketaquid might be represented at this season: a long smooth lake, reflecting the bare willows and button-bushes, the stubble, and the wool grass on its tussock, a muskrat cabin or two, conspicuous on its margin, and a bittern disappearing on undulating wing around a bend.

A conventional approach to painting a word picture of a landscape is to begin with the foreground, move to the midground, and then conclude with the background (Murray 1995). This basic technique mimics how artists might draw a landscape, or how your eye might traverse the setting. Murray suggests looking for texture, color, lines, shadowing, and movement in the same way an artist would. Determine what makes the scene unique, as well as what makes it familiar. Observing and carefully describing a favorite outdoor scene helps the writer understand it in a way that a passive visit does not. If it is not possible to go outdoors, looking out a window or choosing a photograph or painting of a landscape, plant, or animal can provide a useful topic for painting a word picture (Box 6.5).

Journal-keeping activities provide an opportunity for learners to record their observations and sensory experiences, and probe their reactions to readings and landscapes (Brew 2003). Rous (2000) suggests several topics for use with upper school students. She has students write about their favorite place. Students are encouraged to close their eyes and think about a place where they enjoy spending time or where they go for refuge. Then students imagine their feelings as if they were currently in that place. What sounds, tastes, smells, sights, and textures do they associate with the place? Students then write a description (in class or at home) of the favorite place with enough detail that readers can place themselves in the scene and appreciate it along with the writer. Rous has the students read their descriptions in class. Discussion centers on the range of places, and what "favorite" places have in common. This provides the opportunity to share the importance of nature to students' lives and to analyze good writing techniques.

To help students become more aware of what their senses perceive and ways to describe these sensations, Rous has her students write "sensory monologues." Students are asked to sit quietly and describe, rather than name, every sound that occurs (e.g. "the increasing then fading buzz of a hummingbird as it flies by," not "a bird passes"). In another exercise aimed at describing sensation, a bag of objects is passed around and students have to describe the texture, rather than naming the object. In a similar exercise students are blindfolded. Then containers of various foods to taste or aromas to smell are passed around for the students to describe.

For a semester-long course, keeping a journal about a specific piece of land over the duration of several months or longer, offers learners the opportunity to develop a sense of place and facilitates a feeling of stewardship toward the land. Thoreau's (1854) *Walden*, describing his life at Walden Pond, set a standard for writers to share insights about their natural surroundings with a broad audience. Students should choose a site that is easy to access regularly. It can be a small plot, which enhances detailed observation or a larger area that provides more variety.

A number of different journal-styles can serve as a model for a journal-keeping class or private exercise. Examples range from scientific observation, to a literary approach, or an eclectic style that combines prose with visual graphics.

A "Day Book" style records daily and seasonal observations of the plants, animals, weather, and natural phenomenon of a particular place. Over time it can reveal patterns in nature, such as the monthly cycles of the moon or the annual returns of migratory birds. Students are encouraged to notice small things like the order in which flowers bloom or insect numbers increase. Students also can set up monitoring projects and check changes in water quality in a lake or stream with simple equipment or note changes in bird species in a yard.

This approach can be expanded into a collection of exploratory field notes. Charles Darwin's "The Voyage of the Beagle," which led to his theory of evolution, exemplifies this style. Darwin's diary combines detailed scientific observation with personal reflection and discovery. The descriptions of travel in land foreign to him and his boundless enthusiasm help make the prose engaging.

A final approach might combine verbal observations with sketches and other visual mementos of the land. Field journals may juxtapose vivid written descriptions, scientific field observations, or poetry with drawings, maps, photos, designs, or even cartoons (Figure 6.5). This type of journal can inspire new insight in learner's understanding of the land and may help spark creativity.

6.6.3 Evaluation

In the classroom, specific assignments and use of styles can be evaluated based on criteria of appropriate language and writing skills. In the semester-long journal class taught by Rous (2000), students turn in their journals once a week for comments and feedback by the teacher. Rous records her comments on a separate paper, not writing directly on the student's journal to respect the personal nature of the assignment. Rous uses her comments to open a dialogue with the student about their special place. She asks specific questions about the events recorded about the site, and notes which parts of their writing she found interesting, enjoyable, surprising, or unclear.

If the class agrees to this ahead of time, students can swap journals and read and comment on each other's entries. Sharing journals in class gives students new examples of writing ideas and styles. Rous encourages self-evaluation. Half-way through the semester, she asks the students to write in their journals what they have

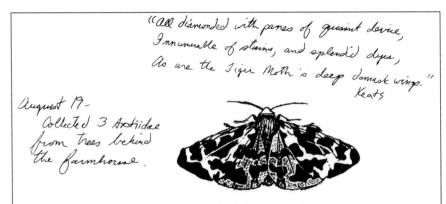

"all diamonded with panes of quaint device,
Innumerable of stains, and splendid dyes,
As are the Tiger Moth's deep damask wings."
Keats

August 19 –
    Collected 3 Arctiidae
from trees behind
the farmhouse.

Fig. 6.5 Field journals may combine written descriptions with sketches, poetry, and other personal expressions.

learned so far about themselves, their places, and journal keeping, and to set goals for the remainder of the semester. She asks them to consider what approaches have worked, what new approaches they would like to try, and what else they want to know about the place. At the end of the course, students write a final entry to critique whether they have accomplished their stated goals and what they have learned.

Rous grades the journals midway through the semester as a progress report and again at the end of the semester. She bases her grades on the consistency and quality of effort. Criteria include full entries turned in each week, entries demonstrating in-depth observation and reflection, and efforts to try a variety of journal-keeping approaches.

Students also give an oral presentation about their site at the end of the course. Their presentations include:

- Reflections from the self-evaluation of the journal-keeping process.
- Excerpts from their journals.
- Information they have collected from documents or interviews about soils, plants, animals, and history.
- Environmental problems associated with their site and possible solutions,
- Visual representations using photos, drawings, maps, or videotapes of their land (Rous 2000).

From samples of student journals and their reflections in their final entries, Rous has found ample evidence of how affection for a piece of land starts and grows through the journal-keeping process.

6.7 Environmental theater

Dramatic performances appeal to people's emotions and use the traditional technique of a story to engage an audience. They also can incorporate music and dance into the performance to appeal to a broad range of people.

The theater group Human Nature has produced two musical comedies to help audiences examine the relationship between humans and nature. They designed their first play, *Queen Salmon*, to address the demise of king salmon in their Pacific Northwest watershed, where habitat has been destroyed by shortsighted forestry, fishing, and planning policies. The play became a rallying point for their local community and instigated the formation of the Mattole Watershed Alliance, charged with changing logging standards and sport fishing regulations. David Simpson, the playwright, said *Queen Salmon* was supposed to entertain, yet, at the same time, he notes, "The dramatic form is a great vehicle for presenting the many conflicts and points of view of people in our community interacting with the natural resources. Theater can create a scenario to show people how they can live and work and even laugh together" (Jacobson 1999).

A theater performance was used in Oregon, to introduce high school students to the complexity of natural resource issues. The play, *Saving Eden Creek*, allows audiences to view—through the safe environment of storytelling—fictional characters espousing a variety of land ethics. Loggers, environmental activists, developers, homeowners, and a threatened squirrel are represented. Students can either read or act the script and afterwards audience members and players can join in a conversation about the differing perspectives presented (Creighton *et al.* 2004). In experiencing such a performance, people come to a better understanding of their own values, beliefs, and actions in relation to a complexity of environmental issues.

6.7.1 Planning

Staging a performance is a major undertaking. Environmental educators may enlist the help of a local drama club or they may engage an entire community. The goal for the performance, time, budget, and expertise available, and the intended audience will dictate the scope of the program. A dramatic performance entitled *The Plague and the Moonflower* was composed in the United Kingdom by Richard Harvey and Ralph Steadman. It was staged in Armidale, Australia, to raise awareness about environmental degradation and the need for people to behave in more sustainable ways (Curtis 2003b). The pageant combined music, drama, and poetry to portray a story of the destruction and protection of nature.

The planning for staging *The Plague and the Moonflower* in the small community of Armidale took over a year and involved 170 performers and crew. The Armidale symphony orchestra, Armidale Musical Society Choir, a folk band, and two dance groups (which included dancers from local high schools) participated in the event. The involvement of so many groups and individuals facilitated advertising the performance throughout the community by using the networks of the many organizations.

6.7.2 Implementation

Careful planning and community participation resulted in a well-attended event. Two performances played to a total of 1000 people.

The Plague and the Moonflower additionally incorporated other art forms into the event to enhance the audience's experience. At the beginning of the performance, the audience was confronted with images of people's discarded rubbish. They walked through a foyer with litter and plastic lining the walls and part of the floor. For contrast, an art exhibit celebrating the beauty of the natural environment was on display next to the foyer. During the performance, images and photographs of the environment were projected above the orchestra. The performance itself included a variety of musical styles, dances, and drama. Collectively, the work extolled the beauty of nature, reflected on the impacts of industrial society, and raised the hope of living in a sustainable manner in the future.

A portion of the ticket sales was donated to the Armidale Tree Group to fund their woodland education center. The education center would later be the site for displaying the concert's art exhibition. Additional art contributions were made by school groups that, with help from the local office of Greening Australia, created illustrations of native flowering plants to depict native biodiversity (Curtis 2003b).

6.7.3 Evaluation

Assessment of environmental performances must consider audience numbers, their satisfaction, and potential shifts in attitudes or increased knowledge. Curtis (2003b) conducted in-depth interviews with participants and members of the audience at varying times after the performance. He asked why they came to the performance, which performances they saw, what was most memorable for them, what messages they gleaned from the event, how they felt immediately after the performance and at the time of the interview, whether they would attend again, what effects the event had on their beliefs or behaviors, and how they would improve the production.

Results indicated that audience members were strongly moved by the performance and identified with the environmental message of the work. Although the numbers interviewed did not allow extrapolation to the larger audience, the results provided evidence that the drama seemed to engender a strong feeling for the environment and appreciation of their community. The event expanded the audience for environmental awareness. Many audience members were regular concertgoers, not people necessarily attracted to environmental content. The musicians and choir members reported being exhilarated by the performance. As a result of performing the work, participants' refined their opinions about environmental issues (Curtis 2003b).

The goals of the performance were to increase respect and appreciation for the natural world and engender disapproval of environmental degradation. Many interviewees mentioned being inspired and uplifted by the performance. Specific behavioral changes were not advocated by the play so it was not possible to measure direct impacts on subsequent activities of audience members. Some respondents mentioned, however, that they were reminded to reexamine how their consumerism and waste was affecting their environment.

Many communities are divided by economic and philosophical differences. Families may find participating in theatrical activities less controversial and more appealing than joining an environmental group. In this way, theatre can help bridge social or intergenerational divisions and bring people together to work on environmental problems.

6.8 Music and the environment

Music can play a role in conservation education because of its ability to attract the attention of an audience and to foster a positive attitude toward the environment (Morgan 2001; Ramsey 2002). Music can connect people with nature. An event organized by the Archie's Creek Afforestation Group in Gippsland, Australia, uses music to attract their target audience: teenagers. The musical event was attended by hundreds of youth volunteers who planted corridors of trees on farmlands to the accompaniment of loud dance music. In this way, music was able to attract a large audience, and create a fun, social event to provide a lot of hard labor for the land (Curtis 2003b).

Educators can introduce music through multimedia presentations, musical performances, and by enjoying natural sounds in the outdoors. Music can be used to explore issues associated with preservation or destruction of the environment. People can learn through melodies and lyrics (Turner and Freedman 2004). As music producer Darryl Cherney said, "Music is sugar-coating for the truth pill. It reaches the heart, and usually the heart is more open to new ideas than the mind" (Jacobson 1999).

Music can invoke emotional responses. Children have a natural affinity for music and rhythm, suggesting it could be a powerful technique when incorporated into educational programming about the environment. Educators have successfully engaged students with a musical experience and piggybacked the learning of an environmental topic on this successful hook. Researchers embedded learning objectives within a musical activity that youngsters were already physically and often emotionally invested in, and found they were more likely to remember the content of the lesson (Lenton 2002). A group of Canadian teachers and musicians created a CD entitled *Passengers*, with songs about environmental stewardship. The lyrics were written to encompass the elementary environmental science curriculum. This integration of science themes with music allowed students to have fun singing while rehearsing the learner outcomes specified in their curriculum. Teachers were happy to check off their program requirements; students derived additional benefits of a community singing experience, such as boosted self-esteem, increased confidence, and social connection with others (Lenton 2002).

Musical intelligence is one component of Gardner's Theory of Multiple Intelligences (Chapter 2). In his book, *The Disciplined Mind*, Gardner (1999) suggests that teachers incorporate more integrated skill sets, including musical and kinesthetic intelligences, into learning activities. Using complementary intelligences in conservation education would foster environmental literacy for a greater variety of participants. The emotional benefits would help learners internalize lifestyle approaches to lessen their impact on the earth (Lenton 2002).

A number of different techniques exist for integrating music into a classroom or outreach program. Many popular singers have recorded songs that describe environmental problems, such as singer Billy Joel's "Downeaster Alexa." His lyrics describe the declining fishery in the Northeast United States and its impact on the life of a commercial fisherman. The song is the hook that engages students' interest in an aquatic resources education activity.

A variety of songs can be played from a CD or tape and the audience can listen for key words, or can follow along with a lyrics sheet. Teachers may ask students to act out particular parts of a song. For example, in a song that explores marine mammal natural history, kids can listen for the word "whale" and then act out a whale spouting by shooting their arms up with a whooshing sound. Children listen with focused attention for their next cue. The audience can accompany the music with physical actions and sound effects. After the song, a variety of language arts exercises can reinforce the concepts from the song. The emotional and creative activity of listening to, singing, or acting/reacting to the song helps stimulate the audience to learn (Lenton 2002).

Participants can write additional lyrics to songs, or can be encouraged to write their own songs or music to explore environmental topics and better reflect on their own understanding and feelings about the issues (Orleans 2004). Students can make a spoken-word recording of a poem or song they have composed with a simple percussion background. More musical students can collaborate to create a school ecology song for World Environment Day or some other celebration. Local musicians may be willing to participate in these activities and add their talent to songs that celebrate everything from biodiversity to recycling and watershed protection.

Songs are a traditional way in many cultures to pass on folk tales. They have been used to build awareness about a variety of environmental problems. In Tripoli, Lebanon, a campaign to improve the collection and disposal of solid waste made use of a catchy jingle. Tripoli, with a population of a half million, has only 225 workers for cleaning and waste collection in the city. A preliminary planning assessment revealed that people throw their waste on the streets and out their windows. The municipality wanted residents to bag their waste and place it in cans for collection. A song addressed to housewives, who often listen to the radio while doing housework, was broadcast on local and national radio stations (Mehers 2000). The song repeated simple slogans to prompt compliance, such as:

"Good morning Madame, the cleanliness of your house goes well with the cleanliness of your street."
"Don't throw your waste out of the window. Put your waste in well-closed bags."
"Respect the hours fixed by the municipality."

A catchy melody made the messages hard to forget.

6.8.1 Singing a conservation message

The Bahamas National Trust launched a campaign to protect the several thousand remaining Bahamas Parrots that survive on two islands in the Bahamas. A primary goal was to establish a national park to protect the Caribbean pine forest where the parrots nest on Abaco Island. The campaign had many facets, from a human-size parrot named Quincy that visited schools to the use of songs to attract a wider audience and to carry the conservation message out into the community.

6.8.2 Planning

During the early months of the Bahama Parrot campaign, coordinators Monique Clarke Sweeting and Lynn Gape and the Bahama Parrot Conservation Committee met with a number of local musicians. Musicians were invited to lend support to the program by donating their time and talents. Several songs were produced including a children's song for use in schools, and a rap song and music video for broadcast on radio and television for a broader audience.

Clarke worked with the Ministry of Education to plan visits to schools throughout the targeted islands. They designed a parrot costume of their mascot, Quincy the Parrot, for the coordinator to wear for school visits. Funding to record musicians singing the rap song and to develop the music video was obtained through donations from a regional bank, other businesses, and the Audubon Society.

6.8.3 Implementation

The project coordinators visited over 60 primary schools reaching more than 26,000 students on the four target islands—New Providence, Grand Bahama, Abaco, and Great Inagua. The parrot song was printed on sheets of posterboard and pinned to a blackboard at each school. Children were asked to read the words out loud. The song's first few verses were played using a cassette recorder and the kids were invited to listen to the music to get its beat. The children then joined in singing the parrot song with the tape. The coordinators encouraged the children to sing louder with the challenge that students sang with more enthusiasm at the previous school visited. Once students could sing the words, they were invited to clap and do a "parrot dance" as well as sing. Quincy the Parrot danced to the song with the students and teachers, much to their amusement. By singing the tune over several times, the information about the parrot was easily remembered by the child through an enjoyable format. During the presentation, Quincy also presented parrot badges to students who answered questions correctly at the end of the presentation. This reward kept the students attentive and provided a tangible reminder of the experience.

Project coordinators Sweeting and Gape additionally wrote the lyrics for a rap song about the Bahama Parrot (Box 6.6). This was a popular form of music in the Bahamas and provided a vehicle for including a lot of information about the parrots in an enjoyable, entertaining manner. Local rappers TMC rallied to the cause, providing music and talent for a recording. An inaugural version of this song was

Box 6.6 A campaign to protect the Bahama Parrot included the creation and production of several songs. This rap song, written by Bahamas National Trust project coordinators Lynn Gape and Monique Sweeting was recorded by a well known Bahamian rapper.

Quincy Rap

Save our parrot, it's only just
Our Bahama Parrots, we must we must
Save our parrot, we must we must
Our Bahama Parrot, it's only just

The Bahama Parrot, Quincy needs a park
Quincy needs protection
Give it from your heart
The Bahama Parrot, Quincy needs a park
Quincy needs protection
Give it from your heart

Our national bird, we all know
Pink and pretty Flamingo
Bahamians tell me have you heard
The squawks of our emerald green bird
Bright and beautiful watch 'em fly
Junkanoo Parade up in the sky

Bahama Parrot I tell you so
Been here from long long time ago
Bahamas! Help our parrots please
Don't cut down our pine forest trees
Bahama Parrots help them to survive
We must we must keep them alive

Columbus 1492 so many parrots
Now so few
Lucayans had no emeralds green
Bahama Parrots were given to the Queen
Today our parrots only seen
Abaco, Inagua in the trees they preen
If you were a parrot how would you feel
Chased from your home by a quick land deal?
Quincy keep on flying!

Parrots in Inagua live in trees
Mahogany our own species
Abaco Parrots don't like heights
They nest underground, out of sight
Bahama Parrot mates for life
The pair's devoted there is no strife
Chicks are born featherless and blind
Without parents no food they find

Bahama Parrots, Quincy needs a park
Quincy needs protection, give from your heart
Bahama Parrot, Quincy needs a park
Bahama Parrot, Quincy needs a park
Give, give, give from the heart
Quincy, Quincy give Quincy a park

Before the birds are fully grown
People sell what they don't own
Hunters shoot them people eat them
All around, we mistreat them
Our parrots decline, numbers few
Less than 3,000 that's true
A real treasure in the sun
Save them before there is none

Put a national park on Abaco
Like Great Inagua, that's how to go
Keep our parrots flying free
For unborn future eyes to see
All must pay attention please
Guard this endangered species
Stop illegal parrot-taking activities
Help your National Trust
And Land Surveys-Forestry

Save parrots' home for our children
Their pine trees give us oxygen
Have compassion, show concern
Parrots depend on what we learn
Everybody on this land
Come and give a helping hand
Bahama Parrots help them survive
Or soon there'll be none left alive

produced by Tecnol Studios and was used in local discotheques. The song was re-recorded by Tony (The Obeah Man) McKay—a well-known musician and recording artist. He recorded the Bahama Parrot Rap and featured it on an album. McKay additionally agreed to donate part of the royalties received from this album to the conservation of the Bahama Parrot.

The lyrics skillfully included many pertinent facts about the Bahama Parrot, including the coloring of the parrot, its habitat requirements, its uniqueness, status, and the threats it is facing. The song also prompted the audience with a specific message about the solution to the conservation of the parrots—the establishment of a national park on Abaco Island.

The Bahamas National Trust provided taped copies of the rap song to radio stations in New Providence and Grand Bahama. It also was used in the production of a music video. With sponsorship from local businesses, a professional production company filmed the music video over a 5-day period and included many scenes with McKay, Bahamian school children, and some captive Bahama Parrots. Copies of the video were sent to schools, and the video appeared as a filler on local television channels as well as a feature on several Bahamian TV-magazine shows. Local movie theaters aired it before the main presentation.

6.8.4 Evaluation

The role of music was not evaluated as a separate technique from the overall outreach campaign to save the parrots. However, the coordinators found the songs and music tremendously popular among every group they worked with. The campaign resulted in almost a thousand school children writing to the Minister of Agriculture to encourage the establishment of the Abaco National Park for the protection of the Bahama Parrots. The legislation was approved a year later. A new park encompassing 8200 hectares of Caribbean pine forest was established for the Bahama Parrot.

6.9 Summary

The arts—including painting, photography, literature, theater, and music—offer an emotional connection to people. Using art for conservation can help attract new audiences, increase understanding, introduce new perspectives, and create a dialogue among diverse people. The arts can be used to inspire people to take action. Visual arts have helped bring the loss of wilderness and the beauty or destruction of nature to public attention. Photographs of the Arctic National Wildlife Refuge were used in a campaign to highlight the tremendous biodiversity of this seldom-seen landscape. The pictures helped spur policy-makers to attempt to protect the area from oil development.

Art can attract new audiences to conservation. A series of field trips combined with simple art projects attracted a new audience to a program in New York City sponsored by The Nature Conservancy. Participants gained a new understanding of nature in their urban environment as a result.

Some art increases concern for the environment. It can illustrate or describe environmental problems in ways that arouse emotions. The dramatic production *The Plague and the Moonflower* stirred Australian audiences to contemplate their consumption patterns and the amount of solid waste they produced. Songs about the plight of the endangered Bahamas Parrot awakened school children's concern for its conservation through simple lyrics and an engaging presentation.

Art can open people's eyes to different perspectives. An art exhibit on forests brought contentious forest management issues to public attention in the Pacific Northwest United States. Audiences at the shows reported that it increased their awareness about the complexity of forest issues. Sponsors found that art helped to attract new and more people to learn about forests, in ways that their typical outreach programs did not.

Planning art activities requires reaching out to artists and the art community, audiences with whom scientists and educators may seldom interact. Implementing programs involving the arts may take a cast of one, dozens, or hundreds. Although participating in an art event or singing songs may not directly stimulate changes in environmental behaviors, it can make participants more open to information or engage their positive feelings in support of a cause or organization. Conservation problems require creative solutions. It makes sense to access multiple ways of knowing and caring about the world in order to take care of it.

Further reading

The Arts And Restoration: A Fertile Partnership? by David Curtis (2003). *Ecological Management and Restoration*, 4 (3), 163–169.

Teaching With The Brain In Mind, by Eric Jensen (1998). Association for Supervision and Curriculum Development, Alexandria, VA.

The Sierra Club Nature Writing Handbook: A Creative Guide, by John Murray (1995). Sierra Club Books. San Francisco, CA.

Literature and the Land: Reading and Writing for Environmental Literacy, by Emma Rous (2000). Boynton/Cook Publishers, Portsmouth, NH.

How Emotions Affect Learning, by Robert Sylwester (1994). *Educational Leadership*, 52 (2), 60–66.

Connecting classes and communities with conservation

Ultimately, success in conservation education and outreach will be measured by the biodiversity conserved worldwide, yet success hinges on our ability to link environmental conservation with the quality of life of individuals, groups, and institutions in our communities. This chapter highlights techniques to help conservation educators use real issues in a community to achieve conservation goals. Conservation education techniques, such as service-learning, issue investigation, and project-based learning were designed to involve students, teachers, and community members in finding creative solutions to issues such as backyard habitat restoration and solid waste management.

Other techniques such as community-based research, citizen science, and mapping target adults, as well as audiences in schools. Community-based research has resulted in concrete environmental and social changes, such as safe drinking water for a community in Kentucky,. Citizen science involves public participation in research, like field monitoring, such as the Christmas Bird Counts, which have monitored bird populations worldwide since 1900. Local communities from Botswana to Bolivia have used mapping to discern gender differences in access and use of natural resources. All six techniques *can* apply to both students and adults if the goal is to impact both learning and conservation in the community.

The techniques in this chapter share many similar characteristics and often overlap (Box 7.1). Students immersed in project-based learning could focus on a service-learning project. Fitting existing conservation education programs into neat descriptive categories can be difficult because some programs reflect several techniques described in this chapter. For example, the environmental education program, GREEN, (Global Rivers Environmental Education Network) is an action-oriented program in which students from more than 130 countries monitor the water quality in their watershed (Stapp *et al.* 1996). Initiated by educators, GREEN has been described as "action research" (a foundation of community-based research) and "community-based problem-solving" (often a source for project-based learning). Recognizing the intended characteristics and goals of your conservation education program is more important than making your program reflect a single technique.

Box 7.1 Techniques that help conservation educators use real issues to connect schools and communities to conservation.

Technique	Purpose
Service-learning	Applies academic learning to community needs by involving learners in planning, service, reflection, and celebration of environmental action
Issue investigation	Involves students in defining, investigating, and analyzing an environmental issue and developing an action plan
Project-based learning	Uses projects as an organizing framework to engage learners in addressing a real-world problem with no known solution and presenting their findings in groups
Community-based research	Involves research conducted by and for communities with a goal of social and environmental change
Citizen science	Uses the general public, teachers, or students in any or all steps of a scientific research project
Mapping	Allows individuals or groups to create visual representations of resources, community, and regions. Can document either (1) knowledge and perceptions or (2) scientific data

Although similar, these techniques differ in their historical roots, variation in degrees of community participation, and goals (Box 7.2). The thread that holds the techniques in this chapter together is the capacity to increase attitudes, knowledge, and skills and affect environmental and social change. With these strategies, conservation education *can* make a difference in people's lives and their environments.

7.1 Service-learning

Service-learning is a form of experiential education where learning occurs through action and reflection as students apply what they are learning in a classroom to community needs (Eyler and Giles 1999). Service-learning has also been defined as a "teaching and learning approach that integrates community service with academic study to enrich learning, teach civic responsibility, and strengthen community" (Fiske 2002). A critical characteristic of service-learning is that the service impacts both the community and the students.

As opposed to a volunteer program or a service requirement for graduation, service-learning connects objectives for *both* service and learning. Effective service-learning involves four key elements based on the experiential learning cycle (Chapter 2): planning, service, reflection, and celebration (Box 7.3). The planning and reflection is what makes service-learning different from volunteering (Easton and Monroe 2001).

For example, in one college biology class, students participate in volunteer training at a local nature center and then work as volunteers, while also receiving

Box 7.2 Historical roots, goals, and initiators of the techniques in this chapter.

Technique	What are the historical roots?	What are the primary goals?	Who usually instigates?
Service-learning	Experiential education Community service	Impact on community needs and student learning	Teacher Community group Students
Issue investigation	Goals for curriculum development in environmental education	Student learning	Teacher
Project-based learning	Experiential education Neuroscience and Psychology Constructivism	Student learning	Teacher Students
Community-based research	Paulo Freire's work Participatory research Action research Service learning	Collaborative research Meeting community-defined needs Social and environmental change	Community members Researchers
Citizen science	Field research Public participation in science	Collaborative research	Researchers
Mapping	Participatory research methods	Data collection	Researchers Facilitators

certification in several environmental education curricula (Haines 2003). At another high school, students collect trash from a streambed, a service that aids the community. But they also analyze the sources of trash and present the findings to neighborhood residents. Their service-learning experience builds skills in analyzing data, understanding of water pollution, and communicating science to the public (NSLC 2004).

In summary, service-learning:

- Connects to academic content and standards.
- Involves students in helping to identify and address community needs.
- Benefits the community and the students by combining service and learning.
- Can be used with any subject area.
- Works with all ages.

Service-learning presents a win–win opportunity for conservation education organizations, students, teachers, schools, and communities (Figure 7.1). Many pressing community needs reflect environmental problems, such as air and water quality, environmental justice issues, habitat degradation, and urban sprawl. The

---

Box 7.3 Effective Service-learning includes four steps of planning, service, reflection, and celebration (Adapted from Cairn, R. (2003). *Partner Power and Service Learning: Manual for Community-based Organizations to Work with Schools.* ServeMinnesota, Minneapolis, MN.)

Planning (Orientation and Training)	Meaningful service	Reflection	Celebration and demonstration of learning
• Mission, goals, philosophy, history of organization • Team-building among participants • Skills needed to perform service • Needs of those served, including their social or environmental context • Problems-solving skills • Guidelines for safety and confidentiality	• Service meets real human or environmental needs • Organizations, educators, and students all involved • Partner organizations and educators state clear purpose and goals • Service engages and challenges students • Experience values diversity • Partners evaluate program and work to improve it	• Occurs before, during and after service • Includes description, analysis of situation, and possible future actions • Explores solutions to problems that arise during service • Examines social/environmental causes and solutions • Requires students to apply classroom learning • Ties assignment into curriculm	• Students receive concrete feedback from site supervisors, teachers, and community members if possible • Students report what they learned • Students receive acknowledgement for their level of contribution • Partners celebrate the outcomes and impacts on the environment

---

majority of conservation groups and community-based organizations lack adequate staff or resources to address their priorities. Service-learning can help meet these needs while enhancing student learning about the environment.

7.1.1 Planning

Why plan a service-learning project for a conservation organization or school? Identifying the benefits of service-learning is an important piece of planning, as well as looking at potential constraints, such as funding, transportation, connection to the curriculum, and time. The benefits of service-learning are unique to each of the partners involved. Through service-learning, students experience the hands-on application of classroom learning and networking with conservation staff, while schools benefit through access to community resources and enthusiasm from students (Cairn 2003). Conservation organizations and agencies can gain service from volunteers, positive publicity in the community, and access to resources of schools and universities.

Fig. 7.1 Service-learning projects, such as the removal of invasive species from community parks, can involve mutual benefits to conservation organizations, students, teachers, schools, and communities (photo by the Service Learning Office at Warren Wilson College, Ashevile, NC).

The components of planning a service-learning project include: (1) exploring community needs, (2) identifying partners for your school or conservation organization, (3) setting objectives for learning and service, and (4) organizing logistics. The details of these steps will differ slightly for staff members with a conservation organization or for teachers and students initiating the planning process. This section addresses strategies for service-learning from the perspective of both schools and conservation groups.

Exploring needs

Conducting a community needs assessment will help identify feasible projects in the local area. As students are citizens of their community, they have a unique vantage point to help determine the environmental needs in a neighborhood or community. With support from teachers or conservation educators, students can conduct observations and interview local residents for their input on issues including household wastes, wildlife, transportation, water quality, air quality, trees, greenspaces, gardens, and natural disasters (Dobbins and Pitman 2001). Students can consider issues of feasibility, environmental benefit, educational value, available resources, and connection to academic content. The youth action guide entitled, *Give Forests a Hand* (Easton *et al.* 1996), gives students the leading role in exploring needs and planning the service-learning project.

For conservation organizations, a needs assessment can explore ways that students could help address organizational and community needs, such as the development of a docent program that has been pushed down the "to do" list for years at a local nature center. Staff can help identify activities that have been neglected due to lack of human resources or time, and brainstorm roles for students to take part in these activities.

Identifying your partners

Partners to schools should include organizations or individuals that can provide financial, educational, or technical support. For example, in Mississippi four local schools worked on a service-learning project to provide more greenspaces and play areas at school sites. Their partners included the Mississippi Department of Forestry, Mississippi Project Learning Tree (PLT), the federal Bureau of Land Management, the Jaycees (a professional group), the Parent Teacher Organization, a garden club, the Plant-a-Tree foundation, and a hardware store (Dobbins and Pitman 2001). Like the Mississippi project, schools can partner with conservation groups, governmental agencies, departments of natural resources, forestry, wildlife, service clubs, television networks and radio shows, businesses, neighborhood associations, youth groups, and educational organizations in their area.

After identifying a potential project for the schools, write a letter to possible partners with an outline of a plan, goals, and what type of support you are seeking. In your letter, ask to schedule a meeting to discuss the collaboration. The organizing teacher and several students should attend the initial meeting. If students and teachers are brainstorming a project with a partner, consider asking the representative to visit the classroom and present his or her ideas for projects. Be sure to write a letter of thanks after your meeting and follow-up by e-mail or telephone to secure the details of the partnership (Dobbins and Pittman 2001).

As a conservation agency or nonprofit, your first contact with a school may be a teacher or students. Think about individual contacts you have that could get your foot in the door. Brainstorm how the service-learning project could benefit the teacher and students as they cover the required academic standards.

Setting objectives for learning and service

Define objectives for both learning and service with a clear connection between the two. Remember that the service experience allows students to apply academic content in real-life settings. As an educator, communicate these objectives to your partners, so they can give the students feedback during evaluation. One challenge is that service-learning projects that start with good intentions can become mere volunteer programs without these defined objectives. Communication of these objectives from the beginning can overcome this barrier.

For conservation professionals working with schools, set clear expectations for students engaged in service-learning. Be specific about the number of hours expected. Research shows that service-learning programs that involve students in 30 h or more have an increased impact on student learning (Cairn 2003). Also insist on regular oral and written reflection, one of the key elements of effective service-learning. If possible, develop regular monthly or yearly projects to involve classrooms over time and build a sustainable service-learning program.

Organizing logistics

Advance planning of logistics increases the success and helps overcome inherent challenges of the service-learning experience. Begin by establishing a timeline for the project. Who will do what and when? Students can plan major milestones and tasks for the service-learning project, in collaboration with partners. Identify specific roles and outcomes and put them in writing, so expectations are clear. Gain support from any decision-makers in the participating organization or school by involving them early in the project. Often presentations by students are effective methods for harnessing support. Also identify methods of transportation for students at this stage in planning.

Develop a list of potential sponsors that could donate time, money, or publicity to the cause. Next, strategize about channels to advertise the service-learning project, through news releases, public service announcements, posters, or a mayor's proclamation. Spread the positive news about your plan, and you may find other institutions that want to support you. Lastly, develop a budget and raise any funding necessary for the service-learning experience (Dobbins and Pittman 2001). Granting opportunities through Project Learning Tree's GreenWorks! fund environmental education projects that combine service-learning with community action and forestry in the United States.

From the standpoint of conservation agencies, you may be responsible for planning the training of students. Consider who will supervise the students, how you will ensure their safety, and the methods for communicating standards of dress, behavior, and confidentiality (Cairn 2003).

7.1.2 Implementation

As you implement the project, adjust the timeline as needed and make sure to document the process by keeping a journal, taking pictures, and recording reactions from students, partners, and community members. Also continue to seek media coverage by inviting government officials like the mayor or city council members to the project or asking environmental experts to speak to your group. If you organized an on-going project, like a stream restoration, take pictures of the location before the project and invite the local media to a celebration at the completion of the project.

Working with students and schools may present new challenges for conservation organizations. Remember that teachers spend most of their day in the classroom and have little time to communicate about the project, so be persistent as you establish a relationship with teachers and the principal. Determine the best way to communicate with teachers, whether through e-mail, school phone, home phone, or notes in the school mail box. Emphasize to teachers how the service-learning project can help students, since that is the first priority for the schools (Cairn 2003). See Chapter 4 for more tips on collaborating with schools. If your conservation organization is working with youth for the first time, remember that treating young people with respect is a major step toward earning their respect. When giving students responsibility in service-learning projects, ask for their creative input early, give them space to make decisions, and document their responsibilities (DesMarais *et al.* 2000).

One organization, Campus Compact (2005) provides resources to colleges and universities implementing service-learning. Their Web site includes program models, databases, grants and fellowships, training resources, and service-learning syllabi. One example is "Neighborhoods and Watersheds," a senior capstone course that involved students in a real-life proposal to develop a resource center at a community-supported farm.

During the course, students worked in teams to determine the costs and feasibility of the center, develop a long-term vision, conduct a survey of residents and businesses, design environmental education tours, prepare an exhibit, and initiate a neighborhood fair. They also presented proposals for watershed protection and conservation to business partners and city representatives. The assignments for reflection included individual reflection in journals, group reflection in a team activity log, and class discussions.

Reflection is an important aspect of implementing any service-learning project, as students evaluate their service and its impacts on themselves and the community. Effective reflection is periodic, guided, allows for feedback, links to academic content, and helps clarifies values (Blanchard 2004). Options for reflection include small group discussion, scrapbooks, learning logs, presentations, and research papers (Box 7.4).

The scope of a service-learning project can range from a conservation organization inviting students to help remove invasive species to students planning their own environmental service-learning project in the community. The six models of service-learning reflect the diversity of approaches to integrating service into a curriculum (Box 7.5). Each model involves learners in the same steps of planning, service, reflection, and celebration.

7.1.3 Evaluation

Especially for on-going service-learning, evaluation is a process of collecting data to assess impacts, but also determine directions for improvement. An evaluation of three national service-learning initiatives in the United States compared the impacts of service-learning on middle school and high school participants with a comparison group (Melchoir and Ballis 2002). The evaluations of the Serve-America, Learn and Serve Program, and Active Citizenship Today (ACT) all used the same design of pre- and post-program surveys, content analysis of school

Box 7.4 Methods for effective reflection of service-learning can include speaking, writing, activities, and multimedia/performing arts (Adapted from Cairn, R. and Coble, T.L. (1993). *Learning by Giving: K-8 Service-Learning Curriculum Guide*. National Youth Leadership Council, St Paul, MN. www.nylc.org.)

Speaking

- Individual conferences with teacher
- Group discussions
- Oral reports
- Discussions with local experts or community members
- Public speaking
- Teaching material to other students or groups

Writing

- Reports
- Research paper or essay
- Daily or weekly learning log
- Evaluation of the program
- Self-evaluation
- Published article

Activities

- Research to understand project
- Recruiting other participants in project
- Role-playing
- Training other students or participants
- Planning for future projects

Multimedia/Performing arts

- Web site
- Scrapbook
- Slide show or multimedia presentation
- Painting, drawing, sculpture, collage
- Theatre, music, or dance

records, interviews, and observations. All three programs increased students' confidence in their ability to identify issues, organize and take action, work in groups, and commit to service-learning in the future (Melchoir and Ballis 2002).

Follow-up surveys of 72% ($n = 764$) of the 1052 students (both participants and comparison group) in the Learn and Serve evaluation revealed that one-time involvement in service-learning did not result in long-term impacts on attitudes and behaviors. Yet, students who continued to participate in service-learning projects the year after the program showed significant impacts on measures, such as service-leadership, educational aspirations, school engagement, and reduced alcohol use (Melchoir and Ballis 2002). The findings support the need for on-going service-learning programs, rather than one-time participation in service.

In the past, most evaluations of service-learning focused on impacts of service-learning on students, rather than on multiple stakeholders including teachers, community groups, and school administrators (Driscoll *et al.* 2003; Gelman 2003). The last 5 years, however, have seen an increase in evaluations that examined multiple constituencies of service-learning. These "multi-constituency" evaluations typically have assessed the impact of service-learning on students, teachers, the school, and the community. Students need evaluations from the community organization about their work, and the community groups need feedback from students about the efficacy of the service-learning project.

Box 7.5 Service-learning projects vary in scope and scale, reflected by these six models of service-learning. (Adapted from Heffernan, K. (2001). *Fundamentals of Service-Learning Course Construction.* Campus Compact, Providence, RI.)

---

1. *'Pure' service-learning.* These courses and projects send students into the community to serve without connecting the course to a specific academic discipline. The intellectual core of the course is service to communities.

2. *Discipline-based service-learning.* Students have a presence in the community and reflect on their experience using specific course content as a foundation for analysis.

3. *Problem-based service-learning.* Students work with community members or a community organization to understand a specific community need. Students work as "consultants" in the community. For example, botany students might identify invasive species and suggest removal strategies.

4. *Capstone courses.* Capstone courses require students to use knowledge gained throughout their studies in a discipline and combine it with service in the community. These courses establish professional contacts and synthesize knowledge and skills in the real-world context.

5. *Service internships.* Students work as many as 10–20 h per week in the community with on-going reflection. Unlike traditional internships, the service internship reflects a mutual benefit between the student and the organization or community.

6. *Community-based action research.* Students work closely with teachers or faculty members to learn research methods while exploring and promoting solutions to community issues.

One strategy for organizing your evaluation is a matrix that addresses: (1) What will we look for? (2) What will we measure? (3) What methods will we use to measure? and (4) What are our information sources? (Gelman 2003). For example, the matrix in Box 7.6 reflects an evaluation of the benefit of service-learning to students in a Portland State University project in Washington. Their evaluation matrix included data collection methods of survey, focus group, interviews, and reflective journals for the evaluation of the students' commitment to community service and personal and professional development. Students can help develop evaluation tools and collect data from different stakeholder groups for an evaluation as well.

If the original objectives of the project included an improvement in the local environment, document any changes through tools such as pre- and post-biological surveys and photographs. If community members were involved in the service, use interviews or surveys to solicit their input. Perhaps the service-learning project aimed to change local environmental behaviors, such as the number of households

Box 7.6 Sample matrix for evaluating impacts of service-learning on students (Adapted from Gelmon, S. (2003). How do we know that our work makes a difference? Assessment strategies for service-learning and civic engagement. In Campus Compact (eds). (pp. 231–240). *Introduction to Service-Learning Toolkit: Readings and Resources for Faculty.* 2nd Edition. Campus Compact, Providence, RI).

What will we look for? (concepts)	What will we measure? (indicators)	How will it be measured? (methods)	Who will provide the information? (sources)
Commitment to community service	Attitude toward involvement	Survey	Students
	Level of participation over time	Focus group	Community patners
	Plans for future service	Interviews Reflective journal	Faculty Service-learning coordinator
Personal and professional development	Changes in awareness of personal skills	Interviews	Students
	Communication skills	Observations	Faculty
	Self-confidence	Focus groups	Service-learning
	Leadership activities	Reflective journal	coordinator
	Impacts on environment and organization		

involved in backyard composting or urban recycling. Again, use pre- and post-project tallies to show evidence of an impact.

As a formative and summative evaluation tool for your service-learning program, consider the use of a published self-assessment tool to help document success and identify areas of improvement. Developed at the University of Minnesota, "Shumer's Self-Assessment for Service-Learning" (Shumer *et al.* 2000) allows you to conduct a quick assessment, in-depth analysis, and action plan for your service-learning initiative. The data from evaluation will help you improve the efficacy of service-learning for the benefit of students, the community, and the local environment.

7.2 Issue investigation

Conservation issues are rarely one-sided scenarios with simple solutions. Typically, issues, such as global warming, waste management, pollution, habitat degradation, environmental racism, and biodiversity conservation involve multiple stakeholders with diverse perspectives. Learning how to address environmental issues involves the ability to define, investigate, and analyze environmental issues and possible actions, as well as communicate with people who have conflicting values from your own.

The structured technique of issue investigation has been used primarily with middle school, high school, and college students to build knowledge and skills that can contribute to positive environmental behavior (Ramsey *et al.* 1981; Ramsey and Hungerford 1989). The framework for issue investigation and action training is based on six modules in the curriculum guide, *Investigating and Evaluating Environmental Issues and Actions: Skills Development Program* (Hungerford *et al.* 2003).

Students participating in an issue investigation begin by analyzing given environmental issues and their own environmental beliefs and values. They then identify local environmental issues, develop research questions, and conduct background research. After developing data collection tools, students collect and interpret data and develop strategies for environmental action. (Students using issue investigation develop strategies and skills for action, but do not always implement these actions).

Fourth- and fifth-grade students on the island of Molokai in Hawaii were researching solid waste management issues in their local area when the local waste dump announced a significant fee increase for waste pick-up and disposal (Winther 2001). The fee increase represented a financial burden for these Native Hawaiian families, and the students began investigating the issue of a bottle bill to reduce solid waste in their neighborhoods.

The students conducted background research and discovered the legislature had recently defeated a bottle bill. They spoke with state legislators and then developed their own research questions and a survey for community members. Their results from the survey revealed that the majority of residents supported a bottle bill. The students then presented their results to the state legislators, who reintroduced the bill and heard testimony from three of the students (Winther 2001). Ultimately, the bottle bill was passed in Molokai.

7.2.1 Planning

A first step in planning issue investigation involves reading the curriculum guide (e.g. Hungerford *et al.* 2003) and collecting articles and resources about environmental issues. Before their own original investigations of local environmental issues, students should build a set of skills, and practice these skills using examples and scenarios. You can find relevant examples using newspaper and magazine articles, Web sites, and your own knowledge of past environmental problems and issues.

A problem is any situation in which something valuable is at risk. An issue arises when two or more parties or "players" disagree about the solution to the problem (Winther 2001). To prepare students for identifying the players in an issue and the values that shape their positions and beliefs, consider the following values that may influence a person's choices about environmental solutions (Hungerford *et al.* 2003):

- Ecological: the maintenance of natural biological systems
- Religious: the use of belief systems based on faith
- Social: shared human empathy, feelings, and status

- Egocentric: a focus on self-centered needs and fulfillments
- Legal: national, state, or local laws
- Economic: the use and exchange of money, materials, and services
- Ethical/moral: present and future human responsibilities, rights, and wrongs, ethical standards

Given an environmental issue, such as logging in a protected area in Uganda, students would analyze the players and their positions, their belief statements, and underlying values. One stakeholder, or interested party, in this issue is the logging company whose position is that loggers should be able to harvest trees in the protected area. Their belief statement would be that timber has economic importance to both the companies and the local communities, so the underlying value is economic.

As a supplement to planning, workshops also are available for teachers or nonformal educators by certified trainers in "Investigating and Evaluating Environmental Issues and Actions" in countries like Argentina, Canada, and the United States. Educators also can attend an annual conference, hosted by the Center for Instruction, Staff Development, and Evaluation, such as the 2005 conference located in Molokai, Hawaii. Case studies (Chapter 5) for issue investigation and action training also have been developed for teachers and tested in classrooms. One such case study uses fishery issues and human impacts along the southeastern Gulf Coast and Florida Peninsula (Culen *et al.* 2000).

7.2.2 Implementation

These initial steps in implementing issue investigation are called "issue analysis," as these skills set the stage for the students' own investigations of local issues. Students should practice identifying and analyzing an issue by naming the players, their positions, beliefs, and values in the scenarios developed during planning. These scenarios become case studies (Chapter 5) for analysis by the students. Ensure that students also analyze their own beliefs and values about specific environmental issues, such as population control and hunting. A careful look at their own beliefs and values will enhance students' ability to perceive the values of others in real environmental issues.

Next, have students brainstorm local environmental issues they are interested in investigating. Use your collected resources, such as newspaper articles and Web sites, as material for students to read when developing their lists of issues. Students can conduct the issue analysis individually or in cooperative groups. Practice comparing the perspectives of information sources, perhaps by reviewing articles on different Web pages or newspaper articles written with conflicting political slants.

When students have identified an environmental issue, work with them to develop research questions. For example, what are the beliefs and values about the lack of public transportation in the province? What is the extent and location of illegal dumping sites in their county? Research questions should be open-ended, indicate a population or geographic area, specify the variables for measurement and any relationship between the variables, and involve a relevant environmental issue (Hungerford *et al.* 2003). The research questions then will direct the populations or area for sampling and the sampling method. Students may combine surveying a geographic area, such as the location of illegal dumping sites, with a questionnaire of knowledge and attitudes of government officials and residents.

Developing the data collection instruments involves consultations with the teacher or facilitator and a review of existing surveys and questionnaires for models. After collecting data, students should summarize their results and present the information in graphs or summary tables. From these analyses, student can draw conclusions and make final recommendations.

The data collected by students and their interpretations of the findings are used to inform their action plans. To begin, have students study environmental actions taken by both individuals and groups in local, regional, national, and global contexts. Four methods of taking action include (Hungerford *et al.* 2003):

1. Persuasion: used when someone or a group of people try to convince others that a certain action is correct.
2. Consumerism: involves buying or not buying something based on your philosophy.
3. Political action: refers to any action that brings pressure on political or government agencies or individuals. Political action can mean supporting political candidates or influencing officials through letters, petitions, e-mails, and phone calls.
4. Environmental action: involves responsible physical action taken with respect to the environment.

You should also discuss legal action with students, although this method requires adult involvement.

Next, help students develop an action plan and decide if they will execute the action and if so, on a local, regional, or national scale. One example of a student action includes informing state officials about public support for reintroduction of the timber wolf in central Wisconsin by fourth-grade students (Winther 2001). An important point is that students implement these action plans on their own volition, rather than the influence of a teacher or conservation educator. Again, students must develop their own action plan but are not required to implement the action strategies.

Some questions to guide students as they consider environmental action include (Hungerford *et al.* 2003):

- Is there enough evidence to pursue action?
- Are there legal, social, economic, or environmental consequences of this action?
- Do my personal values support this action?
- Do I understand the beliefs and values of other stakeholders about this issue?
- Do I know the procedures necessary to take this action?
- Do I have the skills and resources for this action?

Finally, have students present their findings and recommendations to the class and if possible, to a larger group, such as parents or the city council.

7.2.3 Evaluation

Each component of issue investigation should include performance objectives, which form the basis for evaluating the progress of students. For example, the curriculum guide asks students to analyze issues presented by identifying players, positions, beliefs, and values. An assessment could ask students to apply those skills to a new issue, such as a recent newspaper article about a local environmental issue. This formative evaluation measures progress of students while learning the content. The summative evaluation, on the other hand, assesses the original issue investigation and action plan implemented by students. An effective tool for assessment, a portfolio is a collection of student work throughout the issue investigation, which corresponds to the instructional objectives. Students can document their portfolio in Web-based format, a three-ring binder, or even a file folder (Hungerford *et al.* 2003).

To monitor the progress of students, use a task checklist as a formative assessment tool (Box 7.7). Another useful evaluation tool is a contract between the students and facilitator that documents the number of points and criteria for evaluating the students' work, including the quality of the research question, techniques used, quality of data collected, accuracy of conclusions and inferences, accuracy and organization of final report, and quality of presentation (Hungerford *et al.* 2003).

In addition to evaluating individual students, you should assess the overall effectiveness of your instruction. Did the students accomplish the objectives? Does the

---

Box 7.7 This task checklist is a formative assessment tool for issue investigation. (Adapted from Hungerford *et al.* (2003). *Investigating and Evaluating Environmental Issues and Actions: Skills Development Program*. Stipes Publishing, Champaign, IL.)

Investigation task	Due date	Teacher's approval
1. Selection of research topic	—	—
2. Search for secondary sources	—	—
3. Research question	—	—
4. Letters for information	—	—
5. Interview questions/format	—	—
6. Data collection plan	—	—
7. Instrument development	—	—
8. Data collection completed	—	—
9. Charts, tables, graphs	—	—
10. Conclusions, inferences, and recommendations	—	—
11. Action plan	—	—
12. Final report	—	—

instruction need modification? How motivated were the students? You can use reflective journals during the issue investigation for both the students and instructor to collect data.

Researchers also have used quasi-experimental designs to assess the impact of issue investigation and action training on variables influencing environmental behavior (e.g. Ramsey *et al.* 1981; Ramsey and Hungerford 1989; and Ramsey 1993). For example, a study using a modified pretest–posttest design revealed that eighth-grade students who participated in the training showed significantly higher knowledge about resolving environmental issues and stronger beliefs about their ability to affect the outcomes of such issues (Ramsey 1993). A similar study of seventh-grade students revealed that issue investigation and action training promoted responsible environmental behaviors (Ramsey and Hungerford 1989).

7.3 Project-based learning

Humans learn best when they perceive a *need* to learn (Newell 2003). In our jobs, families, and communities, we learn best when a task, situation, or project demands or attracts our intellectual and physical involvement. For example, a couple that buys a new home learns to maintain a garden using xeroscaping techniques. Similarly, wildfires that threaten a neighborhood incite residents to learn how to clear underbrush around their homes to prevent future damage or promote prescribed burning in natural areas.

Project-based learning draws on this innate ability of humans to learn when immersed in a real-life task of interest (Figure 7.2). Project-based learning uses projects as a primary means to learn basic skills, learn how to learn, and learn about interacting with the world (Newell 2003). The projects often emerge from a real-life context, addressing issues faced by students, the school, or the community.

Students involved in project-based learning typically (Rogers and Andres 2004):

- Plan their own project
- Face a problem without a known solution
- Design a process for reaching a solution
- Access and manage the information gathered
- Reflect regularly on the process and outcomes
- Help define criteria for evaluation
- Present findings from their project
- Use technology, if available, as an organizational and multimedia tool

The terms project-based learning and problem-based learning are sometimes used interchangeably, as both techniques immerse learners in concrete issues to build content knowledge and problem-solving skills. However, problem-based learning more frequently uses scenarios and role-plays in prescribed problems, such as those used in medical schools (Markham *et al.* 2003). These scenarios reflect the case study approach also used in law and business schools (Chapter 5).

Fig. 7.2 Project-based learning immerses participants in addressing a problem or question without a known solution (photo by U.S. National Park Service).

As a student-centered approach, project-based learning shifts the role of teacher to resource provider and facilitator. At Minnesota New Country School, teachers are called "advisors," and classrooms are set up like offices with workstations. A group of students visited a local nature center near the school to get ideas for projects in environmental studies and biology. The students found several deformed frogs on the grounds of the nature center, which became the basis for a class project. Upon returning to school, they put the information on the school Web page. They worked with the Minnesota Department of Natural Resources staff and university researchers to secure funding for environmental research, which continued with school collaboration for 6 years. Students at this school may have three or four projects underway at one time (Newell 2003).

Most schoolchildren have experienced the ubiquitous science fair project. Project-based learning, however, is *not* this type of add-on to the curriculum, but an organizing framework for learning. Project-based learning has its roots in John Dewey's experiential education (Chapter 2), as well as research in neuroscience and psychology that reveals how learners construct their own knowledge through past experiences, culture, and community (Markham *et al.* 2003). The project question or issue creates the need to know what drives learning.

The current need for adult learners who can interact in a rapidly changing world has influenced the growth of project-based learning. For many adults starting new jobs, the ability to work in teams and manage multistep projects is an indicator of successful job performance. Studies have shown that students engaged in project-based learning show a deeper knowledge of the academic content, increased motivation, and improved problem-solving skills, when compared with traditional instruction (GLEF 2001). Such skills prepare students as effective adult members of their workplaces and communities.

For conservation educators collaborating with schools, project-based learning presents a technique for engaging students in relevant projects that affect their daily lives and the environment. For teachers, project-based learning is a way to excite students about environmental issues of interest to them. While project-based learning often focuses on the schools, adults also are more effective learners when engaged in real-life projects. Conservation organizations can engage adults, such as members of a Kiwanis Club, in projects and enhance their learning about conservation strategies.

7.3.1 Planning

At Avalon Charter School in Minnesota, a student met several content standards in biology through a project restoring her backyard to its native habitat. She wrote a project proposal that included the tasks and activities, resources needed, and assessment rubric. She researched prairie habitats and then planted native plants and flowers (Newell 2003). At Monteverde Friends School in Costa Rica, a teacher involved children as young as first and second grade in project-based learning. She used the local cloud forest as a context for projects that taught skills ranging from mathematics to literacy.

Each individual educator may use a slightly different sequence of steps for planning project-based learning, but these guidelines from the *Project-based Learning Handbook* (Markham *et al.* 2003) provide a foundation for planning. Remember to involve the learners in these planning steps:

- Summarize the theme or main ideas for the project.
- Identify the academic standards students will learn from the project.
- Identify key skills students will gain from the project.
- Craft the "driving question"—the essential question or problem statement for the project.
- Plan the assessment by defining the products and artifacts of the project and stating the criteria for exemplary performance. You should define products for the beginning, midpoint, and end of the project.
- "Map" the project. Look at one major outcome for the project and analyze the tasks needed to produce a high-quality product. What do students need to know and be able to do to complete the tasks successfully? How and when will they gain that knowledge and skills? Draw a storyboard or a map of the project, with activities, resources, and timelines.
- Consider any modifications needed for special-needs students.
- Meet with other students, teachers, and resource people to refine the project design.

The planning process for New Country School involves a project proposal form completed by students and signed by parents and the teacher (Newell 2003). The proposal requires them to identify the title of the project, the topic of investigation, three questions they would like to answer, the importance of the project to the community or world, an outline of the project, a timeline of tasks, three different types of resources, and the educational standards that will be addressed. The students must review the proposal with a friend, parent, and teacher, and then the project planning team.

7.3.2 Implementation

Implementing project-based learning is when the fun begins, as students learn knowledge and skills through real-world interactions, rather than traditional lectures and memorization. As with project management for a job, project-based learning for students requires time- and task-management tools as organizational aids. Help the learners by providing them with a project checklist or weekly planning sheets to enhance efficiency and record progress (Box 7.8).

At the beginning of each week, ask students to complete a weekly planning sheet that documents what products and investigations the student will work on. The sheet also includes space for reflection at the end of the week. What did I learn this week from the project? For the entire project, the students should have their own copy of an implementation sheet to document the focus on their project, necessary tasks and due dates, resources needed, and how they will demonstrate learning (What? How? Who? and Where?).

Decide on methods for students to organize the data they collect from research, such as a project research log with citations, names of persons interviewed, Web sites and descriptions of information gained from each secondary and primary source. These tools can be simple research journals or a standard worksheet for recording data. The key to implementation is to manage the information collected in an efficient manner. Helping students organize their data using spreadsheets on the computer and project Web pages will aid them in the process (Markham *et al.* 2003).

In addition to research logs, you may decide to use learning logs and time logs during implementation. In a learning log, students document their specific goals or tasks, what they accomplished, their next steps, concerns or problems, and major learnings. A time log is simply a documentation of how much time students spent on the project and on which tasks. Students also can create products, such as Web pages, a journal of volunteer work, a demonstration or model, as a result of their investigations during implementation. For example, at Mountlake Terrace High School in Washington, teams of students in a high-school geometry class used project-based learning to design a state-of-the art energy-efficient high school for the year 2050. The students made architectural drawings and a model, created a budget, wrote a report and presented their work to architects who "judged" the projects.

Presenting the products, findings, and reflections is an important component of project-based learning. Before the presentation, students should outline what they expect the audience to learn from the presentation, their responsibilities during the

Box 7.8 Sample weekly planning sheet and project milestone sheet to use with students in project-based learning (Adapted with permission from Markham, T., Larmer, J., and Ravitz, J. (2003). *Project-Based Learning Handbook*. (2nd edition) Buck Institute for Education, Novato, CA.)

IMPLEMENTATION TOOLS

Student Weekly Planning Sheet

Project:_____ Student:_____ Date:_____

This week I will work on the following products:
1._____ Begin by myself Continue with _____ Complete with _____ 2._____ Begin by myself Continue with _____ Complete with _____
This week I will conduct the following investigations: 1._____ Begin by myself Continue with _____ Complete with _____ 2._____ Begin by myself Continue with _____ Complete with _____
Reflections at the end of the week: What did I learn?

PROJECT MILESTONES

Project: _____ Student:_____ Date:_____

Milestone	Due date	Completed
		❏
		❏
		❏
		❏
		❏
		❏
		❏

presentation, a plan for preparing for the presentation, what they expect to learn from giving the presentation, and what technology or visual aids they need (Markham *et al.* 2003).

7.3.3 Evaluation

Working with students to create assessment rubrics turns evaluation into a participatory process rather than the standard top-down grading process. Since the criteria for evaluation are transparent, assessment becomes less about teachers judging students and more about accountability and improvement. The skills you want to document in an evaluation rubric will depend on the mission and goals of your organization or school, as well as the educational standards.

At the Minnesota New Country School, the evaluation rubric covers three key areas: basic project skills (documentation of time and learning, tasks, project assessment, project quality, resources, ownership, and task completion); critical thinking skills (comprehension, competency, and context); and life performance skills (a set of skills ranging from mediation to organization) (Newell 2003). The rubric documents criteria for each competency. At this school, an assessment team sits with the student to quiz them at the completion of the project and evaluate the project using the rubric.

Other skills to document in evaluation rubrics include: accessing information, selecting information, processing information, composing a presentation, making a presentation, individual task management, individual time management, group task and time management, and group process (Markham *et al.* 2003). The assessments by teachers or facilitators are often combined with a student self-evaluation (Box 7.9).

7.4 Community-based research

Community-based research (CBR) is a partnership between researchers and community members who collaborate on a research project to address social and environmental problems and affect change (Stoecker 2001). CBR is research conducted by, with, and for communities (Sclove *et al.* 1998). Institutional support for CBR can range from a single research project in an academic department or conservation agency to the community research centers that support multiple projects in countries such as Canada, Israel, and South Korea. In the Netherlands, for example, universities conduct CBR to address specific research questions posed by grassroots groups, unions, and local agencies (Sclove *et al.* 1998).

Unlike traditional research, CBR is tied directly to the participation of community groups with a vested interest in the practical applications of research findings to their environment and quality of life. The Policy Research Action Group in Chicago, Illinois used CBR in energy conservation retrofits of more than 10,000 low-income housing units in the city (Sclove *et al.* 1998). The goal of a connected corridor, called Y2Y, from Yellowstone National Park in the United States to the

Box 7.9 This evaluation rubric for project-based learning assesses individual time management with sample descriptive criteria for the first element of the rubric. Add your own descriptors depicting expectations of what students should be able to do as they develop skills in time management. (Adapted from Markham, T., Larmer, J., and Ravitz, J. (2003). Project-based Learning Handbook. (2nd Edition.) Buck Institute for Education, Novato, CA.)

	Limited	Developing	Proficient	Advanced	Exemplary
Uses time effectively	Does not take action to use time efficiently		Uses time efficiently and completes most work within given time constraints		Prioritizes tasks, recognizes time constraints, meets deadlines, and uses time effectively
Estimates time realistically					
Establishes a schedule for completing work					
Stays on schedule					
Completes tasks in a timely manner					

Yukon in Canada is a transnational endeavor between scientists and surrounding communities. For the project, Canadian and American biologists have joined environmental organizations in CBR to promote ecosystem-based land-use planning and species protection (Krajnc 2002).

Community-based research in the Philippines addressed the crisis of a decreasing supply of quality water in the Manupali River watershed. Community members received training in water quality monitoring and analyzed the results of thousands of water quality samples, which revealed clear connections between water supply degradation and clearing of land for roads and agriculture. Local governments incorporated the community findings and recommendations into their natural resource management plan. With these findings, the community members took the lead and formed a non-governmental organization (NGO), whose president serves on the Natural Resource Management Council of the municipality (UN Department of Economic and Social Affairs 2003).

Community-based research is often described by other terms such as action research, participatory research, empowerment research, and popular education. Diverse historical and educational movements have influenced the growth of CBR, including:

- The popular education work of Paulo Freire's (1970) *Pedagogy of the Oppressed* that advocates education as a tool to affect social change.
- The participatory research and participatory action research arising from struggles in India, Africa, and South America and critiques of Western research.
- Action research by industrial psychologist Kurt Lewin who used research to increase worker satisfaction and productivity.
- Service-learning that sought more civic involvement in schools and universities (Stoecker 2002; Strand *et al.* 2003a, b).

For conservation educators, CBR is an exciting, innovative technique to build networks within a community, connect people's quality of life with the environment, and use research to create change. CBR reflects many of the same steps as citizen science described in this chapter. However, the focus on social and environmental justice and the higher level of public participation in CBR differentiates the two techniques.

7.4.1 Planning

The partners involved in planning CBR should have a clear understanding of the three principles of CBR (Strand *et al.* 2003a):

1. Collaboration between researchers and community members. The "researchers" may include faculty, scientists, or students. The community may include a diverse range of stakeholders, from business associations to women's groups.
2. Validation of multiple sources of knowledge and the use of multiple methods of discovery and dissemination.
3. A goal of action to achieve social and environmental change and social/environmental justice.

Given these principles, planning CBR involves more than simply developing a research plan. It involves organizing meetings, building relationships in the community, having informal discussions with community members, creating trust between partners, and brainstorming strategies for lobbying and building networks for change with research results. Community participation also drives the traditional steps of choosing a research question and identifying methods (Stoecker 2002).

Choosing a research question

Unlike traditional research, the researcher in CBR does not define the research question. Rather, building relationships with a community group or community members and understanding their goals for social and environmental change is integral to choosing a research question. Some CBR centers hold focus groups with community organizations or call for proposals to identify local research needs (Stoecker 2002). Individual faculty or researchers may find that a community group approaches them due to their expertise in issues such as water quality or wildlife management.

In Tennessee the Highlander Research and Education Center collaborated with the Yellow Creek Concerned Citizens (YCCC) of Kentucky, a group who challenged Middlesboro Tanning Company's disposal of hazardous chemicals. The community group's concern was the driver for the research question investigating contamination of the water source and a possible connection to the health of residents. With the support of researchers at the Highlander Center, residents conducted a health survey, videotaped illegal dumping, and tested the water quality for toxic chemicals from the tannery. The project resulted in a safe water line for the community, as well as a settlement with the tanning company 10 years later that invested $11 million in a community health fund (Sclove *et al.* 1998).

Identifying the methods

Choosing the data collection methods will depend on the research question, the involvement of community members, and the fit between the researchers' expertise and the situational needs. Research questions involving environmental health issues may combine social science methods, such as interviews, surveys, focus groups, and review of records, with quantitative measurements of soil, water, air, and biodiversity. Identifying methodology also requires consideration of the stakeholders involved. What methods are most likely to result in rich, valid data while retaining the trust of community members? Data collection methods, such as a lengthy phone surveys might do more harm than good in communities where residents have learned to distrust bureaucratic processes. Identifying the methodology and building trust with input from your partners will help ensure a fit between your methods and the local context.

In Benin, a United Nations program uses CBR with local fishing communities to compare the effectiveness of two fishing methods. The FAO's Sustainable Fisheries Livelihood Program sponsored this study to compare the effectiveness of 2-in. mesh that leaves undersized fish behind with the current 1-in. mesh that catches juvenile fish that would otherwise mature (afro1news 2003). In this case, CBR addresses the challenge of convincing small-scale fishing communities faced with extreme poverty to switch to the new nets. The methods chosen for the research were to enlist one fishing group to test the new nets for 18 months while a technician lived in the community to measure the size and value of the catch. With more room in the seine for larger fish, the nets catch more higher-value fish.

Preliminary results in the Aido Beach community showed that the group using the new nets caught 24 tons of fish in 9 outings (US$140), while the control group using the small mesh pulled in 30 tons in 9 outings, but only earned US$75.

Representatives of the experimental group say the difference convinced them firsthand of the value of the new nets (afrol News 2003).

7.4.2 Implementation

The steps in implementing CBR follow the traditional research process of conducting the research, analyzing data, and reporting results—but with the caveat of community involvement from start to finish (Stoecker 2002). These differences between traditional academic research and community-based research influence the implementation of the research steps.

Conducting the research

Conducting the research often involves data collection by students and community members, so training in research methods is critical. The principal investigators for the research should oversee data collection to ensure validity and precision of measurement. With appropriate training, however, students and community groups often become enthusiastic, persistent partners in the research, as they gain insight into their community and hands-on experience from the CBR (Figure 7.3).

Analyzing the data

With CBR, analysis of data often is done by researchers, but the bulk of interpretation occurs with all partners involved. For example, a social scientist might categorize and code data from interviews, but faculty, students, and community members

Fig. 7.3 Community-based research involves data collection by community members, which requires appropriate training in methods and protocols (photo by Martin Hutten).

would discuss and interpret the data together in an open meeting. Or a toxicologist would analyze data from local water sources and hair samples from residents and present the findings to the stakeholders who would decide on recommendations and subsequent steps.

Reporting the results

Outlets for reporting the results in CBR are as diverse as the stakeholders involved. Since a goal of CBR is the use of research results, the strategies for reporting often target specific stakeholder groups. Avenues for reporting results can include Web sites, listservs, community meetings, presentations at city council meetings, demonstrations, newspaper articles, and public radio broadcasts—in addition to the standard journal article. Decisions about disseminating results usually occur with the involvement of all partners.

One CBR project, which involved exposure of migrant farmers to harmful pesticides, used a workshop to disseminate findings to relevant stakeholders. The research involved scientists from Wake Forest University and Oregon Health Science University, farmworkers, community organizations, and agency representatives in a workshop to discuss research results and the environmental health risks of pesticide exposure to farmworkers. The workshop approach for reporting results also allowed the stakeholders to help identify directions for future research and actions (Arcury *et al.* 2000).

Challenges of CBR

Despite the potential for social and environmental change, CBR poses real challenges as academics and community members tread new ground in conservation research. Challenges include balancing or blending the needs of faculty, the curriculum, and the community; the interdisciplinary nature of real-world problems and the narrow focus of academics; the "activist" role of researchers; cultural differences between a community and researcher; the quality of data and ownership of results; identifying which community group or "community" to work with; and the issue of volunteer time of community members versus paid time of faculty (Stoecker 2002). Awareness of these challenges can help members of the partnership address them if and when they arise.

7.4.3 Evaluation

With its focus on community needs, CBR differs from the majority of conventional research and development focused on the interests of business, military, industry, pharmaceutical, government, and the scientific and academic community. In the United States, traditional research and development costs $70 billion per year with an average of $50,000 to $1 million spent per research project. In comparison, an average CBR project costs $10,000 (Sclove *et al.* 1998).

Community-based research projects have documented outcomes including: new social relationships, increased environmental quality, improved public health in the community, enhanced learning among students in terms of research skills, public speaking, and environmental understanding, and application of academic research to real-world change (Sclove *et al.* 1998; Stoecker 2001, Strand 2003a; 2002). But the need exists for increased evaluation of the efficacy of CBR. A study of 12 organizations conducting CBR found that none implemented systematic evaluation of the impact of their research (Sclove *et al.* 1998).

To design your evaluation of CBR, consider the impact of the research on multiple stakeholders, including the environment. A multi-constituency matrix, described in the service-learning section of this chapter, is useful for evaluating CBR. Work with your partners to identify the concepts you want to evaluate, the indicators of success, data collection methods, and sources of information. For example, concepts to evaluate would include:

- Quality and use of research results
- Pedagogy/teaching and learning of participants
- Social and environmental improvement
- Partnership between community members and academics
- Knowledge, skills, and empowerment of citizens involved in the research

Identify data collection methods, such as interviews, reflective journals, document review, observation, and focus groups to gather information about these indicators and choose the appropriate sources of information. As always, use the results of any evaluation to improve future research projects, share lessons learned, and gain support for CBR in your community.

7.5 Citizen science

Public participation in research has been called citizen science, volunteer environmental monitoring, student field research, research-based education, and collaborative research. The term "citizen science" refers to an organized project involving the general public, teachers, or students in any or all steps of scientific research (Prysby 2001).

For example, the Audubon Society recruits volunteers to work with scientists monitoring 265 nests of bald eagles in Florida. The volunteers locate an average of six new nests a year and rescue injured birds. The Cornell Lab of Ornithology involves both youth and adults in several citizen science projects, from Project PigeonWatch in urban centers to a study of uncommon birds in forested landscapes. With its emphasis on public involvement, citizen science shares some common characteristics with community-based research (CBR), but in CBR, citizens typically are involved in more steps in the research process.

Citizen science projects either start with scientists who have a research question and want to work with non-scientists or with non-scientists who want to collect useful data to answer a scientific research question (Prysby and Super 2006). "In practice, most citizen science projects have begun with a lead scientist defining the research question and protocols," says Michelle Prysby, citizen science director for the Great Smoky Mountains Institute in North Carolina. "From an educational standpoint, it's always better to involve the public, but in citizen science to date, the public is typically involved in data collection, as opposed to defining the research question. With community-based research, however, the community must be involved in defining the research question, because the research stems from community needs."

Some of the earliest examples of citizen research include the Christmas Bird Count in the United States, which started in 1900 and wildlife monitoring programs, such as the Bird Banding Program begun in 1920. However, monitoring projects aided by citizen volunteers seem to have originated in England where they are called recording schemes. Recording schemes have covered a wide variety of taxa and produced databases guiding conservation in England (Stevenson 2004). In Canada, citizen monitoring is overseen by the Ecological Monitoring and Assessment Network (EMAN) and the Canadian Community Monitoring Network (Stevenson 2004).

Citizen science in conservation can serve a range of uses including monitoring wildlife populations over time, monitoring genetic diversity, conducting inventories of biodiversity, gathering natural history data, and measuring water and air quality (Prysby 2001). Some of the key principles of citizen science are maximizing access to information, including stakeholders as researchers, building opportunities for participation, and identifying stakeholder needs in the research (Coastal Zone Australia 2004).

7.5.1 Planning

One leader in citizen science is the Appalachian Highlands Science Learning Center at Purchase Knob in the Great Smoky Mountains National Park. The science learning center offers research opportunities for middle school students, high school, and college students, as well as training for teachers. One project involves seventh-grade students in the observation of an ozone bio-monitoring garden planted with species sensitive to ozone. Through their research, students use estimation techniques to determine the amount of damage to the leaves.

In the eighth-grade program, students collect water quality data and information for a salamander mark and recapture study. They also visit a permanent monitoring plot to sample for spiders. All programs are correlated with the National Science Education Standards.

To share lessons learned from such citizen science programs, 25 individuals representing environmental education centers and National Park Service learning centers across the United States gathered in 2003 at the Great Smoky Mountains Institute at Tremont. Roundtable discussions at this Citizen Science Forum later resulted in a collaborative document entitled, "Best Practices in Citizen Science for Environmental Learning Centers" (Prysby and Super 2006). The best practices include the following steps for starting a citizen science project. The recommendations below assume that a conservation group or agency is developing a research project to involve community members, students, teachers, or other members of the general public (Prysby and Super 2006).

Research question

Define your research question with input from stakeholders, if possible, and determine the data needed to answer the question. Send your question to a specialist for review, and if possible, have a professional scientist as the Principal Investigator for your study. Before implementing your research questions, make sure you can answer the following questions: Why is this research question important? How does the project fit the mission of your organization?

Project design

Consider the following questions in the design of your project (Prysby and Super 2006):

- Who is the intended audience? Identify potential partners, such as researchers, schools and universities, clubs, landowners, and funding organization. (Read more about creating partnerships in Chapter 8).
- Will the results be used for decision-making?
- What are safety concerns? Think about potential risks and any safety training your staff or participants may need, such as first aid. Consider your liabilities and procedures for dealing with any injuries.
- Is this project feasible with volunteers/citizen scientists? Be clear with your volunteers about the level of commitment. Identify ways to support and acknowledge your participants, and ensure they receive adequate training. If working with students, distribute permission forms for the parents or guardians to sign. As participants gain more experience, give them additional responsibilities.
- What is the funding for the project? Identify sources of funding, and keep track of all income and in-kind donations to use as matching funds in grant proposals.
- What are the protocols for your research? Check the scientific literature to follow protocols if they exist for your type of research question. Are the citizen scientists going to do all data collection? Who is going to enter and analyze the data? Involve citizen scientists in development of protocols and go over data sheets and database entry in your training. Consider making online data entry available to your participants. On-line systems are used by programs such as GREEN (Global Rivers Environmental Education Network) and the Cornell Lab of Ornithology. Develop methods to ensure quality of data, such as data entry by teams. Test the protocols and data sheets before any training.
- Will your project have a strong education component? Decide on educational products for your project, such as class visits, news articles, Web pages, and public lectures. If you are working with students, collaborate with local teachers to design projects that consider both the academic standards and logistical factors, such as transportation. Think about issues of accessibility and special needs of your participants in the project design. Develop pre- and post-visit

lessons for students, or pay teachers to help you design these lessons. Try to incorporate opportunities for students/citizen scientists to test hypotheses during their field visits.

- What is the time frame of the project? How will you promote the sustainability of your research project? Help maintain the longevity of your project by documenting all your decisions, and identifying strategies for keeping your pool of citizen scientists involved over time.
- What will you do with the data once you have it? Determine how you will publicize your results, such as news media, public meetings, a newsletter, and Web site. Identify contacts at local newspapers and TV stations to promote coverage of your project (Prysby and Super 2006). (See Chapter 10 for tips on writing news releases).

Review the steps for planning with an example of an existing citizen science project at the Great Smoky Mountains National Park (Box 7.10, Figure 7.4).

Box 7.10 This citizen science project in-progress illustrates the use of the planning steps. (Adapted from Prysby, M.D. and Super, P. (2006). *Best Practices in Citizen Science for Environmental Learning Centers*. A collaborative document produced from the Citizen Science Forum, November 13–16 2003, Great Smoky Mountains Institute at Tremont).

Research question: The Smoky Mountains National Park is taking an inventory of all species of plant, animal, and fungus. Since many visitors to the park have an interest and experience in identifying the fungi, a mycologist at a local university decided to use visitors to document the seasonal and geographic distribution of species of fungi that are easily identified in the field. The research question was:

At what time of year and in what locations of the park do the target species of fungi produce mushrooms? In what microhabitats are they found? With what tree species are they associated?

Project statement: Park staff adopted the name FungiMap for the project, after discovering a similar project by this name in Australia. They planned to analyze the data with GIS and compare it to existing datasets containing soils, geology, topography, and habitat data. They hoped to produce probability maps to indicate the probability of finding each species at a given location. A staff member was assigned as Principal Investigator (PI) for the project, and the park's GIS expert would conduct the analysis. No end date was established, although the general inventory had a timeline of 10 years.

Partners: Partners included a local mycologist, the non-governmental organization (NGO) working with the park on the inventory, the park's environmental education center, and a local amateur mushroom club.

Citizen scientists: The initial citizen scientists included the local mushroom club, some of the groups attending the EE center, volunteers with the NGO, and park employees.

Funding needs and sources: This project did not need initial start-up funds, since supplies came from the general operation of the park. The greatest cost was printing a color key for identifying target species.

Protocols: The mycologist and the mushroom club independently compiled lists of 50 target species. The mushroom club produced data sheets for their records, and the park's data management expert worked with the PI to design the database.

When the PI tested the data sheets, the protocols seemed to work. But once the mushroom club started collecting data, some problems emerged. The park wanted all records to have location coordinates recorded in UTM units from a Global Positioning System (GPS). Some volunteers with the mushroom club did not have GPS units or preferred to record latitude and longitude units for location. Park staff determined that latitude and longitude data could be used, and they added a verbal description of the location to the database.

Education: The education plan consists of a training program for citizen scientists, as well as the picture key. When the program opens to the general public, the park plans to publish the picture key and information about the research project.

Sustainability: The park staff in charge has communicated information about the project and location of key files to another employee. At present, the project does not have a sustainability plan.

Publicity and publications: The park's Web site now includes the picture key, protocols, and data sheet. The PI has made presentations to the mushroom club.

Evaluation: Meetings between the mycologist, the PI, and the mushroom club provide feedback on the project. Each data record is evaluated by the PI at the time of entry into the database.

Feedback after first year of research: Two problems were that some of the species on the list were not as easy to identify as hoped, and few records were being reported to the PI. Based on this feedback, the PI worked with the mushroom club and mycologist to revise the list and also increased involvement with the club, the NGO, and park staff to increase participation.

7.5.2 Implementation

The exciting aspect of implementation is watching the collaboration and learning that occurs between participants like students, teachers, researchers, and other community members. The GLOBE (Global Learning Observations to Benefit the Environment) Program, for example, is a worldwide, hands-on program that involves students in taking valid measurements of water, soils, land cover, and the atmosphere; reporting their data online, analyzing the data on an interactive Web site, and collaborating with scientists and other students in GLOBE. Workshops

Fig. 7.4 Citizen scientists from the local mushroom club record the distribution of species in the Great Smoky Mountains National Park (photo by Great Smoky Mountains Institute at Tremont).

and training for teachers help to disseminate the protocols for participation in GLOBE.

Implementation of field research and citizen science does not require a global on-line system and extensive funding. Some of the most successful projects have involved one group of students or adults helping scientists monitor the quality of soils or water in a creek in their watershed. Local non-profits like the Riverkeepers can help your group get the necessary training for research, such as water quality monitoring. In Canada, EMAN distributes peer-reviewed standardized monitoring protocols in an Environmental Monitoring Tool-Kit. The tool-kit also includes features, such as an introduction to EMAN, a needs assessment guide, quick methods for analysis of water, soil, and air quality in your community, and a list of government offices, nonprofits, and community groups. (The tool-kit is available by contacting environmental.network@smu.ca).

The most important factor in implementation is collection of data about the process and any needed changes in protocols, study sites, and equipment (Prysby and Super 2006). Even an extensive planning checklist cannot anticipate challenges during the first months of implementation. Collect these data and make changes as needed.

If you have funding from an external source, ensure you meet all reporting requirements. Follow your plans for data analysis, publication, and reporting to partners. Ensure that decision-makers have access to your data analysis and recommendations for action. Your results can affect positive change for the environment. Document the process to share with others and congratulate your group!

7.5.3 Evaluation

Your evaluation can range from a simple survey with participants to a pre- and post-visit evaluation with follow-up surveys a year and more after the program. Your evaluation also should assess the process of data collection to ensure data quality at all stages from collection through analysis. Your evaluation tools should address your educational objectives for your participants. Issues your evaluation could consider include (Coastal Zone Australia 2004):

- How well were the educational and research objectives met?
- Which target groups participated in the research?
- How satisfied were the citizen scientists and the researchers with the process?
- What aspects of the project would they change? What would they keep?
- How useful was the data collected? How well were the results disseminated?
- Did the data influence conservation decision-making?
- How cost-effective was the project? Is it sustainable?
- How adequate was the timing, funding, and other resources?
- How effective was the training of participants?

The science learning center at Purchase Knob uses multiple evaluation tools, such as pre- and post-program tests, questionnaires, journals, interviews, photos, and concept mapping. Concept maps are visual representations of a learner's knowledge on a given topic. Teachers often use concept maps to understand the connections students are making among concepts, as well as any misconceptions. Staff at Purchase Knob asked high school interns to draw concept maps for the topic "Great Smoky Mountains National Park," before and after their internship at this center for citizen science. Analysis of the pre- and post-internship maps shows a change from general knowledge of the park and its resources to an in-depth understanding of connections between multiple resource groups (Figure 7.5a, b).

7.6 Mapping

People have used maps throughout history to record the past, and chart decisions for the future. Mapping is a structured activity that allows individuals or groups to create visual representations of their resources, communities, region, country, and world (Feuerstein 1986). Mapping can reveal differences in perceptions and uses of resources among different groups of people in a community, as well as changes over time in land use, resources, and social structures. Imagine asking a group of senior citizens to draw maps documenting open spaces or clean water sources in the community during their youth and sharing those results with students in the schools or members of a city planning commission.

Now, you get to create your own concept map on the main topic of *Great Smoky Munains National Park*.

(a) Pre-program map

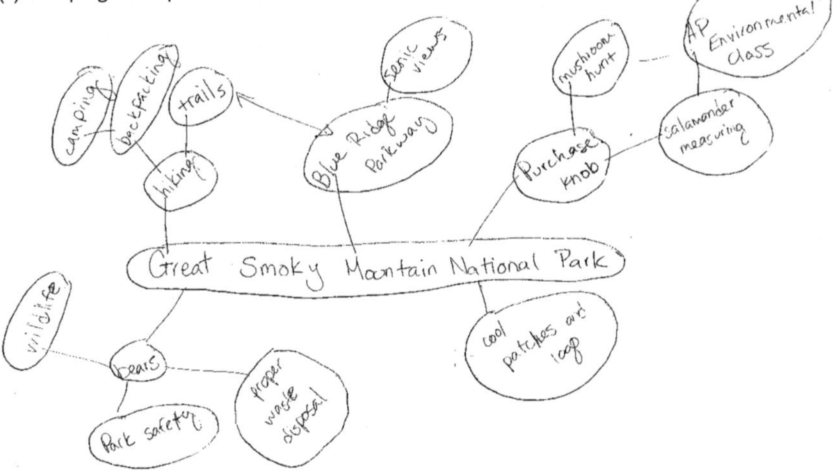

(b) Post-program map

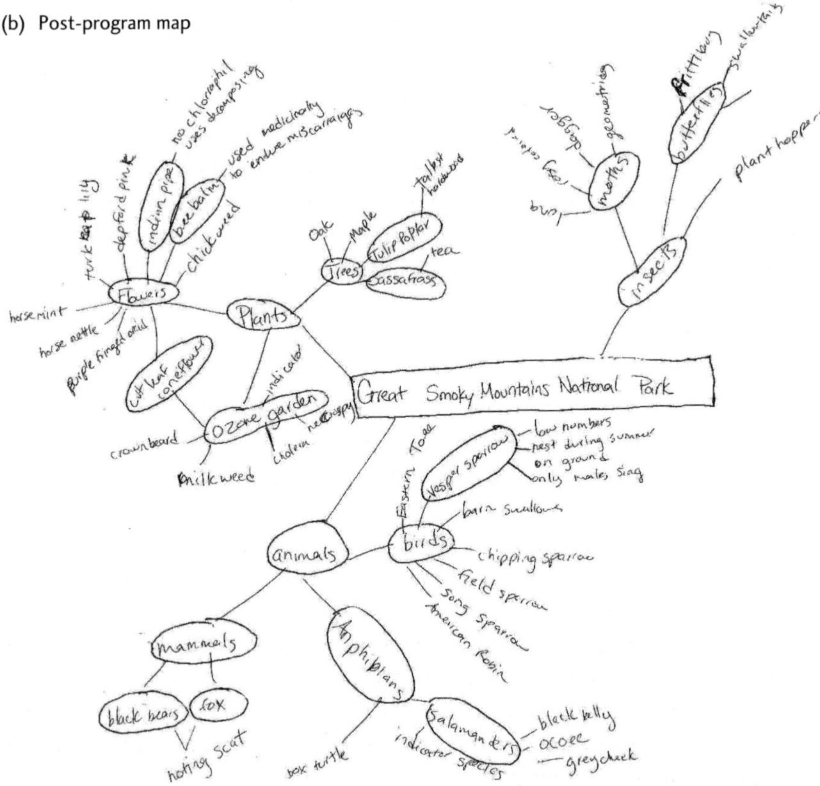

Fig.7.5 (a),(b) A comparison of the pre- and post-program concept maps, used as a tool to assess impacts of internships at a citizen science center in the Great Smoky Mountains National Park (photo by U.S. National Park Service).

When working with multiple stakeholder groups, mapping can become a springboard for discussing resource use, understanding environmental issues in a community, and building support for conservation initiatives. Mapping can be important for developing mutual understanding among groups, such as local land users, administrators, politicians, and elders—residents who share resources but may not share perceptions or priorities.

In a mapping activity, members of the community usually create the maps but can use existing maps of scientific data to generate discussion about conservation issues and actions in the past and future. Because mapping is a visual activity, you can use this tool with participants who do not read or write, due to age or language barriers. Paper and pens are not even required for mapping, as community members can create maps on the ground using sticks for drawing and rocks, leaves, and grasses as symbols.

7.6.1 Planning

Mapping can achieve many different objectives, from identifying priorities for land use and wildlife conservation with adults to developing a sense of place with children. The first task is to identify the objectives and participants of your mapping session. If you are using mapping to compare perspectives, group participants to reflect the comparisons, such as dividing into groups of men and women or elders and youth.

Mapping generally falls into two categories: (1) mapping of knowledge and perceptions and (2) scientific maps. When you use mapping to document perceptions about natural resources, a map is not right or wrong; rather, the facilitator uses the map to understand different perceptions. Scientific maps, on the other hand, document the identified location of natural resource features in a community and are often used to inform decision-making.

The objectives of a mapping session can include:

- Demonstrating that different people have different mental maps of where they live, depending on their information and perspective.
- Showing that different groups of people, such as men and women, use and value different resources in a community.
- Developing a sense of place among children and adults.
- Reflecting changes, such as land use, social structure, wildlife diversity, and resources over time.
- Revealing changes in knowledge, such as comparisons of participants before and after a conservation education program.
- Increasing knowledge and access to scientific data about conservation issues, such as maps of hydrology, wildlife, development pressures, air quality, and deforestation.
- Using environmental and social data to influence environmental and land-use planning (Pretty *et al.* 1995; Slocum *et al.* 1995; Sobel 1998).

Content of a mapping session

The content of your maps will stem from the objectives of the mapping session. Maps can help show:

- Social structure, such as location of neighborhoods, ethnic groupings, social centers, leadership.
- Communication and transportation networks, including roads, bridges, trains, paths.
- Geographical features, such as rivers, oceans, mountains, forests, deserts.
- Physical features, including the size of the town, location of houses, schools, factories, farms.
- Natural resources like wildlife, trees, plants, pastures.
- Health and wellness features, such as hospitals, water sources, sanitation facilities.
- Changes over time, which can be shown with overlays to compare changes with the original map (Feuerstein 1986).

The complexity of maps can range from maps drawn in the dirt to maps created using GIS (Geographical Information Systems). With GIS, you can create digital maps showing different features and overlay these "layers" to create a rich picture of an area. The Wombat State Forest in Victoria became Australia's first community managed forest, resulting in the development of issue-based working groups to tackle management issues, such as invasive exotic species. Prior to the establishment of working groups, the region's Blackwood community had struggled with obtaining assistance to control invasive weeds for 10 years. The newly established Weed and Pest Animal Working Group (WPAWG) helped the community obtain university assistance in developing maps showing the distribution of weed species. Community members walked the bush in difficult terrain to map the location and densities of weed species. The resulting GIS maps were used as evidence in the Blackwood Weed Management Strategy, a funding proposal to the government (Baral *et al.* 2004).

7.6.2 Implementation

This section presents four examples of implementation, including mapping the neighborhood, mapping the town, mapping to compare stakeholder perceptions, and mapping with youth.

Mapping the neighborhood

One strategy for beginning a mapping session is to give each participant the opportunity to create their own map of the neighborhood using paper and pen. Ask participants to think about where they live and draw major landmarks and natural resources of importance to them. After 10 min, post the maps on the wall and allow everyone to observe differences in the maps. Ask follow-up questions such as,

"What natural resources did you draw? What symbols did you use to represent features on the map? Who has lived here for more than 10 years? How does that knowledge of the natural resources compare with someone who just moved here? What changes in natural resources have occurred over time in this neighborhood? What conservation measures, if any, are needed to protect the resources portrayed in the maps?" (Pretty *et al.* 1995).

Mapping the town

Another strategy for implementation is to divide your participants by the relevant demographic variable in the group. For example, you could divide the group into men and women, farmers and city dwellers, community members who grew up in the town and those who recently moved to the area. Give each group a large sheet of paper and pens, and ensure that each person has access to a pen if you are using paper. You can photograph the drawing for documentation.

Ask the participants to create a map of the important natural resources in the town or city and give them about 45 min for their drawing. Emphasize that each group member should be involved in the drawing. Ask each group to present their map and explain the process of creating the drawing. Reflect on the maps using questions such as "What did your group agree upon? What natural resources did you include? What did you exclude? What did you emphasize? What did you not emphasize?" Ask the community members to analyze what the maps reveal about the natural resources and conservation issues of the area (Pretty *et al.* 1995).

Comparing perceptions using maps

Maps can reveal perceptions about natural resource management between different stakeholder groups in a community. In Abaco National Park in the Bahamas, mapping was used with six different stakeholder groups to gather input and assess current and desired uses of the park for recreational management planning (Marks *et al.* 2004). Facilitators of the mapping session distributed topographical maps to all participants who used colored pencils to identify areas for different activities in the park. After the individual mapping session, groups of stakeholders used a large topographic map to identify their preferred location of zoned activities, such as hunting and ecotourism, by gluing colored squares on the large map (Figure. 7.6). For the majority of groups, the maps revealed a combination of ecotourism and protection as most important for the park's future.

In Nepal, a participatory video workshop sponsored by USAID's GreenCOM brought community forest user groups together to compile a community video letter (Chapter 11) as a strategy for presenting their concerns about forest issues to government officials. Since the workshop included ethnic groups who rarely spoke to each other, facilitators used mapping as a tool to open dialogue with these disparate groups who share the same forests (Greiser 2000). The participants

Fig.7.6 A group mapping exercise in Abaco National Park in the Bahamas provided input for recreational management planning (photo by S. Jacobson).

produced a map using red mud and objects to mark boundaries, rivers, forests, and houses. The creation of this collective map sparked discussions among all stakeholders about forest issues that affected their community. Many participants noted that the map provided a context for conservation and was more interactive and fun than a simple discussion (Greiser 2000).

Another method for comparing perceptions is called gendered resource mapping, which reflects gender differences in resource control and use (Slocum *et al.* 1995). Begin by listing the major classes of vegetation, land use, and tenure, such as forest, woodland, savannah, perennial crops, annual crops, conservation reserves, houses, public property, roads, and, rivers. Label the land cover and land-use categories in the dialect of your participants.

Next have participants identify the land user groups, such as men and women, children and elders. On another sheet, ask participants to sketch the distribution of land use and land cover types in the community. Note who controls and uses these land use/land cover types (Slocum *et al.* 1995). With the participants, make detailed sketches and inventories of the community, with a matrix of who controls and uses each product and place. Such tables and sketches can guide discussions for natural resource management with community groups, integrate the interests of both men and women, and increase commitment to solutions if all involved are heard.

Mapping with youth

Mapping with children can build a sense of place and a spatial perspective of the natural resources in a young person's community. Young children should begin mapping exercises by representing their home and then move to their neighborhood, corresponding to their developmental levels (Chapter 2). As students get older, their maps can explore their country and even the world (Sobel 1998).

High school students in Fairhope, Alabama mapped the growth and development in their community. They first drew maps of their neighborhood and then the entire town. Next, pairs of students interviewed elderly residents who grew up in their town. The students created maps with the senior citizens, drawing from their memories of the town. Finally, the students compared historical maps of the town with the ones drawn by themselves and their elders. As a final mapping activity, the students drew a visioning map of how they wanted to see their town in the future. This mapping exercise revealed the changes in land use—from rural to suburban—and the need for land-use planning to protect their natural resources.

You also can use mapping in combination with another technique, such as action projects (Chapter 4), service-learning, or issue investigation, described in this chapter. For example, the service-learning program called ACT (Active Citizenship Today) uses mapping exercises to begin this program that connects the study of civics with community involvement. Students first draw their community, and the maps highlight community issues that become a focus for their research and service-learning project (Loney 2000).

7.6.3 Evaluation

Maps drawn by students or adults can show gains in knowledge as a result of time or an educational program. The urban program VINE (Volunteer-led Investigations of Neighborhood Ecology) asked youth participants to draw and label what they had seen living in their schoolyard, before and after the program (Figure 7.7). The pre- and post-program drawings were scored based on change in three criteria: application of information presented in a VINE activity, organization and context, and complexity (Hollweg 1997). Thus, mapping can be an important evaluation tool for your conservation education initiative. (One constraint is that developmental changes may account for some observed changes in the post-program drawings.)

When evaluating a mapping activity itself, observe your participants and document the level of participation. Is everyone involved in the mapping and decision-making? Who is taking the lead in drawing and discussions? Who is holding back? At the end of the mapping session, ask participants to reflect on the process. What did they learn from creating a map with others? What differences did they observe from different maps?

If your participants create maps for land-use planning and management, follow the group or community to document how often the maps are used in decision-making. Are the maps sitting on a shelf or posted on the walls for planning

Fig. 7.7 Mapping can show pre-and post-program gains in knowledge, as seen in these maps of schoolyard ecosystems drawn by children in the VINE (Volunteer-led Investigations of Neighborhood Ecology) program.

(For details describing this technique and suggestions for enhancing its use, see pages 53–62 in Hollweg, K. (1997). *Are We Making a Difference? Lessons Learned from VINE Program Evaluations*. NAAEE, (NAAEE.org), Washington, DC.).

purposes? City planner Elizabeth Teague in Fletcher, North Carolina posted brightly colored maps of a proposed greenway around the town hall during community visioning sessions. She invited community members to post their comments about the maps, which became focal points for a stakeholder-based planning process. "The maps were central to our planning," she said, "because the visuals gave people a reference point for describing the town they wanted to see in the future."

7.7 Summary

Effective conservation aims to integrate, rather than compete, with the needs of the human communities that share the landscape with biological communities. This chapter includes six techniques to connect classrooms and communities with conservation. Techniques, such as service-learning, issue investigation, and project-based learning were developed to connect conservation with classrooms. Although techniques such as service-learning originated in an academic setting, conservation educators can use service-learning with any group of learners, from teachers to senior citizens, if the goal is to impact learning and conservation in the community.

Community-based research, citizen science, and mapping are techniques developed to work with either community groups or students. Choosing which technique works best for your conservation education program depends primarily on your overall goal. Do you have a field research project that would benefit from community involvement? Citizen science could provide an educational technique to help your program. Do you want to connect the work of your conservation organization with academic learning in local classrooms? Service-learning enhances learning while involving students in conservation work. The tips for each technique in this chapter offer important tools for building that connection with conservation. No matter what your objective, you can align it with one of the above techniques to bring conservation issues to the forefront of communities and classrooms and ultimately help achieve your conservation goals.

Further reading

Partner Power and Service Learning: Manual for Community-based Organizations to Work with Schools, by Rich Cairn (2003). ServeMinnesota! and Minnesota Department of Education, Minneapolis, MN.

Give Forests a Hand Youth Action Guide: A Youth Program for Environmental Action and Community Service, by Janice Easton and Martha Monroe (1996). Cooperative Extension Service, University of Florida, Institute of Food and Agricultural Sciences, Gainesville, FL.

Partners in Evaluation: Evaluating Development and Community Programs with Participants, by Marie Feuerstein (1986). MacMillan Education Ltd, Hong Kong.

Investigating and Evaluating Environmental Issues and Actions: Skill Development Program, by Harold Hungerford, Trudi Volk, John Ramsey, Ralph Litherland, and R. Ben Peyton (2003). Stipes Publishing, Champaign, IL.

Community-based Research and Higher Education, by Kerry Strand, Sam Murullo, Nicholas Cutforth, Randy Stoecker, and Patrick Donohue (2003). Jossey-Bass, San Francisco, CA.

Project-Based Learning Handbook, 2nd edition, by Thom Markham, John Larmer, and Jason Ravitz (2003). Buck Institute for Education, Novato, CA.

8

Networking for conservation

If you have ever moved to a new town, you know what it means to network. You start by meeting new neighbors. If you have children, you meet new teachers and other parents. You learn the locations of and faces at grocery stores, markets, public transportation routes, medical facilities, and the library. At work, you become familiar with your responsibilities and colleagues. In short, you become adept at introducing yourself to others and learning about them. You are building relationships to integrate yourself into a new community.

Effective conservation demands these same skills to build relationships in a community. People make decisions in relationship to other people. Therefore decisions about conservation often hinge on our ability to network with other individuals and institutions—especially with unlikely partners we may perceive as the opposition.

This chapter highlights examples of conservation networking. The Nature Conservancy, World Wide Fund for Nature, and Conservation International, for example, joined forces to create the Conservation Measures Network. On an international level, organizations such as the Natural Step International Gateway work with representatives in 12 countries through a network of scientists, universities, and businesses to promote sustainability in communities across the globe.

Networking for conservation allows you and your partners to become advocates for each other and find common ground between your interests. In the business community, networking has been described as "linking together individuals who, through relationship building become walking, talking advertisements for each other" (Speisman 2005). Effective conservation, such as landing a sale or a new job, requires building relationships through effective networking. We in the conservation community need skills for promoting our message and reaching out to audiences beyond the limited circle of conservationists.

Networks can create synergy between groups, galvanize resources, expand your support base, and promote your conservation objectives (Box 8.1). The techniques for networking in this chapter begin with organizing environmental clubs, conducting workshops and seminars, giving public presentations, developing professional posters and information booths, and hosting conferences and special events. Partnerships are vital for networking, and the chapter includes tips for identifying potential partners, cementing strong partnerships in the community, and assessing the effectiveness of that partnership. Maintaining strong alliances also involves

Box 8.1 Networking for conservation involves techniques to build relationships and alliances and thus promote conservation goals.

Technique	Purpose
Environmental clubs	Mobilize individuals into a group that shares a common interest or stake in a conservation issue.
Workshops and seminars	Provide a structured forum where people come together to increase knowledge and skills, work on a common task, build consensus, and solve problems. Workshops use strategies such as discussions, small and large group activities, and reflection
Public presentations	Address a public audience to present latest scientific data, influence attitudes and behaviors, respond to questions in a community, or clarify public misconceptions
Information booths	Provide information at a booth designed to raise awareness about a conservation organization or issue to people gathered at an event, such as a festival
Professional posters	Convey information in a visual format in order to share research results and network with professionals typically gathered at a conference
Conferences	Bring together like-minded people or members of an organization to share information, network, conduct business, and build skills
Special events	Provide an event, such as a festival or concert, centered on a conservation theme with goals such as raising funds, recruiting members, building awareness, and attracting publicity
Partnerships	Create an institutional arrangement between two or more organizations that provides mutual benefits and helps achieve conservation goals
Conflict resolution	Uses strategies such as negotiation skills to resolve conflicts between stakeholder groups by looking for mutual gains and resolving issues on their merits, rather than the positions of conflicting groups

solid skills in negotiation and conflict resolution, a critical technique in the conservation arena. One take-home message from this chapter is that conservation techniques for networking can be creative and productive. Networking for conservation can be an adventure with positive outcomes, just like moving to a new town!

8.1 Environmental clubs and groups

In Kenya, 50 students gathered in 1968 at Kagumo High School to create a national conservation organization, the Wildlife Clubs of Kenya (WCK). The students realized that tourists, rather than Kenyan citizens, had primary access to the national parks and wildlife that were symbols to the world of their country's resources. These environmental clubs lobbied and gained free entry into the

Fig. 8.1 The Wildlife Clubs of Kenya is the largest grassroots conservation organization for youth in Africa (photo by M. McDuff).

national parks for their members. In the 1970s, the clubs rallied 5000 youth to protest elephant poaching and hunting. Club members have marched in the streets of downtown Nairobi to encourage the use of baskets for shopping, as opposed to the now ubiquitous plastic bags. The formation of WCK, the largest conservation organization for youth in Africa, prompted the growth of similar clubs that promote wildlife conservation across Africa, Latin America, Asia, and Europe (Figure 8.1) (McDuff 2000).

Environmental clubs are a technique to mobilize individuals with a common interest or stake in conservation issues. Environmental organizations for youth can provide significant life experiences that help develop environmental interests and actions (James 1993; Palmer 1993; Chawla 1999). Conservation groups build motivation and skills to conserve natural resources by:

- providing practical experiences in outdoor settings
- developing hands-on conservation experiences
- increasing knowledge of environmental issues
- identifying roles for youth in community environmental action
- developing environmental responsibility among youth (Voorrdouw 1987)

Environmental clubs are not only for youth. In Ireland, a group called Invest Northern Ireland organized the Environmental Performance Improvement Club. Developed for the local construction industry, the club encourages small companies to adopt environmental management practices to improve their business and environmental performance.

A club can be a rallying point for a conservation cause, a meeting place for people interested in environmental action, and a social group with a passion to make a difference. A Jewish environmental club, the Mosaic Outdoor Club sponsors environmental events such as an island cleanup and service for the Jewish Earth Day, Tu Bishvat. An active club at the Nanyang Technological University (NTU) in Singapore is Earthlink NTU, whose goal is to promote environmental awareness among students. Activities of students include reforestation projects, a summer EcoCamp, and field trips such as the treetop canopy walk in Central Catchment Nature Reserve. Another group, The Kosovo Young Ecologists, is involved in environmental activities such as raising environmental awareness, research, and capacity building.

8.1.1 Planning

In a survey of 75 school environmental clubs in Wisconsin, club advisors gave a range of reasons their members started the club, including:

- increase positive attitudes and activities involving the environment
- promote awareness of ways to care for the earth and its resources
- enhance knowledge about the surrounding school and community
- empower students to make environmental change
- create awareness of local environmental issues
- build participation in environmental action (Schulz 2000).

You may decide to start your own club or use existing organizations, such as Girl Guides, Boy and Girl Scouts, Rotary Clubs, and Kiwanis Clubs to address a conservation issue. Collaborating with clubs that already have a strong membership base saves resources and uses the existing infrastructure in your community. Many clubs have monthly speakers, so volunteer to present a program on your conservation issue or involve members in a special event for conservation. Wildlife conservation agencies such as the US Fish and Wildlife Service have worked with Boy and Girl Scouts to correlate activities in the Project WILD curriculum with the badges earned by Scouts. Environmental educators in Florida have used 4-H youth groups to conduct conservation projects. Often common interests in outdoor activities drive the creation of recreational clubs, such as a kayaking club or garden club. Recreational groups such as canoe and kayak clubs depend on natural resources such as rivers and estuaries, so their members can become strong advocates and partners for conservation.

If you decide to start an environmental club, brainstorm objectives for your group. Does the need exist for increasing awareness among members or will your club lobby for a specific environmental action, such as restoration of habitat or conservation of public lands? (Box 8.2) Established by primatologist Dr. Jane Goodall, Roots and Shoots Clubs promote conservation in countries as diverse as Germany, Uganda, Tanzania, the United Kingdom, and China. The International School of Beijing Roots and Shoots group works to end "bear farming," a practice where Asiatic black bears are kept in small cages, while the Shanghai Roots and Shoots raises money to create an improved enclosure for chimps at the Shanghai Zoo.

Box 8.2 Six steps to planning an environmental club. (Adapted from Schulz, C. (2000). *School Environmental Clubs in Wisconsin: 2000 and Beyond.* Wisconsin Center for Environmental Education, Stevens Point, WI)

1. Find a core group of people who want to help plan the start-up of the club.
2. Brainstorm objectives for your club or organization.
3. Conduct an informal needs assessment of your school or community:
   a. How are other groups organized?
   b. Do you have a solid group of potential members? Word-of-mouth is often an effective way to assess this need.
   c. Are there other environmental groups with whom you could partner and learn?
   d. Do you have potential sources or ideas for funding?
   e. What are the major environmental issues, problems, or needed vision in your school or community?
4. If working with a school, write a proposal for your environmental club to outline the purpose, activities, and needs from the administration. Identify an advisor for the club as well.
5. Work with your planning team to develop a draft organizational structure to present to the larger group of members at the first meeting. The entire group should decide the organizational structure by a decision-making process, such as consensus, majority voting, etc.
6. Recruit new members. Ask yourself why would someone want to join this group? Plan and publicize your kick-off meeting as a recruiting tool.

8.1.2 Implementation

With a new club, the first step is to initiate a kick-off meeting, the perfect time to build excitement among members. Plan a time when you think most of your potential members can attend. If you are working with an existing club, contact key members to ensure that people can attend your presentation or event. Reserve a room that you think will be a bit small, as a full room for the first meeting can create positive energy (Schulz 2000). Plan to serve food at the meeting as an incentive for participation. People will want to stay involved if the meetings are both fun and focused. Post the agenda for the meeting so everyone can see.

Give an introduction to the group and ask for participants to introduce themselves. Plan an icebreaker such as a partner introduction to allow people to relax and get to know each other. Provide an overview of the proposed organizational structure and ideas for activities. Let the group brainstorm and take notes on flipchart paper of their vision for the club and ideas for activities. When working with another club, you can ask members to brainstorm avenues for collaborating with conservation objectives. At the end of the meeting, discuss a time schedule for future meetings. To end on a fun note, consider giving door prizes, like a gift certificate from an outdoor store. Make sure to include some discussion and decision-making about organizational structure in the agenda for the next meeting.

Promote the environmental club through advertising that gives the group visibility and increases awareness and knowledge among students or community members. Send press releases to the local newspaper to advertise club events and initiatives. Make contacts with a reporter who tends to focus on environmental news. Unless you have outside funding, your club may need to plan fund-raising events for your activities. If you are using an existing club, advertise the environmental work of the group in the local media.

One local environmental group, PARC, People Advocating REAL Conservancy, formed in direct opposition to the proposed sale of a downtown park in North Carolina, US to an upscale resort hotel. The city proposed to sell part of the downtown park to the hotel owners who intended to build a 10-story condo, but there was little press coverage and no public announcement about the proposal. The city council then voted to sell the land, and the mayor ignored 1500 signatures on a petition opposing the sale due to the loss of green space. Several months later, the local newspaper published a poll showing that 82% of readers thought the hotel would build the high-rise, despite the public opposition.

But PARC prepared leaflets, signs, displays, and a scale model of the proposed building. Their Web site showed a drawing of the building superimposed on the park. PARC sent e-mail messages to members of the local environmental community, and word spread of the public opposition. A League of Women Voters forum on the sale of the park attracted a crowd of 400 people who confronted the president of the hotel. With such strong support, PARC proposed a boycott of the hotel. Shortly after, the hotel called off the development due to the 'financial feasibility' of the project. By targeting a specific objective for their group, PARC achieved a conservation victory despite the initial wishes of both the city government and a development project.

Other clubs such as Friends of Nairobi National Park in Kenya and Friends of the Rocky Mountain National Park in the United States have lobbied successfully for funding and increased protection of the parks they aim to protect. The Friends of Sherwood Island State Park in Westport, Connecticut, US have planned and implemented a butterfly garden, written and received grants to fund projects, co-sponsored community cleanups, purchased equipment, and helped with ecological renovation. The current goal of the Friends is to raise the funds necessary to build a nature center.

8.1.3 Evaluation

Many environmental clubs or groups experience large membership at their initiation, with participation stabilizing or dwindling over time due to the departure of a charismatic leader or the end of a galvanizing issue. Monitoring and evaluation can help identify trends in membership and factors associated with these trends. Tools you can use to evaluate the effectiveness of your club include:

- periodic verbal feedback sessions or focus groups with members
- surveys to gather data from members and nonmembers

- documentation of success stories and failures—document all projects to learn from the past activities
- accurate records of members to track trends in membership.
- Interviews of members who left the club to gain insight into factors influencing the members.
- Interviews of community leaders to assess the effectiveness and support for your environmental actions.

The WCK records showed that 1800 clubs had registered since 1968, but only 25% had regularly renewed their membership 30 years later. The national office had begun to focus on a target audience of adults and had accumulated substantial debts. In the 1990s, the office created an endowment fund and refocused on the local clubs with yearly themes such as young women and biodiversity conservation. The WCK also conducted a participatory evaluation to involve its stakeholders in evaluation of the program. The evaluation created criteria for assessing success and established membership registration as a priority area for the organization.

8.2 Workshops and seminars

Workshops and seminars are a structured event where people come together for a common task to increase knowledge and skills, resolve problems, and build consensus for action. An effective workshop can enhance small group communication, promote ownership of outcomes, and generate feedback from participants on alternative solutions or actions (Citizen Science Toolbox 2004a). Workshops use strategies such as large- and small-group activities, discussions, reflection, role-plays, and case studies with participants representing either a variety of stakeholders on an issue or a specific group, such as teachers or environmental planners.

Workshops can bring together decision-makers who do not typically share information or network. The Appalachian Sustainable Agriculture Project in the Southern Appalachian Mountains (US) hosted a farm-to-school workshop that brought together 50 stakeholders including nutrition directors, teachers, farmers, cooperative extension staff, environmental educators, and health department officials. The goal of the workshop was to take steps to improve childhood nutrition and conservation of farmlands by providing local schools with foods from local farms. Participants of the workshop even got to taste these delicious foods—herbed goat cheese and homemade bread, fresh salads, and quiche made from local eggs—during breaks between sessions.

After 1 h of a set agenda, the participants started asking each other questions, eager for information from the other stakeholders. The school nutrition directors explained their requirements for accepting local produce from the farmers, and the farmers explained the foods available during each season. The workshop resulted in a proposal for funding field trips for the school nutritionists to visit the farms, as well as talking points for explaining the concept of farm-to-school to school boards and county commissioners.

8.2.1 Planning

Proper planning for a workshop includes three phases—building support, designing the workshop, and selecting instructional strategies.

Building support

You need to build support from two groups: (1) the people with whom you will work, such as directors of your organization or school, supervisors, and donors, and (2) the participants of your workshop. Begin by identifying the funding, personnel, or resource materials you could receive from your partners or sponsors. Consider a broad range of sources for support, including for-profit, nonprofit, private, and governmental agencies. Think about issues such as the philosophy of the supporting organization, a clear understanding of the goals of your workshop, and support for incentives for participants (Braus and Monroe 1994).

Second, you must assess your audience. Conducting a needs assessment will give you information to match the organization of the workshop to the needs and interests of your participants. You can gather this information from written or e-mail surveys, participant interviews, or questions on the registration form. These data will also help you refine your agenda and develop your goals and objectives for the workshop. Most participants appreciate the chance to offer feedback prior to the workshop. Ensure that you gather information from a representative sample of attendees.

Seven steps to designing an experiential workshop

This seven-step process builds on the experiential learning cycle (Chapter 2) that involves interactive experiences, reflection, generalization, and application of learning. Use this model for designing a 2-h session or a 1-week course (adapted from Braus and Monroe 1994).

1. *Set the climate.* Begin with an activity that excites participants about the workshop. The opening also can explain the relevance of the subject to participants. During this time, make introductions, conduct an icebreaker, or use a demonstration to pull people into the setting. Think about the impact of room arrangement on learning and interaction. You can arrange chairs in a circle with areas for break-out groups or have four to five people at each table, angled so everyone can see the front.
2. *Review goals and objectives.* Have the workshop goals and objectives written on flipchart paper and state the knowledge and skills you want participants to gain by the end of the workshop. Discuss how you used information from the needs assessment to develop the agenda. Give everyone a chance to ask questions. It is often helpful to ask participants to list their expectations on a flip chart, so you can explain which ones will be met during the workshop.

3. *Conduct the activity.* During a workshop session, engage participants in an activity that gives them the opportunity to "experience" a situation relevant to the topic of the workshop. Common examples include role-plays, case studies, field experiences, and small group activities.

4. *Process the experience.* Allow participants to share their reactions to the activity. Ask questions such as "What happened in this activity? What worked well? What was challenging for you or your group? What would you change?"

5. *Generalize.* Try to identify key generalizations about the experience, so participants can see how the activity relates to their own lives. Ask questions such as "What insights did you get from the experience? What was the most important lesson for you and why?"

6. *Apply.* Help participants identify and share how they plan to incorporate these insights and lessons into their lives and work. Questions such as "Now what?" and "How can I use what I learned?" can help begin applying lessons learned.

7. *Reach closure.* Summarize and make connections to the goals and original expectations. Also be sure to conduct workshop evaluation to see if you achieved your objectives and gather reactions from participants.

If you are scheduling a 1-day workshop, set aside at least 45 min for lunch and include opening and closing times. Include 15-min breaks in the morning and afternoon, which leaves three or four blocks of teaching time. If your workshop is less than 1 day, be sure to focus your objectives on a reasonable amount of material. Avoid the mistake of including too much material for a short workshop. For workshops longer than 1 day, you can include out-of-class assignments, fieldwork, and more opportunities to practice new skills and get feedback from peers. The National Conservation Training Center in West Virginia, US, for example, offers a 4-day workshop on Education Program Evaluation attended by many environmental educators. The 4-day format allows participants to learn the skills of evaluation and then apply those skills by evaluating a local environmental education program. The participants receive guidance and feedback on their evaluation from the facilitators during the week.

Selecting the instructional strategy

A variety of teaching techniques exist to accomplish your workshop goals. Role-plays, lectures, small group discussions, field trips, case studies, hands-on activities, and many other techniques (Chapter 5) can involve your participants in the material (Figure 8.2). Think about your overall workshop design, and integrate multiple approaches to keep attention and enthusiasm high.

For small group activities, decide how you want the large group divided, how long they will work together, the roles of people in each group, and how small groups will report to the large group. You can divide your large group into pairs, threes, or larger groups. In general, groups of about five participants result in the most interaction.

Fig. 8.2 A workshop for teachers included hands-on experience using ozone bio-monitoring gardens to assess the impact of ozone on plant species (photo by U.S. National Park Service).

8.2.1.4 Logistics

A large part of planning a workshop is logistics, including the food, venue, audio-visual equipment, nametags, transportation, and back-up plans in case of bad weather. A checklist for pulling off the perfect workshop gives a concise overview of the logistical arrangements to consider (Box 8.3).

8.2.2 Implementation

After designing the workshop and planning logistics, turn your attention to facilitation skills. An effective facilitator can put the participants at ease and help them clarify their thinking and build their understandings. Four key facilitation skills can help implement an effective workshop (adapted from Braus and Monroe 1994):

1. *Asking questions*: The ability to ask good questions is critical to guiding any discussion. The best questions are open-ended and help participants reflect on activities. Write down questions and practice, such as "What are some different ways this activity could be used?"
2. *Paraphrasing*: This skill is important for clarifying and highlighting a comment. With paraphrasing, you rephrase what someone else says. For

Box 8.3 Checklist for conducting the perfect workshop or seminar (Adapted from Braus, J.A. and Monroe, M. (1994). *EE Toolbox: Workshop Resource Manual: Designing Effective Workshops.* Kendall Hunt Publishing, Dubuque, IA.)

Initial planning

- Discuss the workshop with administrators, colleagues, students, donors, and other partners who should be involved from the start.
- Determine sources of funding for the workshop. Will participants pay? Will you have an outside sponsor?
- Identify your audience and conduct a needs assessment.

Workshop design

- Develop workshop goals and objectives, incorporating results of the needs assessment.
- Decide if you need a co-facilitator. Plan the agenda and workshop sessions, including objectives for each session. Include a variety of techniques and activities to hold the interest of the group and appeal to different learning styles.
- Choose the date, time, and location, and make sure these work for your target audience.
- If inviting special guests, make arrangements for their participation, including their roles and compensation.
- Determine the materials you will distribute during the workshops and who will gather, develop them, and make copies.

Logistics

- Reserve the room or area appropriate for the workshop. It should have enough floor and wall space, outlets, tables, comfortable chairs, ventilation, etc. The area also should provide access to the outdoors. Organize the room in a way that best suits your needs.
- Determine what incentives, if any, you will offer participants.
- Decide how you will publicize the workshop.
- Determine how you will evaluate the workshop. What type of follow-up efforts will you ask of participants?
- Decide what kinds of food and drinks you will provide. Have them delivered to a separate room at least 30 min before you need them.
- Order workshop supplies, such as pencils, resources, notebooks, curriculum packets, at least two months before workshop.
- Determine which audiovisual equipment you will need (flipcharts, VCR, slide carousel, laptop computer and projector, extension cords) and reserve them well in advance. Make field trip arrangements. Check on costs, special clothing, transportation. Pack a first aid kit and have an alternate plan in case of bad weather or other complications.

- Send introductory materials to your participants—a map of the site, travel directions, parking information, phone number, agenda, and items they should bring. Decide if you want to include a pre-workshop survey to gather additional information from confirmed participants.

Last-minute reminders

- Set up early so you can talk with participants when they arrive. Display posters, charts, and other materials.
- Put signs along the route and in the building so participants can locate the site.
- See that food and drinks are ready. Set out name tags with markers and wear one yourself.
- Make sure goals and objectives are written on flipcharts or overheads before you start. Write small group tasks on flipcharts to post for individual sessions.
- If you are using flipcharts, tear off masking tape in advance and stick the pieces along the easel for easy access.
- Write down questions you want to ask participants. You also can write reminders or notes to yourself on prepared flipcharts using a pencil—you'll be the only one who can see them.
- Before everyone arrives, make sure all equipment works. Make sure all handouts are ready to go, including the evaluations.

During the workshop

- Greet participants warmly when they arrive. Point to name tags and handouts as you mingle with the group.
- Start on time. After your welcome and icebreaker, give an overview of the entire workshop, including the agenda and goals. Highlight breaks, meals, restroom locations. Introduce yourself and allow participants to do the same.
- Stay on schedule, give breaks as indicated, and go over the scheduled ending time of workshop.
- Leave time in each session for asking generalization and application questions and for closure.
- Leave enough time for evaluation at the end. Collect all evaluations before participants leave.
- Pass out a sign-up sheet early to record participants' names, addresses, phone numbers, e-mails. If possible, make copies and distribute the list before everyone leaves. If you already have this list, late registrants can add their names.

After the workshop

- If you are working with co-facilitators, meet at the end of the day to review what worked well, what didn't, and what needs revision for the next day or next workshop. Keep debriefing sessions brief, upbeat, and focused.
- Know how you plan to follow-up with participants. Remind them of any post-workshop assignments.
- Give yourself time to clean up room at the end after you've answered questions and seen your guests to the door. Congratulations, you're finished!!

example, you might say, "Did I understand that you said ..." or "I hear you saying that..." Avoid overusing paraphrasing though, so you are not constantly repeating every comment.

3. *Summarizing*: Summarizing allows you to guide a discussion, draw similarities and differences between participants' comments, or transition to the next session. You can note, "It sounds like we all agree that having green space in our town is important, but we differ on how to get there. Is that correct? Does anyone want to add anything?"

4. *Offering encouragement*: You can use many techniques to encourage discussion. Maintain eye contact with participants. Use encouraging body language by paying attention when individuals are talking. Give people time to think by waiting 5 s before calling on respondents. Call people by name.

The checklist includes other helpful hints for implementing the workshop and ensuring smooth facilitation. During the workshop, make sure you stick to the schedule, especially ending on time. And be sure to say, "I don't know" when appropriate.

Building ongoing support

A concern about short-term workshops is the long-term value for the sponsor and participants. People often need more than a 1-day program to make a difference in their behaviors or practices. However strategies exist to build support and ensure that the workshop is just one step in a sustained initiative for change. Ongoing support may include funding for new programs, resources, newsletters, e-mail correspondence, and other networking opportunities. Think about how you plan to interact with the participants after the workshop, and how they can interact with each other.

One strategy is to register participants in pairs and design the workshop so they can continue to work together at their site or via e-mail. Give partners an assignment to do after the workshop with publicity of their efforts in local newspapers, and offer incentives for participants who complete action plans (Braus and Monroe 1994).

To provide ongoing information and reminders, schedule a follow-up meeting or phone conference to check on their progress. Stay in touch with e-mail. Submit a grant and involve participants in editing and next steps. Have participants write a postcard to themselves reminding them of their commitments and mail it in several months. The Leopold Education Project, an environmental education program focused on the writings of Aldo Leopold, conducts 10-h workshops for educators. The workshops include a homework component where participants must complete an assignment such as teaching a lesson or correlating a lesson to the state education standards. Such simple incentives ensure that participants do not leave their new skills and resources behind when they leave the workshop site.

8.2.3 Evaluation

Evaluation is a critical part of any conservation education workshop to provide feedback about what works and what needs improvement. One strategy breaks down the evaluation of a workshop into four categories (Kirkpatrick 1998):

Reaction: How much did the participants like the workshop?

Learning: What principles, skills, facts, and techniques did they learn?

Behavior: What changes in participants' behavior, if any, resulted from the workshop? (Behavior changes may not be an objective of your workshop.)

Results: What were the tangible results of the workshop in terms of learning, collaboration, networking, or environmental conservation?

To assess reactions to the workshop format, ask participants: What were your most favorite parts of the workshop? What were your least favorite parts? What improvements would you make to the workshop? You can also use written evaluations or small group interviews to collect reactions.

A typical strategy to document changes in learning is a pre-test/post-test design or a comparison with a control group. Other tools include performance assessments, focus groups, or interviews. To evaluate changes in behavior, you can observe the participants or ask them to measure and report their own behavior. If participants include teachers, have them document how often they incorporated the conservation education lessons into their teaching in the two months after the workshop. Include incentives for reporting these behaviors.

Lastly, to evaluate results, conduct a survey or interviews 3–6 months after the workshop. Schedule interviews with a representative group of participants on the phone. Make sure you keep good records of paperwork, lesson plans, evaluations, and your own debriefing of the workshop. These records will prove invaluable as you prepare for future facilitation!

8.3 Public presentations

Public presentations are a part of most workshops, seminars, conferences, and public meetings. With conservation issues, public presentations offer the chance to present the latest scientific data, address public misconceptions, influence attitudes and behaviors, and respond to questions affecting the lives of community members. If your audience has a stake in the issue, a public presentation can speak to the heart of their concerns. In Rocky Mountain National Park (US), researchers funded by the EPA spent 2 years assessing the possible effects of climate change on factors such as the local economy, hydrology, and wildlife. Public presentations to local stakeholders on the findings gave community members insights into future environmental scenarios in their hometown.

When audience members are divided on a conservation issue, public presentations can become lightning rods for opposing political stands. In cities across the world, opponents and supporters of proposed developments often convene at public presentations. During these forums, speakers may present the costs and benefits of construction of large shopping areas in terms of destruction of habitat for wildlife, growth of local economy, and the demise of small business. Often the stakes are high in terms of conservation of land and small businesses, and the public presentations touch the core of community concerns. Whether you are giving a presentation to a group of 20 at a local environmental organization or talking to a packed city council meeting of hundreds, concrete tips for public presentations will enhance your performance.

8.3.1 Planning

Singer Patty Scailafa, a member of Bruce Springsten's E Street Band, overcame her fear of performing in front of thousands by focusing on singing to one person in a crowded auditorium or stadium. Public speaking can evoke similar fears, but time-tested strategies exist for delivering effective public presentations.

The key to planning is appealing to the needs and concerns of your audience. Think about what makes you pay close attention to a friend or family member talking at a loud, crowded party. Typically, a good story with a personal connection can make you tune out background noise and focus solely on the details of the tale. With public presentations, good stories also hook the audience by weaving the key points of the talk into meaningful anecdotes, supported by substantive data. Remember that effective presentation styles are individualistic and natural. An audience can tell if you are affecting mannerisms that do not fit your speaking style.

When you begin practicing your talk, construct notes that will cue you, rather than deliver your lines word for word. You can write out your talk but then construct notes in outline form. Nothing bores an audience more than being read to. Practice your talk aloud and in front of a small group of colleagues and videotape yourself if possible.

Publicity is another key to planning a public presentation to ensure you have an audience. For some talks, you will be the invited speaker on a predetermined agenda, such as an Audubon Society meeting or the local zoning board. But if you are organizing your presentation to attract a variety of stakeholders, publicize the presentation through multiple channels used by your audience, including newspaper, flyers, e-mail lists, and word-of-mouth.

Organization

Presentations generally use the structure of an introduction, main body, and conclusion. The three-step mantra repeated in many high school debate teams holds true today: tell them what you are going to tell them; tell them; then, tell them what you told them. The introduction should draw the audience into the talk through revealing anecdotes or surprising facts. Highlight the key points of your talk during the introduction and present a concise thesis statement.

The body includes the key points that support your thesis statement. Limit the body of your talk to the most significant points related to your primary topic and support your points with evidence such as personal experience, research, expert testimony, and historical data (Meany and Shuster 2002). Make sure you rehearse the transitions between the main points. The National Park Service in the US trains its interpretive staff to develop audience connections to both tangible and intangible resources to make the presentation relevant to the audience. Tangible resources are those that can be seen, such as props, photographs, or charts. Intangible resources are universal and apply to diverse cultures, such as values like love, safety, and family. Interpretive staff are trained that every presentation should include both tangible and intangible connections to support the main points of the talk (National Park Service 2005).

The conclusion should summarize the main points and create a memorable message to leave with the audience. After concluding the talk, plan to take questions from the audience as well.

Audiovisual aids

Presentation aids—from a flip chart to computer-projected images—can enhance a presentation using the following tips. Remember though that audiovisual aids do not substitute for developing an effective talk!

- Visit and set up the room or outdoor space prior to your talk. Consider all equipment you need, such as VCR/DVD player, pointer, flipchart stand, and computer projector. Practice with the equipment you plan to use. Check for electric outlets or bring an extension cord.
- If you are using overheads, make sure you number the overheads in the order of your presentation. If you drop the overheads on the floor, you can easily pause and put them in the correct order.
- Set up chairs so all audience members can see the audiovisual aids. Consider arrangements for the hearing and visually impaired, if relevant.
- Evaluate the lighting in the room. If using slides or computer-projected images, do not plan to plunge your audience into darkness at the beginning of your talk. Address the audience before dimming the lights.

The use of graphic software such as Power Point has become ubiquitous in presentations. There are many advantages and conveniences of these tools, but you should heed some guidelines to ensure your slides enhance, rather than dominate your talk.

- Plan for one graphic slide per 15 s, but vary the time from a few seconds to a minute (Jacobson 1999).
- Limit the amount of information on a slide, so the text does not overwhelm the audience. Aim for key phrases rather than complete sentences. Some presenters only use images such as photographs to illustrate talking points without including any text on the slide.
- Prepare text and graphic images that contrast with the background. Dark text on a light background is easy to see, such as black on ivory. Light text on dark background is also easy to read. Use conservative colors such as blue and green. Avoid using reds.

- Select clear and simple fonts. Limit the variety of fonts.
- Eliminate distracting and unnecessary words, graphics, or transitions between screens.
- Have a back-up plan, such as outline notes or text handouts, in case the technology fails you.

8.3.2 Implementation

Practice, practice, and more practice before your actual presentation will boost your confidence during your talk. This practice also will help you eliminate any distracting mannerisms you have, such as saying "uhm" or "you know," and using vague and imprecise words such as "these things." During your talk, use expressive body language, dress appropriately, and maintain eye contact with your audience. Most people have heard a presentation where the speaker faced the flipchart or screen more than the audience. Avoid this trap by using notecards or an outline of speaking points. You may want to choose one or two members of the audience in different places in the room to focus on during the presentation.

Use your normal conversation or speaking style, articulate clearly so the audience can understand you, and vary the pitch of your voice. Try to practice "word economy" as a speaker, using a minimum of words to express what you mean (Meany and Schuster 2002).

At the beginning of your talk, do not worry if you are nervous. You can use that nervous energy to appear excited and focused for your presentation. Another strategy for reducing nervousness is to anticipate "presentation disasters." Try visualizing how you will react to "worst-case" scenarios such as losing your train of thought or arriving at the talk to find ten times more people than you expected (Box 8.4).

Also remember that you are probably the expert on the subject, so try to enjoy conveying your message to the audience. Keep presentations to a larger group to 30–45 min with time for questions and discussion at the end. Always repeat a question asked since other members of the audience may not have been able to hear.

Avoid staying in one fixed position during the presentation, such as behind a keyboard or podium. Rather, engage the audience by stepping away from the podium at key points. People process written information approximately four times faster than when they listen to the same information (Meany and Shuster 2002). For this reason, use your slides as back-up information or illustration of your own speech, rather than talking while your audience reads text from your slides. The slides are reference tools that keep pace with your talk and illustrate your key points. Remember that you can use blank slides during transitions as well.

8.3.3 Evaluation

One concrete strategy to evaluate your presentation is to videotape the actual talk and watch it to critique your own presentation style. Many speakers also prepare an evaluation form for their audience to solicit performance feedback immediately after the talk. Have a stack of business cards available or your contact information on a handout so people can get in touch with you after the event. Your presentation

Box 8.4 How to anticipate and handle presentation disasters (Adapted from Meany, J. and Shuster, K. (2002). *An Introductory Guide to Effective Public Presentations.* Retrieved February 26, 2005, from http://etsportal. mckenna.edu/FITness/Features/docs/meany_presentation_guide.pdf)

How do you deal with a disaster during your presentation? Most disasters are not life-and-death situations, and you can deal with them by following a few helpful hints. Above all, try to keep smiling and calm. So what should you do if...

You find out you have 15 min instead of the 45 min you had planned.

Decide quickly what percentage of the 15 min each part (introduction, main body, and conclusion) should take. Pay attention to your watch and limit yourself.

Someone asks a question about an issue you plan to discuss later.

Answer briefly and say you will go into detail later. If you are presenting in a meeting, do not ask people to hold their questions until the end, as that request suggests you are not confident enough to handle interruptions.

You expect to talk to 5 people but 120 people are in the audience.

If you were planning on working from a set of handouts, change strategies unless you can make copies quickly. Ask for a flipchart and make a sketch of key points for everyone to see. Remember, you are the expert!

You plan to talk through a handout page by page, but people are reading at their own pace.

If you give people handouts at the beginning of the talk, they usually read at their own pace. If possible, give handouts at the end of the presentation. But tell the audience a handout will be distributed, so they do not become frustrated by taking notes.

Several people start a side conversation while you are speaking.

Try these strategies in this order: Ask if there are any questions. Ask if you can do anything to clarify. If they will not stop talking, continue your presentation but move nearer to them. Lower your voice or pause. If absolutely necessary, ask them to continue their conversation elsewhere or ask the group if they want to schedule a new meeting time.

is an opportunity to open dialogue about the topic, and comments after the talk can shed light on your performance. Give colleagues a feedback sheet and ask them to meet with you after the presentation to share their opinions (Box 8.5).

8.4 Information booths

Information booths provide information and raise awareness about your conservation organization or issue to people gathered at an event, such as a festival, fair, concert,

Box 8.5 This sample feedback form can give you reactions from audience members on your performance during public presentations.

Presentation Evaluation

Name _____

Please rate the following:

1. Organization of presentation	Poor	Fair	Good	Excellent
2. Logical sequence of presentation	Poor	Fair	Good	Excellent
3. Knowledge of presenter	Poor	Fair	Good	Excellent
4. Ability to answer questions	Poor	Fair	Good	Excellent
5. Visuals	Poor	Fair	Good	Excellent
6. Professionalism	Poor	Fair	Good	Excellent
7. Overall presentation	Poor	Fair	Good	Excellent

8. What were the strengths of this presentation? Please explain.

9. What were the weaknesses? Please explain.

10. What suggestions do you have for improvement?

exhibition, or conference. Information booths also may be set up at a central location frequented by a specific target audience, such as a grocery store or university commons. Many environmental organizations use information booths, from World Wide Fund for Nature with 5 million supporters worldwide to Earthaven Ecovillage, an environmental community of 55 adults and 17 children.

Information booths used for events typically include a portable exhibit such as a display board that sits on the back of the table, leaving room for publications, sign-up sheets, and demonstrations at the front. An advantage of the information booth is that staff or volunteers can engage people, answer questions, and direct individuals to specific resources.

The Land of Sky Regional Council has an information booth to promote its annual Strive Not to Drive Week (Figure 8.3). The organization places the booth at local fairs and festivals. Volunteers also target an organic grocery store where many shoppers make the commitment to reduce their driving by carpooling, biking, or taking public transit during that week. During a 2-h time slot, for example, volunteers registered 70 people for the campaign to reduce their driving.

Fig. 8.3 An information booth for the Strive not to Drive campaign seeks commitments from the public to use alternative transportation (photo by Land of Sky Regional Council).

The display includes publications about air pollution, local bus schedules, events during Strive Not to Drive week, and an interactive Web site where people can type in the model of their car and find out its emissions. Volunteers also enter the names of people who make the commitment to reduce their driving in a drawing to win prizes such as a new bike.

Information booths can focus attention on one issue, increase awareness, and allow for different levels of sharing information with the public. If people are very interested, they may talk with you for 30 minutes about the issue, while those not as interested may just take a brochure. Challenges of using information booths however are that the staff or volunteer at the booth must encourage involvement from passerbys. Also, the public must be motivated to attend the event where you have set up the booth (Citizen Science Toolbox 2004b).

8.4.1 Planning

Planning for an information booth includes the design of your display, as well as the research and preparation for interacting with visitors. People view the average exhibit for less than 60 seconds. Often less than one percent of people read the whole text of an exhibit (Fazio and Gilbert 2000). Therefore, the display for an information booth should be attractive, brief, and clear, and address the design principles of good exhibits (Chapter 12).

Design

Themes. Your display should communicate a theme to the viewers. By stating the theme in the title, you increase the probability that even visitors who just glance at your information booth will take away your central message. A theme is the key point of any display (Ham 1992). Visitors to an information booth should be able to capture the essence of your organization with just one sentence, such as "Riverlink protects the French Broad River." Avoid displays that include only topics, such as "Riverlink," the name of the watershed protection organization.

Remember that topics are only the subject matter, while themes are the key message. Moreover, people remember and understand themes better than topics (Ham 1992). Think about that one sentence you would like visitors to remember from the booth, such as "Our community depends on us to conserve natural resources" or "Healthy human communities depend on healthy natural communities." Design the display to support this theme and four to five supporting points. Solicit feedback from others on the theme, messages, and design of your display. Chapter 12 includes more details on developing themes.

Attracting visitors. The design is important in attracting people to your information booth. Displays can include opportunities to interact and get feedback from visitors, such as models, posters, visual representations of community issues, "post-it" idea boards, maps for people to mark their most and least favorite green spaces, and demonstrations of proposed community projects (Citizen Science Toolbox 2004b).

Designing opportunities for people to interact with the resources of concern to your organization can attract young and old to your booth. The nonprofit Riverlink, for example, has an information booth at Bele Chere, one of the largest music and art festivals in the Southeastern United States. The Riverlink booth is always one of the most popular sites at the entire festival. What is the secret? Riverlink secured permission from city officials to use the large downtown fountain for "whitewater" raft rides for kids. In partnership with rafting outfitters, Riverlink uses trained rafting guides who guide the boats right into the spray of fountains, to the delight of children during the hot month of July. While parents wait for their kids, they often pick up pamphlets and other information about Riverlink and conservation of the area's creeks and rivers.

Additional strategies for attracting people to your booth include:

- Giveaways, such as tree seedlings, local food, or coloring books.
- Raffle items or contests.
- Movement from videos, mobiles, running water, or any demonstration.
- Lights, such as ones that flash or twinkle.
- Sound, from music to a recorded animal howl.
- Giant objects that draw attention due to their scale.
- Participation by pushing buttons, turning knobs, or anything that involves people in the display (Fazio and Gilbert 2000).

Remember that any technique you use to draw attention must have a direct connection to your theme. Otherwise the strategies will come across as gaudy gimmicks, rather than fun complements to your booth!

Publications and resources. Your information booth should include publications, pamphlets, magazines and other resources for people. Include a sign-up list for people who want more information or have questions.

Preparation of staff and volunteers

The personnel at an information booth are on display as much as the exhibit (Fazio and Gilbert 2000). If you are choosing the staff or volunteers to work at the booth, select these people carefully with the goal of creating positive interactions between your personnel and the audience. Many organizations make the mistake of sending untrained volunteers or staff without much investment in the organization (Fazio and Gilbert 2000). Avoid this mistake by training volunteers and providing them with adequate resources to answer the questions you receive most frequently. The last thing you want is to send someone to oversee a booth and have him or her sit behind the booth reading the newspaper and avoiding questions during the entire event!

Training for volunteers before they work at an information booth should present an overview of the organization, its successes and challenges, and familiarize volunteers with the publications. Provide a take-home tool, such as a handout of frequently asked questions with responses and tips for volunteers working information booths. You can use and adapt the helpful hints from this list to any conservation group (Block 2003):

1. If you are a volunteer, learn about the organization or conservation issue by exploring the website and reading all brochures and handouts. If you have questions, call the staff or volunteer coordinator.
2. Think ahead of a few good opening questions to start conversations with people who stop to look at the display board or publications. For example:
   - Do you have concerns about water quality (or your conservation issue) where you live?
   - Have you heard about our organization?
   - Do you have questions about the organization?
   - Can you find where you live on this map? Here is a marker—make a dot.
3. Practice a short answer to the questions: "So what does your organization do?"
   - Example: "We are a statewide nonprofit that works with communities on environmental issues that affect human health. We especially support low-income communities and communities of people of color that may not have the resources they need" (Block 2003).
4. Learn about the festival. What is the theme? The purpose?
5. Anticipate your audience. Whom do you expect to attend? What ages? What languages will people speak? What types of questions might they ask? For

water quality issues, a rural audience may ask about septic tanks and agricultural runoff. An urban audience might ask about construction site runoff and industrial facilities.

6. If you are picking up supplies for the festival, make arrangements with the staff at least one week before the event.
   - *Standard supplies*: box of brochures, organization banner, display board, maps, tablecloth, pens, nametags, mailing list sheets, duct tape.
   - *Possible additional supplies*: table, tent, chairs
7. Get directions, passes, and any special instructions from staff. Exchange your contact information with staff in case of weather changes or last-minute emergencies.
8. Allow extra time when going to an unfamiliar location. Plan to arrive on time or early. It is often helpful to arrive early so you can quickly check out the content of other booths and the demographic of attendees.

8.4.2 Implementation

You arrive at the site, hear the sounds of bands warming up, and watch as festival organizers point left and right directing exhibitors to their assigned places. Now the fun begins, as you get to relax and interact with people about your passion. For most events, individuals will sign up for 2–3-h time blocks. If you are the staff in charge or the volunteer doing setup, plan to give yourself 30 min extra time to find the location of your booth and arrange the display and publications. When you transition with volunteers for the next shift, make sure you discuss any issues that emerged, such as publications that are in short supply, interesting questions asked, or humorous stories to relate! The following additional guidelines will help implement a successful information booth:

- Before you head off to the festival, be sure you have any necessary parking passes, entrance passes, food tickets, directions, drinking water, site map, and the contact name for the event.
- Make the information booth as attractive and organized as possible. Restock depleted handouts. Put boxes under the table and out of sight. If there is a limited supply of one report, place only one out at a time. Pick up any coffee cups or soda bottles from the table, and hide any valuables such as purses.
- Wear a nametag. Preferably you will have printed/typed nametags, which you can reuse at the next event. If not, ensure the nametag is legible.
- Stand (or sit if necessary) to the side of the table toward the front and look alert and welcoming. Do not stand directly in front of the display or place chairs in front. Try not to pounce on visitors but be ready to engage them with a general opening question. Keep a clicker in your hand to record the number of people you talk to.
- Let visitors take what they want from the table rather than hand everyone a brochure. If they have a specific question, you can recommend a related handout.

- Always ask if people want to be on a mailing list. If so, have them write out their name, address, e-mail, and phone. Ensure that their e-mail address is legible. Indicate if they want to become a volunteer. Know that you won't be able to answer all questions and that's okay! You can respond by saying, "I don't know the answer, but I can take your name and number and have one of our full-time staff give you a call, or you could call the office directly" (Block 2003).

8.4.3 Evaluation

If you are working the information booth, pack up all supplies at the end of the event and thank the organizers. To evaluate the information booth, many organizations use a volunteer evaluation form that allows volunteers and staff to document their concerns, impressions, and questions about the event. What questions did people ask? Were there any misunderstandings or common perceptions and what were they? How many people wanted more information about the organization? Did the volunteers feel they had adequate training for the event?

To evaluate your display, ask visitors or colleagues to give feedback on the design, even before you attend an event. Ask them if they can tell you the main message of the information booth. What were the salient points they got from the display? At the event, choose a representative sample of visitors and ask them for their opinions on your booth. What do they like best? What do they like least? What are suggestions for improvement? Did they learn anything new from the booth?

In their annual reports, many organizations report the number of people reached during festivals and events through information booths. You can also include feedback from people who visited the information booth. Be sure that you make changes to your display based on common recommendations for improvement. For example, if you repeatedly have people tell you that the display has too much text and too few visuals, make those changes immediately before your next event.

8.5 Professional posters

The advancement of knowledge depends on effective communication of research findings to the scientific community as well as the wider public. Poster sessions are an important form of communication at many professional meetings as they allow you to highlight your main message—your research findings. Viewers can study your poster at their own pace and discuss the results with you in an informal and interactive setting. The Society for Conservation Biology meeting held in Brazil in 2005 displayed approximately 100 posters.

Planning a poster follows the usual guidelines of determining your goal, audience, message, and evaluation process. The goal of a poster is to convey scientific information in a visual format to an audience, composed of viewers such as other scientists, extension professionals, and students. Similar to a good oral presentation, an effective poster can help you make professional contacts, establish your reputation in the field, exchange information about research findings with similar researchers, and promote your department or agency. With some planning and preparation you can create an effective poster presentation and make a great impression (Jacobson 2005).

8.5.1 Planning

How do you attract viewers to your poster? The type of meeting you are attending will allow you to determine your audience. You'll be competing with scores of other posters and often a concurrent social gathering. The same principles for designing an effective brochure apply to posters (Chapter 10). Qualities of attractiveness, brevity, clarity, and dynamism are crucial to hook your audience. The audience will want to briskly scan the important points of the your poster to see how it relates to them. An inviting title, enticing graphics and clear layout will encourage your audience to give your poster a closer look. Study other scientific or professional posters that catch your eye before designing your poster (Fig. 8.4).

Content

The conference organizers will often dictate the content of a poster. Posters may be required to include a title, authors and institutions, abstract, introduction, methods, results, discussion, conclusions, and literature cited. A good poster is organized around an introduction, body and conclusion, similar to an oral presentation.

The title should concisely communicate your main message. It should catch the attention of the audience and arouse their interest in your subject. The introduction should briefly indicate why the topic is important, placing it in the context of other scientific literature. It should state the hypotheses and objectives of your project or study and briefly explain your methods. A photograph of your study organism or site helps illustrate this section. If a separate methods section is called for, briefly describe your experimental equipment and techniques and your statistical analysis. Often a flow chart or table can be used to illustrate the research or project design.

Use visuals to present results with using graphs, photos, and other illustrations. Briefly describe if the study supported your hypotheses or if the experiment worked. Figures should have legends that concisely state your major findings. These may be all that many viewers read. The main conclusions and implications should sum up the hypothesis and results, and explain why they are noteworthy relative to other studies and to the real world. Literature citations can be placed at the bottom of the poster, since only dedicated colleagues will want to peruse them.

Keep in mind that most people will spend about 10 s gazing at your poster, so you must catch their interest with your title and graphics. Few will spend more than three minutes. This limits your total text to 200–300 words. It is important to curb the urge to tell the viewer everything about your study. Instead, you must be satisfied if you communicate your messages. Colleagues who are involved in related research will no doubt spend time chatting with you at the poster session, allowing you to describe more details of the study.

Fig. 8.4 Posters present research or program results and stimulate social interactions at meetings (photo by S. Jacobson).

Layout

A poster is made up of three components: text, graphics, and empty space. A rule of thumb is to have equal parts of these three elements. Just as in brochure design, the empty space helps make the poster look inviting and easy to read. Visual material, such as photographs, drawings, diagrams, charts, graphs, and tables will make the poster attractive and help emphasize the main points. Charts and graphs are usually easier to absorb at a glance than tables. These should be simplified to enable the reader to understand quickly the major findings.

To create more empty space, avoid extraneous text and distracting graphics. Colors, graphics and text should work together to make the poster easy to read. The size of the text and the colors selected for text and graphics can help orient the reader and emphasize the main points. Larger text for headings and subheadings will help organize the poster and highlight important points. Choosing a light, unifying color for the background and two or three dark, contrasting colors for your various sized-text and graphics will make your poster stand out and attract attention to your photographs and other visual materials. Color can be used to associate related text and graphics and to separate various sections. But beware, too many colors make the design look busy and confuse the viewer.

Conference organizers will usually specify the size of the posters. A vertical format that is about 120 cm high by 100 cm wide is common, although many posters also use a horizontal layout. The poster title should have lettering large enough to be read from 4 m away to reel in prospective viewers. Poster text should be legible from a distance of 2 m. This requires font sizes of 90–110 pt for the title, 48–72 pt for the section headings, and 24–36 pt for the text. Some scientific posters use a minimum text font size of 18 pt. Remember if you can't make out what it says, no one else can. Only the least important information, such as literature citations should be printed in these smaller fonts.

The layout should follow how people usually view material. A person's first glance at your poster will be at the center. Attract them with a succinct title and effective graphics. English-speakers are used to reading from left to right and from top to bottom. Therefore, orient your poster accordingly. If the allowable poster size is large, think carefully before you fill all the space. Few people will squat down to read your conclusions if they are printed at knee height.

Much of the design elements for posters are common sense. Posters often display text and graphics in three vertical columns because of the size of the text font and the number of words (about 11) that fit across each column and make it easy to read. Lower-case print is more legible than all capital letters. In fact, using all capitals reduces reading comprehension by 25% and reduces speed by 14% (Ham 1992). This means capitalize only the first letters in your title and headings to maximize ease of reading. Bulleted lists are easier to read than full sentences or numbered items.

Many conservation educators and scientists resize PowerPoint files to design professional posters. By adding color, photos, text boxes, and other design features with PowerPoint, it is easy to create and print an attractive product to transport to a meeting or conference. University print shops charge about US $30 for the printing, while copy shops can charge up to $100 to print a high-quality poster designed from computer software. The high cost may be worth it, if you plan to use the poster several times or if you are marketing your program with this tool. If you present a poster in collaboration with other partners, several organizations can share the costs of printing.

If those costs are prohibitive, a cheaper way to make a "poster" is creating PowerPoint slides and printing them with a regular printer, but with a nice piece of construction paper behind each slide as a border. Arrange the slides on a poster-board that you can transport to the meeting.

8.5.2 Implementation

Once you have prepared your poster, you can shift your focus to enjoying your interactions with colleagues at the poster session. Try to arrive early to the poster session to ensure you have a place to mount your poster. Typically, at least one collaborator in the research project should stay by the poster during the session to explain the findings and discuss results with viewers. This is your opportunity to engage colleagues in your research and make additional contacts. Be prepared to explain your work in one succinct sentence. Always have a supply of business cards, the URL of a Web site, or an abstract of your study available to give to viewers if they want more information about your research. Remember to bring thumbtacks or some other mounting equipment in case it is not provided.

8.5.3 Evaluation

Similar to other communication materials, it is imperative to "pilot-test" your poster with members of the target audience. Feedback is critical. Before you actually print your poster, make a small mock-up and ask your colleagues to critique it. Are they confused? Can some verbiage be eliminated? Do they understand your main points? Are your graphics clear? Fix it before you waste time and money printing your final poster. Also ask to see a proof of your poster before printing the final version. You may find that fonts, colors, or shading changes from one computer to another. Then you can attend your meeting and enjoy the benefits of successful outreach. After the poster session, assess how many colleagues interacted with you, read your poster, and discussed the results with you. Keep track of individuals who were especially interested in your research and follow-up with an e-mail if they had questions. During the session, check out the other posters, especially those that attracted many visitors, and get ideas for improving your own poster design (Jacobson 2005).

8.6 Conferences

Have you ever returned from a conference with renewed energy for your work, a creative idea, or a valuable contact with a new colleague? Well-organized conferences can generate enthusiasm, exchange innovations in education and outreach, and create partnerships. A conference usually brings together like-minded people or members of an organization to share information, network, conduct business, build skills, and make decisions together. The scale can range from a 1-day conference with 20 local participants to a 1-week, large-scale international conference with 1000 attendees and a professional conference manager. Conferences can bring together individuals who live and work in different regions of a country or even the world, such as the International Children's Conference on the Environment in Victoria, British Columbia or the International Outdoor Education Research Conference in Victoria, Australia.

The North American Association for Environmental Education (NAAEE) convenes an average of 800 members to its annual conference with varied locations and themes each year. Past locations have included Anchorage, Alaska, US, Cancun, Mexico, and Montreal, Canada. The conference activities each year involve keynote speakers, concurrent presentations, affiliate meetings, poster sessions, a silent auction, exhibits and vendors, an authors' corner, dinners and social events, a service project, entertainment, and field trips highlighting local ecological and conservation education issues. The NAAEE conference in Boston, MA, for example, featured a 1-day workshop on environmental justice that included a field trip to meet local youth working to eradicate environmental racism in their inner city neighborhood.

The setting for a conference should match the philosophy and membership of an organization if feasible. The Gathering, a week-long conference for hikers and supporters of the Appalachian Trail meets in Hanover, New Hampshire, (US) with participants camping on the grounds. Through presentations and informal conversations, attendees exchange information on topics such as lightweight backpacking and conservation efforts affecting the Appalachian Trail, which extends from Georgia to Maine, US.

The location also should consider factors for inclusiveness such as access for the disabled, childcare, distance from transit centers, and affordability. The Association for Experiential Education (AEE) provides childcare at its international conferences and allows children and families at meals and evening entertainment. Such arrangements make conference attendance more feasible for many working parents. While the AEE conference is located at a large hotel or conference center, the marketing materials for the conference include information about affordable lodging, such as youth hostels and campgrounds. If conference attendees cannot convene in one location, videoconferencing is an option that uses technology to bring people together to network and share ideas (Chapter 11).

8.6.1 Planning

Planning a conference is an invitation to get creative, disseminate new information, help members of an organization or cause, and raise money for an organization. The standard menu of options for a conference includes lectures, seminars, poster sessions, hands-on demonstrations, panel discussions, roundtable discussions, social events, plenary sessions, and service projects. Many larger conferences have plenary sessions attended by most participants with concurrent presentations organized by topic. If you are planning a 1-day conference, you might decide to schedule a presentation and large-group discussion in the morning, followed by break-out groups with focused questions in the afternoon. Think about what activities would best serve your audience as you start to brainstorm about the conference, and don't be afraid to try something new! The Association for Experiential Education conference includes "consultations with experts," so new members of the field can sign up for 30 min consultations with a leader in experiential education. The leaders who have been working in the field for years say they gain valuable contacts as well.

Before you begin, decide on a way to organize and document your planning. Most conference organizers have a large three-ring binder with divisions such as: timeline/schedule, initial proposal, contacts, committees, correspondence, budget, contracts, site, marketing, promotional materials, registration, food and reception, theme and program, and audiovisual (Devney 2001). Even if you computerize the majority of tasks such as registration, you will want hard copies of all documents in this binder.

Begin to talk to the people who have organized this conference or a similar conference in years past to get their tips, tricks, and suggestions. Form committees to help you shoulder the big-picture vision and the logistical details of the conference. The administrative committee will include any paid staff and will oversee the other committees. Depending on your number of volunteers and staff, you can combine the committees listed below to cover the necessary tasks (Devney 2001). Remember to build in methods for your committees to communicate with each other and work together, through e-mail updates or periodic meetings.

- Administrative committee: coordinates all the committees, evaluates, and writes final report.
- Business: receives and deposits checks, pays bills, solicits sponsors and exhibitors.
- Facilities: selects and reserves site, diagrams facilities (seating, table arrangements), contacts government officials, obtains permits.
- Publicity: writes press releases, creates posters, designs advertisements, interacts with the media.
- Reservations: tracks reservations, handles reservation desk and packets, greets attendees.
- Kitchen and serving: plans menu, shops, cooks, or selects caterer; coordinates servers.
- Programming: schedules programming, selects keynote speakers, presides over ceremonies, produces printed agenda and program.
- Setup/cleanup: assigns exhibit space, moves furniture, cleans and closes site.
- Theme and decorations: develops theme, decorates site (Devney 2001).

Your next step is deciding the goals of the conference, researching the audience, and deciding on the content and activities. What type of conference appeals to your audience? What are their needs in terms of accessibility, programming, location, and cost?

An important piece of conference planning is outlining the budget and identifying potential sponsors. First outline an estimate of your budget to get an idea of the scale of conference you can afford. Use the attached budget worksheet to assess your income and expenses with categories typical for most conferences (Box 8.6). As you calculate the budget and the estimated number of participants, try to

---

Box 8.6 Budget worksheet for a conference (Adapted from the book *Organizing Special Events and Conferences: A Practical Guide for Busy Volunteers and Staff* by Darcy Campion Devney (2001). Used with permission of Pineapple Press, Inc., Sarasota, FL)

Expenses

Site $_____
   Room and hall fees
   Site staff
   Equipment
   Tables and chairs

Refreshments $_____
   Food
   Alcohol (if included)
   Linens
   Staff
   Catering fee

Decorations $_____
   Flowers
   Candles
   Lighting

Prizes $_____
   Ribbons and plaques

Programming $_____
   Performers
   Travel, lodging

Publicity $_____
   Photocopying
   Postage
   Graphic artist/photographer

Miscellaneous $_____
   Telephone

Paper supplies
Fax
Photocopying
Transportation
Postage

Add 20% for preliminary budget $_____

Total $_____

Income

Admissions/attendees $_____
_____attendees @ $_____ each
_____attendees @ $_____ each

Exhibitors/vendors $_____
_____ booths @ $_____ each

Advertisements/programs $_____
_____ 1/2 pages @ $_____ each
_____ 1/4 pages @ $_____ each

Sale of items $_____
_____ items @ $_____ each
_____ items @ $_____ each

Total $_____

Total profit (or loss)

Total income $_____
   minus

Total expenses $_____

Total $_____

include reduced registration costs or scholarships for students or members with financial need in exchange for volunteer hours at the conference site.

Attendees at the annual AEE conference can choose to work on the service crew on tasks such as registering participants and setting up equipment for speakers. Working eight hours reduces the registration fee by 30%, while volunteering for 24 h results in a 70% reduction. While most participants may not choose to work 24 h, the option sends a strong message that your organization does not want to exclude participation based on financial resources.

Your goals, philosophy, and budget will help determine the conference site. The four basic criteria for site selection include location, cost, size, and facilities (Devney 2001). Calculate your site budget and programming needs and create a list of possible sites, which you can narrow to a smaller list of five to ten. Call these sites to screen for your needs, and then visit at least three sites in person. From this shortlist, you can negotiate facilities, personnel, and cost with the site your committee likes best. Remember that some large organizations require proposals for sites several years in advance. Once you select a site and sign a contract, you can work with the site manager to diagram the seating and tables for each event. More experienced conference planners know to negotiate for deals before they sign a contract. Ask for perks such as free audiovisual equipment, coffee, and complementary rooms.

The advantage of many conservation education organizations is a large pool of dedicated volunteers who can decrease costs by pitching in to help with these planning tasks. A list of conference planning tasks will help keep you organized and on track (Box 8.7).

Box 8.7 Schedule of tasks for a conference (Adapted from the book *Organizing Special Events and Conferences: A Practical Guide for Busy Volunteers and Staff* by Darcy Campion Devney (2001). Used with permission of Pineapple Press, Inc., Sarasota, FL).

When?	What?
5–12 months prior to conference	Recruit and meet with committees Decide purpose, goal, and theme Research audience needs Select site Announce conference theme, date, location Invite exhibitors and vendors
3–5 months prior	Meet monthly with committee Make a preliminary budget Choose caterer or decide menu Outline schedule and program Diagram site (which event goes where) Apply for permits, licenses. Contact entertainers and speakers Send publicity to newsletters
2–3 months prior	Meet monthly with committee Send invitations Select entertainers and speakers Send publicity to external media Send program to printers.
1–2 months prior	Meet monthly with committee Collect reservations Confirm all contracts Collate registration packets Purchase prizes, decorations, etc.
1 week prior	Meet with committee Close reservation list

	Arrange tables and seating
	Print nametags
	Give final head count to caterer
	Purchase nonperishable items
	Order food items
	Borrow/rent any cooking and serving equipment
1 day prior	Purchase perishable food
	Pack up equipment
	Deliver ingredients to cooks
	Pick up keys; transport items to site
	Call all committee chairs to confirm tasks
Conference Day	Put up directional signs
	Meet with site staff
	Unpack equipment
	Set up registration table
	Begin cooking
	Open event officially and begin scheduled activities
	Informally interview participants
	Set tables and serve food
	End scheduled activities and clean
	Pack up supplies
	Check site for lost items
1 day after	Sign out, lock up, and celebrate!
	Take down outdoor signs
	Thank all volunteers
	Clean and return borrowed items
	Make notes on conference evaluation
2–4 weeks after	Meet with committee for evaluation session
	Summarize evaluation techniques
	Finish budget paperwork
	Calculate profit or loss
	Write final conference report

Note: This schedule also applies to planning a special event.

8.6.2 Implementation

The day of the conference culminates months of hard work with the excitement of the actual event. Before people arrive to register, place directional signs on the main routes to the conference site. Hang your conference banner above the site or on the road leading to the conference. Unpack all equipment that you will need for the day, and touch base with the staff at the site.

Your registration committee should set up the registration tables in a configuration that allows a smooth flow for new arrivals. Standing in a long line with irritated newcomers is not the first experience you want folks to have at the conference. For a large conference, some organizations use a registration service that handles reservations, registration, check-ins, bank deposits, and name tags (Devney 2001).

Sometimes personnel from the convention and visitors center will be in charge of reservations for a large event. The registration site is the location of the master list of reservations and payments, information packets, nametags, notices and announcements. Ideally, for a large conference, your registration site will include a separate information/questions desk and registration table, so attendees without complex questions can sail through registration.

Opening day will involve preparing meals and refreshments if you are providing the food. Try to include regional specialties and local foods, but remember to plan for diversity with your meals. Vegetarians visiting cattle country will be unhappy and hungry without some choices. If you are using a caterer, talk with the staff early in the day to ensure timely and sufficient preparation of food. During the conference, you will spend much of your time trouble-shooting. The presenter from the Audubon Society is delayed in the airport... What should we do about her scheduled keynote? With organized planning, you can handle the unforeseen problems that will arise.

8.6.3 Evaluation

Attendance and profit margin are often the first measure many organizations use to assess the success or failure of a conference. Your evaluation must refer to the original conference goals—the drivers in your planning and implementation. As stated, during your planning, review your goals and identify evaluation tools to measure your effectiveness. Evaluate your conference using these methods (Devney 2001):

- During the conference, randomly talk with participants to get informal impressions.
- Have evaluation forms at individual workshops or sessions
- Include an overall evaluation form in the registration packet. Have a box for the evaluation forms at the exit. If possible, provide a small incentive for completing the forms.
- Hold a meeting with your administrative committee soon after the event. Debrief the conference in terms of attendance, registration, goals, programming, and calculate your profits or losses. What worked well? What didn't? What will you do differently next year?
- Photocopy your conference binder for reference by future organizers.
- Write a final report of the workshop to include information such as budget, site selection criteria, program, menu and recipes, sponsor list, committee members, and evaluation summaries.

8.7 Special events

Special events can range from local to national, outdoor to indoor, serious to silly. An event can be anything from a public meeting to promote discussion of a conservation easement to a triathlon to raise awareness of public transit systems. A special event could involve a giant community board game of Monopoly to simulate changes in land use, a parade to celebrate a local harvest, or media events to generate publicity for a conservation cause (Fig. 8.4).

Each August Bat Night festivals occur in countries throughout Europe. In 2004, Bat Night festivals were held in 24 participating countries including Bulgaria, Latvia, and Armenia to celebrate bats and raise awareness about the plight of bats. Festivals feature excursions, exhibits, childrens' workshops, and information about bat conservation.

Earthaven Ecovillage, an ecologically focused community organized a special event called the Permaculture Summer Gathering, which assembled participants to discuss bioregional organizing, permaculture, organic growing, and natural building. The activities included classes, plant walks, tours of Earthaven, display booths, a music and coffee house, and kids' activities. Participants camped on the grounds of the community and ate vegetarian meals prepared in the outdoor kitchen.

In Beijing, China, the World Wide Fund for Nature's Power Switch! Campaign included a Save our Climate festival on World Environment Day that featured the rock band "Toy" debuting a song called "Save Our Climate." And in Ecuador, the village of Ambato celebrates the annual blooming of fruits and flowers with the "Fiestas de las Flores y de las Frutas de Ambato" (Ham 1992). You can create your own special event or connect your conservation cause to an existing celebration, depending on your resources and needs.

8.7.1 Planning

The first step in planning is to decide the type of event appropriate for your goals, audience, and organization. A special event may have the following goals (Devney 2001):

- Raise funds for a cause or place.
- Build spirit among long-term members to solve a problem or initiate a new program.
- Distribute or exchange information.
- Recruit new members.
- Celebrate.
- Attract publicity.

Given these goals, think about the desired interactions between participants at an event, between participants and a natural resource, or between performers/ organizers and attendees. Do you want different stakeholders to share perspectives on a contentious issue? If so, you might use a public meeting. Would you like to bring people together to mingle in a fun atmosphere? Then think about a game where you assign teams randomly. Do you want your target audience to learn about conservation of habitats for local species? How about a birdwatching hike? This planning can help develop the theme for a special event that is appropriate for the desired outcomes. To plan and organize your event, use the budget and timeline of tasks in "Conferences" (Box 8.6 and 8.7).

The organization Riverlink organized the first Swannanoa River Awareness Day with the primary goal of increasing citizen awareness and interaction with the river. "Many people drive on bridges to cross the river, but they have never stood on the banks of this local river," said riverkeeper Phillip Gibson. The event targeted families through fishing lessons for kids, a chef grilling an abundant supply of local trout, river awareness booths, and environmental education games. "We accomplished our objective of increasing access to the river, but we didn't have a large turnout the first year," Gibson said. Scheduled on a Sunday afternoon, the event did not target churches and synogogues its first year. The second year, however, Riverlink staff realized the Sunday date meant they should publicize within these communities, a strategy that increased participation (Figure 8.5).

As with planning conferences, you will want a strong committee structure to facilitate organization. First convene an organizing or administrative committee to oversee planning of the entire event. Then create committees to delegate other tasks essential to the special event, such as committees for the theme, finances, facilities, publicity, food, reservations, serving, programming, and setup/cleanup (see "conferences" for details of committee tasks). The names and responsibilities of your committees will obviously differ depending on the nature of your special event.

Often your theme will drive the decision of the date, site, and logo, as well as your choice of sponsors. Consider connecting to existing celebrations of natural resources to streamline resources and organization. You can find published listings of national and international celebrations appropriate to your country and resource, such as *Chase's Calendar of Events: a Day-to-Day Directory to Special Days, Weeks and Months*, published annually by Contemporary Books. If possible, dovetail outreach efforts with existing community events focusing on nature, such as these holidays and annual festivals celebrated at local, national, and international scales (Ham 1992):

- The town of Tualatin, Oregon hosts an annual Crawfish Festival, a week-long celebration of the crawfish and river resources.
- Costa Rica has an annual "Festival de las Cebollas" "Onion Festival" in the town of Santa Ana, which celebrates the area's agricultural roots.
- Jamaica recognizes National History Week each year with a focus on the historical dependence on nature.
- Countries and cities around the world celebrate Earth Day on April 22 and World Environment Day on June 5.
- Bhutan hosts a Crane Celebration to celebrate this native species.

If you choose an outdoor location, remember to have toilet facilities, drinking water, and a contingency plan for rain. Also with many outdoor recreational events, such as a race or bike-a-thon, you will need a liability form for adult participants and minors.

The details of your planning will depend on the type of event you choose from the diversity of special events (Box 8.8). Your creativity is the only limiting factor in planning a special event that resonates with your target audience and becomes a community tradition!

Fig. 8.5 The organization Riverlink organizes special events such as the Swannanoa River Awareness Day, as well as rafting rides in a downtown fountain during the largest music and arts festival in the southeastern United States (photo by Riverlink).

> **Box 8.8** Creative ideas can connect special events to a conservation theme.
>
> - Fundraisers, such as phone-a-thons, bike-a-thons, dance-a-thons, costume contests, or sales of goods, such as a local crafts fair.
> - Cleanups, such as a beach, city, river, or road cleanup.
> - Games, including live board games, Twister, word games, or improv theatre.
> - Sporting events, such as triathlon, running race, fishing rodeo, kayaking race, adventure race, ultimate Frisbee, soccer, golf, or putt-putt golf.
> - Exhibits or demonstrations, such as field days and farm tours.
> - Fairs and festivals, such as an Earth Day festival.
> - Public meetings to bring stakeholders together on a specific issue.
> - Dances and musical performances such as a "Concert for Conservation," "Rock the World," or an "African dance and drums for wildlife."
> - Block parties, such as parties in neighborhoods with a focus on an issue or community campaign.
> - Tours, such as the Girls on the Move Outward Bound cross-country bike tour or Granny D, a 92-year old political advocate who tours countries raising political awareness.

8.7.2 Implementation

Make sure you contact the heads of your committees the day before the special event to ensure that everyone knows their responsibilities for the big day. Remember that you will be the point person for trouble shooting. The morning of the event, hang directional signs on the roads leading to the event and a banner over the site. Post any licenses or required permits for your site, and check in with security or site officials. Bring all your equipment for the day, such as tables, chairs, water, speakers, etc. Have copies of a site map with numbered locations for vendors, exhibits, educational displays, or performers.

As the event progresses, talk to participants about their perceptions for on-the-spot feedback. You may even be able to tweak some details during the activities. Circulate with committee chairs to address unforeseen needs. Does the drink table have enough ice? Are the vendors satisfied with their locations? Did the videographer show up to record the public meeting? Volunteers and staff will appreciate your calm attention to detail and ensurance that they have the necessary supplies and support. Give your committee coordinators cell phones or radios and keep in touch with each other during the event.

At the end of the event, work with the cleanup committee to pack up all supplies and check the site for any missing items. Take down all signage and banners. Now celebrate and sleep!

8.7.3 Evaluation

Keep meticulous records of your planning to assist future committees. Meet with your organizing committee after the event to document success factors, challenges, lessons learned, and unexpected outcomes. Write down remarks from your informal interviewing during the event and summarize your findings. Another useful tool is visitor feedback surveys, which you can distribute during the special event to ask feedback on what worked well, what didn't work, and what could be improved. Use separate feedback surveys for vendors, volunteers, and visitors.

Your evaluation tools, such as observation, interviews, and surveys, should address these common evaluation questions for special events (Devney 2001).

- Did the attendees enjoy the event?
- Were participants engaged in activities at the event?
- Who was the best speaker or what was the best workshop?
- Did participants like the location?
- Did attendees enjoy the food?
- Did participants feel the event was a good use of their time or money?
- Would they attend the event again?
- Was the event at a good time of year, month, week, day?
- Would they recommend the event to other people?
- What was the most important part of the event?
- What part of the event was unnecessary?
- What are suggestions for improving the event?

8.8 Partnerships

To enhance conservation of natural resources, we must build relationships with other groups to become "walking advertisements" for each other's objectives and mutually meet conservation goals. In business, a salesperson might ask, "How can I sell you what you need? Who can help my bottom line?" In conservation, we scan the horizon of other individuals and institutions and ask, "How can we align our interests?"

Typically, environmental groups are regarded as individual organizations with isolated causes, each with its own membership, messages, and lobbyists. Yet we are all working for the same end—conservation of natural resources. Thankfully, success stories exist with impressive outcomes when environmental groups create partnerships within the conservation community and with other institutions, such as faith communities, corporate groups, and the medical community.

A partnership is an institutional arrangement between two or more organizations that provides mutual benefits, which in turn helps achieve conservation goals. It can range in scope from sponsoring a weekend special event to joint lobbying efforts that span decades. Challenges with partnerships often include communicating between partners, defining roles and responsibilities, and maintaining focus on shared goals. Clear communication and documentation can help address these challenges. To address environmental damage in Northern Europe, a partnership developed called the Northern Dimension Environmental Partnership (NDEP). Led by the European Commission, the Russian Federation, and the European Bank for Reconstruction and Development, this network of donors tackles environmental problems and nuclear waste that affect northwest Russia and nearby countries such as Estonia, Poland, and Lithuania.

In another example, SmartWood, the world's first independent forestry certifier, established formal partnerships with nonprofit organizations to provide local expertise and certifications services. A program of the Rainforest Alliance, SmartWood works in Brazil with Instituto de Manejo e Cerificacao Florestal e Agricola, Brazil's leading certifier of sustainable forestry. In Eastern Europe, SmartWood partners with NEPCon—Nature, Ecology, and People Consult—another nonprofit consulting firm. These partnerships give local perspective to a global sustainable forestry program.

8.8.1 Planning

Planning for partnerships first involves identifying the reasons for aligning your interests with another group. Why should you partner in the first place? When you partner for conservation, you can:

- Share resources, strengths, strategies, and constituencies
- Discover mutual interests behind competing positions
- Build credibility within a community
- Increase media coverage
- Gain access to and influence politicians
- Broaden your understanding of different stakeholder groups
- Put your organization in a better position to leverage new funds.

The next step is identifying potential partners (Figure 8.6). Consider the following groups, as well as your stakeholders, and think about how their issues of concern intersect with conservation: faith communities, such as synagogues, churches, mosques, spiritual retreats; civic and youth groups, such as Rotary International and Boy Scouts; corporate groups, including local and even global businesses; hospitals, clinics, and others in the health community; resource extraction industries; other environmental groups; agriculture; the tourism industry; schools; municipalities; politicians; and unlikely partners, or any group that you perceive as the "opposition".

One of the leading pieces of air quality legislation in the United States, the North Carolina Clean Smokestacks Act, was the result of unprecedented partnerships and networking. Brownie Newman, former executive director of the Western North Carolina Alliance, was a leader in the movement to

Fig. 8.6 Identifying your stakeholders and other potential partners is an important first step in planning partnerships (photo by M. McDuff).

bring together unlikely partners, such as Carolina Power and Light (CP&L), the medical community, and the many environmental groups in North Carolina.

"At first, we saw the power companies as our biggest enemy in this struggle that took three years," Newman said. "But we ended up finding a way that consumers could take on the costs of upgrading the coal-fired power plants, which is a major contributor to our air pollution. With the power companies on board, we had some heavy political clout." The environmental community also came together in this battle and aligned their conservation interests with the health and medical field. In rallies and opinion columns in newspapers, medical experts discussed and wrote about the ever-increasing rates of asthma in the mountains where air quality was declining rapidly. "You may end up at the table with partners you would never expect," advises Newman, "and those partnerships make a major difference in your campaign."

Planning a partnership involves learning to focus on the interests of potential partners, not on their positions. A position is something you decided upon. An interest is the need, desire, or concern behind this decision (Fisher and Ury 1991). For example, the power companies first took the position of opposing costly upgrades in their scrubbers to conserve air quality. Western North Carolina Alliance and the other environmental groups maintained the position that these upgrades were necessary for the health of the environment and the citizens.

Yet an analysis of the interests behind the conflicting positions revealed that the power company's concern was financial. Who was going to pay the costs of the upgrades? The interest of the environmental groups was environmental conservation, which didn't have to conflict with a financial interest. When a strategy for paying for the upgrades was resolved, the groups developed an option for mutual gain. These negotiating strategies, called *principled negotiation* are crucial to networking. Another technique discussed in this chapter, conflict resolution, provides details on the skills necessary for negotiation.

Another key skill to networking and partnerships is looking for a quality *conversation*, rather than a sale or a partner. "Quality conversations build trust, and from trust comes opportunities" (Henderson 2005). One trait that great communicators and salespeople have in common is the ability to make a heart to heart connection with others. So when you plan your partnership, think about how you treat others—not just potential partners—when you interact and talk with them. Practice asking people open-ended questions that start with who, what, how, when, where, and why (Henderson 2005). You will start to learn more about the interests of the groups, the individual lives of their constituents, and the intersections with your own conservation cause. More importantly, you will build listening skills, a critical tool for partnerships.

8.8.2 Implementation

Environmental success stories often come when conservation issues intersect with quality of life. Consider the impact of Nobel Peace Prize winner Wangari Matthai's networking at the community level in the Kenyan tree planting movement. The environment surrounds us and affects our everyday lives, from the water we drink to the air we breathe. Effective partnerships can jumpstart the formation of a unified conservation movement by combining interests of different groups. When partnerships increase the scope and scale of an organization's reach, conservation can increase its impact on both human and natural communities.

Implementing a partnership can involve collaboration between nonprofit organizations with similar goals, such as the network between the Nature Conservancy, Conservation International, and World Wide Fund for Nature. Or a partnership can involve organizations with different institutional goals, but overlapping interests, such as partnerships between public agencies and private groups. For example, partnerships in Brazil between protected areas and non-governmental organizations offer benefits to both stakeholder groups, including better infrastructure, improved management, and public involvement (Rocha and Jacobson 1998). The Poco das Antas Biological Reserve, created by the federal government in Brazil to protect the golden lion tamarin, collaborates in a partnership with the Golden Lion Tamarin Association, a non-governmental organization supported by international funds.

The same strategies for networking in business can improve your success in implementing effective environmental partnerships (Speisman 2005):

- Remember that partnerships are about being authentic, building relationships, and learning how you can help others.
- Define your own goals before you set up a meeting with another group. This exercise will help you choose groups efficiently and define their interests.
- Develop a reputation as a powerful resource for others. Other groups and individuals will turn to you for suggestions and ideas when you are known as a resource in the community.
- Be able to articulate your needs, the goals of your group, and how other organizations or people can help you. Develop responses to the question "How can our group help you?"
- Have a concrete understanding of the work of your group, your goals, target audience, and the necessity and uniqueness of this work. Your staff and volunteers should have a clear understanding of your direction and mission before expecting others to sign on with you!

The opportunities for creating partnerships are only limited by your creativity to consider the interests of different groups. In this section, we examine the implementation and outcomes of partnerships with two potential partners: faith communities and the medical community, two groups that have increased their collaborations with conservation in recent years.

Faith communities

The spiritual beliefs of many world religions reflect care and concern for the Earth. Religious imagery is filled with symbols from the natural world such as the Buddha achieving nirvana while sitting by the bodhi tree or the Old Testament story of Noah's ark and the flood. Whether the images reflect destruction or renewal, nature and religion are inextricably linked.

This connection bodes well for environmentalists who form partnerships with faith communities, from temples to meditation groups to Quaker meetings. The Church of England, for example, began discussions about climate change the day after the Kyoto Protocol went into effect. In a report entitled, "Sharing God's Planet," Archbishop of Canterbury Rowan Williams asked each parish to complete an ecological audit and enact green policies at the local level, such as using organic bread and wine for Holy Communion and selling fairly traded products at parish events. In an attack on gas-guzzling SUVs, the bumper sticker "What would Jesus Drive?" used faith to prompt reflection of environmentally responsible choices for transportation (Fig. 8.7).

Traditionally, a disconnect prevailed between Christian followers and environmentalists concerned with the legacy of "human domination of the Earth." Could you be an ardent Christian and an environmental advocate? Some modern

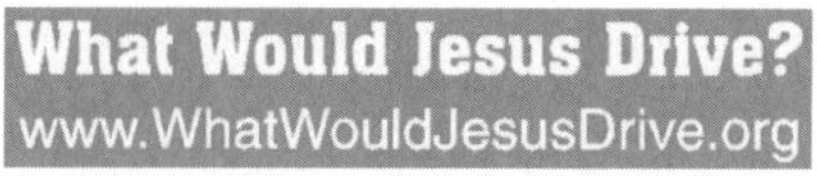

What Would Jesus Drive?
www.WhatWouldJesusDrive.org

Fig. 8.7 The bumper sticker "What would Jesus drive?" united religious and environmental interests to promote fuel-efficient vehicles.

interpretations of the Bible show a strong correlation between conservation of "God's resources" and the faith. Indeed, the Episcopal church has convened conferences on the "Ecology of Clergy." "Green" guides exist such as "Responsible environmental purchasing for faith communities" (Carmichael 2002). Religious leaders can choose conservation education activities from publications such as "To till and to tend: a guide to Jewish environmental study and action" (The Coalition on the Environment and Jewish Life, n.d.).

Even more fundamentalist Christian traditions, such as Evangelical Christians, have created an "Evangelical Environmental Network" (EEN), a group of many organizations such as World Vision, World Relief, InterVarsity, and the International Bible Society. This network published a declaration on "the biblical call to reduce pollution and environmental degradation and the harm they cause to people and the rest of creation" (EEN 2005). The network also sponsors an educational campaign called "Healthy families, healthy environment" to provide families with information to protect their loved ones from environmental threats to their health. Some fundamentalist Christian groups refer to stewardship of the earth as "creation care" to avoid the liberal association with environmentalism.

A partnership called the Sabbath Project involved collaboration between an environmental nonprofit, the Environmental Leadership Center, and churches to address issues such as water quality. Brian Cole, former director of the Sabbath Project, says that water quality issues resonate with clergy and their congregations. "We can talk about the ecology of baptism, since traditionally people were immersed in water for baptism, holy water in rivers," he said. "Now if this holy water is polluted, suddenly a holy act is tainted, and churches have a moral obligation to work toward better water quality."

To that end in 1990, diverse religious leaders and scholars joined the Dali Lama at a "Spirit and Nature" conference at Middlebury College in Vermont, US to advance the idea of conservation of nature as a religious duty. Additionally, a number of books have been published on this subject. Harvard University Press has published 10 books on "Religions of the World and Ecology," with each book exploring the intersections between faith and the environment in major world religions, such as Buddhism, Jainism, Islam, Christianity, Shinto, and Daoism. These intersections between religion and world religions have prompted partnerships such as the annual "Celebrating Sacred Food and Water" conference, jointly sponsored by New Mexico State University Cooperative Extension Service and the three Catholic dioceses in New Mexico.

Medical and health communities

Increasingly, environmental and health concerns parallel each other. A recent US study found up to 167 industrial compounds, pollutants, and other chemicals in participants' blood and urine with an average of 91 chemicals per person. "Of the 167 chemicals found, 76 cause cancer in humans or animals, 94 are toxic to the brain and nervous system, and 79 cause birth defects or abnormal development," according to the Environmental Working Group (2005). Toxic substances in our bodies accumulate from chemicals in the food we eat, skin creams we apply, air we breathe, water we drink, and even plastic containers that hold our food.

Arctic people, such as the Inuit, have the highest concentrations of toxins in their bodies including PCB's, dioxin, DDT, mercury, polybrominated diphenyl ethers or PBDEs (bromine-based flame-retardants), and perfluorooctane sulfonate or PFOS (a type of stain repellent). Since the Inuit diet relies on animals at the top of the food chain, they are getting a hefty dose of these chemicals (Guynup 2004). Many of these chemicals are hormone disrupters and carcinogens. Even additives in food such as high fructose corn syrup put a heavy and detrimental burden on our bodies. The only upside to these alarming statistics is a prime opportunity to work with doctors, medical researchers, public health officials, funding agencies, and families to improve both health and the environment.

Conservation education programs that use youth gardening as a context for learning have discovered success through partnerships with the medical and health community. School gardening programs take students outside to plant and grow food and then eat their produce. Many school-based programs, such as Growing Minds, aim to increase the local, fresh foods served in the school cafeteria.

Since the number of overweight children in the United States has tripled in the past two decades, school gardening programs find their conservation education theme suddenly resonates with the obesity epidemic linked to inactivity of children in "developed" nations. Emily Jackson, director of Growing Minds, says that they have collaborated with health professionals in the schools for their grant proposals. "Our programs address the obesity issues because we're taking students outside, feeding them healthy food instead of junk food, and providing exercise," she said. "And we're teaching them about gardening, plant ecology, and appreciation of the outdoors."

8.8.3 Evaluation

When you begin working with a partnering organization, you should collaborate with your partners to define criteria for judging the effectiveness of the networking and the partnership. Some indicators of an effective partnership include (Strand *et al.* 2003a):

- agreement on goals and strategies
- shared power and mutual trust
- clear communication
- flexibility
- alignment of interests and needs

Convene with members of the partnership on a periodic basis to assess barriers and benefits to a successful partnership. How would you evaluate your progress on each of your criteria for a productive partnership? Brainstorm areas for improvement, make changes when possible, and continue to build a stronger partnership.

A study of the partnerships in Brazil between protected areas and non-governmental organizations used interviews to assess the effectiveness of three such partnerships. The research identified 26 benefits to the partnership and 30 problems such as bureaucracy, lack of legal framework, and unclear definition of roles in the partnership (Rocha and Jacobson 1998). These findings helped frame recommendations, such as increased public involvement, that lead to successful partnerships.

8.9 Conflict resolution

Whenever you bring together different people and organizations, there is the possibility for conflict. By its nature, natural resource conservation involves an increasing population of people competing for decreasing resources (Crowfoot and Wondolleck 1990). Worldwide, different stakeholder groups vie for input into decisions about limited resources such as wildlife, land, water, and forests. A study of 105 ecosystem management projects in the United States found that the largest single obstacle to effective ecosystem management was public opposition due to conflict (Yaffee *et al.* 1996). In contrast, the most important factor in the success of ecosystem management was collaboration.

Conflict is inevitable, and not always negative. Indeed, conflict in institutions and communities can lead to growth, new management strategies, and innovative ways of thinking about issues (Hoban 2000). But escalating, unresolved conflict jeopardizes the most basic conservation efforts. For these cases, strategies exist to resolve conflict and negotiate win–win solutions.

Consider the story of a brother and sister arguing about a bag of ripe oranges their family had received as a holiday gift. Tiring of their argument, the mother finally asked, "Why do each of you need the oranges?" The brother replied that he wanted the pulp to make juice, while the sister needed the rind for baking. They each competed for the entire bag of oranges, but they needed the fruit for different reasons. In the end, they could dovetail these interests and share the same resource.

The story reflects the need to focus on interests, rather than positions, a key strategy in conflict resolution. Conflict results from a disagreement between individuals or groups that may differ in values, needs, attitudes, and beliefs (Hoban 2000). Conflicts about natural resource conservation involve four key factors (Campbell and Floyd 1996):

- *Complex and often uncertain issues.* Conservation involves many complex issues, some of which are not clearly defined or understood. Often with environmental issues, groups such as scientists, the government, and industry disagree about technical issues, and stakeholder groups have strong values about the resources in the conflict.

- *Complicated network of stakeholders.* Conflicts over issues such as public lands involve many stakeholder groups, and new parties often emerge as negotiations start. The decision-making process for different groups varies, such as how indigenous or native community groups make decisions versus government agencies. The level of knowledge, power, and access to resources also varies widely between groups involved in the conflict.
- *Laws and regulations.* The rules that are a part of many natural resource conflicts may contradict each other, or different groups interpret the laws differently. Often an agency maintains that it is following national regulations, while environmental groups assert the opposite.
- *Spiral of unmanaged conflict.* As stakeholders assume sides, their positions often harden and communications come to a halt. If a crisis emerges, decisions may go to the courts, or the decisions are taken out of the hands of the involved parties.

8.9.1 Planning

Conflict involves factors such as needs, perceptions, power, values, and emotions (Hoban 2000). Understanding these ingredients of conflict is essential to planning a strategy for conflict resolution. Analyzing the conflict with the following questions allows you to identify the stakeholders groups involved, the substance of the conflict, and possible strategies for managing the conflict.

Analyzing the conflict

Stakeholder groups. What stakeholder groups are involved in the conflict? How are the groups organized? Do the groups have a history of working for or against each other? Can the groups work together?

The conflict. How did the conflict begin? What are the primary issues? Are any issues negotiable? What information is available and what information is needed? Have the groups taken positions? Do the groups have any common interests?

Strategies. What is the timeline for a decision in the conflict? What external factors are important in the conflict? To what extent will conflict resolution involve the public? Will the media be informed of the process? Will resolution require an outside negotiator or facilitator (Hoban 2000)?

Setting the stage for negotiation

Negotiation is a strategy we use in our everyday lives to resolve conflict. Even when different parties use litigation to resolve conflict, they usually negotiate a settlement agreement. Negotiation in environmental conflict requires several initial steps to increase the probability of success in finding a win–win solution.

First, someone involved in the conflict or an outside facilitator must initiate the negotiation, invite representatives from key stakeholder groups, and identify the time and resources to support the process. At this time, stakeholder groups begin to determine which issues in the conflict are negotiable to them.

Second, the facilitator or the involved groups set rules for communication and the overall objective of the negotiation. The groups can determine an agenda of the issues to cover during the negotiation, as well as what information is needed to help the decision-making. The stakeholder groups must decide what technical or cultural information is known and relevant to the negotiation.

8.9.2 Implementation

As a strategy for conflict resolution, negotiation looks for mutual gains and resolves issues on their merits, rather than the positions of each stakeholder group. Developed by Fisher and Ury (1991), this approach called *principled negotiation* involves four key principles of effective negotiation.

Principle 1: Separate the people from the problem. Every stakeholder has two kinds of interests—one in the substance of the conflict and one in the relationship. In many conflicts, the histories and relationships of the people involved become woven into the fabric of the conflict, making conflict resolution almost impossible. Effective negotiation requires stakeholders deal with the substance of the conflict and try to understand the perspectives of other stakeholders. Often active listening to other stakeholder groups, particularly those with different values, can promote understanding of different perspectives and identification of these values. Stakeholder groups are more likely to agree to proposals consistent with their values (Fisher and Ury 1991).

Principle 2: Focus on interests, not positions. A position is the ultimate goal you want to achieve. An interest is the reason you want to achieve the goal. Negotiation requires reframing each issue in terms of interests, not positions. Usually each stakeholder group will have multiple interests. The most influential ones include basic needs such as health and security. As a facilitator or a group involved in negotiation, this step is critical to avoid a deadlock in negotiations. Groups involved in negotiation may not know what their interests are. Ask each group why they take their particular positions to uncover these unknown interests. As you discover these interests, write them down for everyone to see (Fisher and Ury 1991).

Principle 3: Brainstorm options for mutual gain. Have all groups engage in brainstorming options. At this stage, avoid rejecting an option or looking for the "best" answer. Also try not to focus on your own interests. Look for chances to dovetail differing interests, like those of the siblings arguing over the bag of oranges. The goal during this stage is to look for shared interests for mutual gain.

Principle 4: Develop objective criteria. Stakeholders must develop some objective criteria for choosing an option. Some examples of objective criteria include cost, scientific judgment, or community norms. The development of these criteria must be fair, and groups may use a combination of criteria. The combination of criteria is useful when groups do not agree on the validity of scientific information, such as whether or not a species should be listed as threatened or endangered. As you develop criteria for selecting or combining different options from the brainstorming, you may decide to collect new data and consult with outside experts, community members, or other partners for feedback. At this stage, it is also important to revisit the interests that conflict, because these interests need acknowledgment, even if the solution does not accommodate them (Fisher and Ury 1991).

8.9.3 Evaluation

Reaching an agreement in a negotiation does not signal the end of conflict resolution. Each of the stakeholder groups must adopt the agreement and identify strategies for evaluating the effectiveness of the solution. Ideally, each group involved in a negotiation can help develop a monitoring and evaluation plan to assess the progress of the agreement, resolve problems, and renegotiate if necessary (Hoban 2000). The evaluation plan should also include landmarks for celebrating the success of the agreement. One evaluation strategy is to bring representatives from the stakeholder groups together at periodic times to give feedback on the agreement. Since issues change over time, stakeholder groups may find that an agreement becomes outdated, and the parties need to renegotiate. Continued networking through structures such as advisory committees provide a proactive strategy to anticipate conflict between different groups. Many government agencies, such as wildlife refuges or national parks have advisory committees with representatives from stakeholder groups to help maintain good relationships and review upcoming decisions.

8.10 Summary

Networking involves aligning your interests with other individuals, groups, and communities to increase the success of a conservation effort. If you are talking with festival attendees at an information booth, forming a long-term partnership, or using negotiation skills to resolve a conflict between groups, networking means making connections. Making connections can involve creating or using existing environmental clubs and groups, as well as promoting your conservation objectives through public presentations, workshops, and professional posters. The skills to organize conferences and special events promote networking with a larger audience. Each situation and conservation issue requires something different in order to reach audiences, promote conservation messages, and ultimately achieve conservation objectives. Used alone or to complement one another, the techniques in this chapter provide ample opportunities to network and attain success both within and beyond the conservation community.

Further reading

EE Toolbox: Workshop Resource Manual: Designing Effective Workshops, by Judy Braus and Martha Monroe (1994). Kendall Hunt Publishing, Dubuque, IA.

Organizing Special Events and Conferences: A Practical Guide for Busy Volunteers and Staff. 2nd Edition, by Darcy Devney (2001). Pineapple Press, Inc., Sarasota, FL.

Getting to Yes: Negotiating Agreement without Giving in. 2nd Edition, by Roger Fisher and William Ury (1991). Penguin Books, New York, NY.

Environmental Interpretation: A Practical Guide for People with Big Ideas and Small Budgets, by Sam Ham (1992). North American Press, Golden, CO.

Partnerships for conservation: protected areas and nongovernmental organizations in Brazil, by Ligia de la Rocha and Susan Jacobson (1998). *Wildlife Society Bulletin*, **26** (4), 937–946.

Marketing conservation

A media campaign in Egypt promoted water conservation practices to farmers. An elementary school in Ontario, Canada engaged students and their parents in a litter-free lunchroom. A series of programs in Michigan, US encouraged residents, businesses, and policy-makers to restore the Rouge River.

These education and communication programs focused participants on a particular issue and set of behaviors. Rather than providing experiences that make concepts more meaningful, (such as a field trip to investigate the sources of water pollution) these efforts were designed to elicit a change. They used techniques of social marketing. Social marketing is the application of marketing principles, those techniques that companies use to sell toothpaste and shoes, to issues that will improve society. It has a long history in the health field where it has been used to promote immunizations and oral rehydration therapy for toddlers and discourage cigarette smoking and drunk driving. More recently, conservation educators have applied social marketing principles to environmental issues (Day and Monroe 2000, RARE 2003).

Social marketing hinges on a complete understanding of the audience's motivations and perceptions so that very specific persuasive communication messages can be crafted. Like other education and outreach programs, social marketing techniques use the values and motives of the audience to appeal to their concerns. By knowing how information travels in a community, we can be more efficient about where we invest our outreach materials. It also uses psychological theories to suggest other techniques, such as incentives, feedback, modeling, and commitment.

Social marketing, however, is not just about communication. After exploring the perceptions and barriers to the behavior, a social marketing program may suggest simple modifications to elicit the desired change, such as extending the hours of an office, placing additional waste receptacles at key intersections, or making physical changes in a program. In Sweden, there is a clear correlation between participation in waste recycling and the promotion of waste recycling opportunities. Those waste-reducing activities that are not supported by government policy and local organizations (e.g. repair shops and used clothing collection) are less frequented by the public (Lindén and Carlsson-Kanyama 2003).

Some conservation educators may disagree with this type of programming. They are more comfortable with traditional environmental education programs that aim to provide audiences with the awareness, knowledge, and skills to make their own decisions. In fact, traditional education programs often take great pains to avoid the appearance of advocating one solution. But will the traditional approach result in significant change, quickly enough? Those conservation educators whose ultimate goal is a behavior change may choose to design a program that targets and changes that behavior. Unlike traditional programs, they are less concerned about fairly representing the advantages and disadvantages of all possible outcomes; they are promoting one particular outcome, such as water conservation or carpooling.

It may be helpful to consider situations where social marketing is not appropriate. During a heated controversial issue, for example, it may not be wise for one party to announce and "market" their version of a proper solution. Unless there is some urgency to act quickly, many communities will want to figure it out themselves. Where communities have decided what behavior is appropriate, such as washing hands before eating, educators may be encouraged to change behavior and support the norm. Thus, community approval can be a prerequisite to using social marketing tools.

Many programs combine the two strategies. To reduce a perceived barrier to waste reduction, Toronto's Norway Public School used a variety of efforts to generate enthusiasm and maintain interest in a school-wide challenge to reduce the amount of garbage generated at lunchtime. Initially, the school sent an informative flyer about the waste involved in pre-packed lunch foods and the increased nutritional value of lunches packed from home ingredients (Kassirer 1999). They also advertised to parents and sold reusable plastic containers for student lunches. Older youth made compost piles and monitored the waste collection for decomposable materials. A follow-up survey of parents indicates that the program helped reduce packaging in their children's school lunches by 60% and helped reduce by 40% the amount of packaged products they purchased. They also increased recycling and composting activities at home. In this case, social marketing approaches complemented traditional communication and education techniques to help make the program effective.

Social marketing techniques can be powerful tools for changing human behavior. That alone raises important questions: who decides which behavior should change, what role does the audience have in agreeing to it, and does social marketing conflict with efforts to build civil societies?

These questions are rarely given a second thought when multinational corporations seek to change our behavior and encourage consumption; we recognize it is their job if they want to stay in business. But when natural resource agencies use social marketing tools to conduct their job of protecting endangered species, regulating water, or managing parks, some people are nervous about being told what to do. In an era where public participation is a tool that improves program success, could social marketing work against this goal?

There are several considerations that could help organizers decide whether to use social marketing in conservation: (1) if the goal matches the publicly accepted mission for the agency or organization, (2) if the audience is involved in the design of the social marketing program, or (3) if elected leaders wish to pursue a course of action with urgency. This chapter suggests that community leaders could be engaged in every decision (Andrews *et al.* 2002), although it is possible to implement a social marketing program without public participation. In some crisis situations and with trusted implementers, this may be most expedient.

Where community participation is feasible, consider working with an existing group of community members. If one does not exist, you might develop your own advisory group of well-respected community leaders. Once a relationship is established, there are many opportunities to involve them in the development of a behavior change program. If you represent a different cultural perspective than the community, or if the community consists of several cultures, consider teaming with their representatives. You may even hire and train local people to become your staff (Ady 1994). They can help define the goal, select the behavior, and make the barriers clear. They can work to design and field test messages. They can be featured in posters to promote the new change. By giving community leaders access to the powerful tools of social marketing, conservation educators are partners to the change process, not manipulators (McKenzie-Mohr and Smith 1999).

Social marketing follows a process of asking key questions to understand the audience, develop tools, test those tools with the audience, revise the tools to be most effective, implement the program, and evaluate the outcome. In many ways, it is not different from the development of any other educational program that begins with a needs assessment and uses evaluation techniques to improve the product (Jacobson 1999, Day and Monroe 2000, and Chapter 1).

This chapter begins with an explanation of planning, implementing, and evaluating a social marketing program, and then explores eight techniques common to social marketing programs, offering strategies for planning, implementing, and evaluating each. This information will help conservation educators use the power of social marketing techniques in an honest and reasonable way, to empower communities to conserve resources, and to offer opportunities to design a behavior change strategy.

9.1 Beginning a process for social marketing

The first step in social marketing is to identify the behavior that should be targeted. This may involve meeting with the community, introducing or responding to their concern, and asking for their cooperation in designing a strategy to solve the problem. By starting with the selection of a behavior, you can help focus attention on how people can change a situation or solve a problem. Douglas McKenzie-Mohr, an experienced social marketer and trainer, suggests, "There are multiple activities that could be promoted and it is useful for organizations to think about which of these they are best suited to pursue. Too often organizations blindly jump into program development without thinking across their options." The process begins by listing feasible activities, selecting one that would promote the intended behavior and one that would dissuade an existing behavior. Next consider the barriers and benefits for both activities. What might influence the audience to participate? Finally, try to ascertain the realistic impact this activity would have on the problem. This process may help reveal the most appropriate behavior for your project.

9.1.1 Assessing the audience

The second step is to collect information about why some people practice the intended behavior and why others do not. Initial community meetings may provide some ideas, but preconceived notions can be faulty. Furthermore, ideas and barriers are not the same from issue to issue. For example, knowing how people feel about wildlife is not likely to help design messages about water quality. It is best to conduct interviews, focus groups, or a small survey to better understand the reasons behind the action and inaction. Theories about human behavior (Chapter 3) suggest some key questions that may assist this process:

- What do people know about the problem?
- How important is this problem and how does it affect their goals for the future?
- What motivates people to care about this problem or solution?
- Do people believe any change is likely to resolve the problem?
- Do people have the opportunity to make the change?
- Who is already conducting this behavior and are they respected?
- What do people think that others (particularly respected and important others) will think of them participating in this behavior?
- What barriers stand in the way of making the change and which are most important?

A series of questions like these were used to understand how residents in Quito, Ecuador perceived their waste and a new municipal recycling program (Hernández 2000). In addition to asking residents to define waste and how they dispose of each type of waste, the survey included questions such as:

- How many places to dispose of waste are there in your house?
- Who is in charge of handling waste?
- What happens to the waste you dispose of? Where is it taken?
- Why would one separate waste?
- What advantages do you see in separating waste?
- Who has a say in how you dispose of the waste generated in your house? Who else?
- Who would approve of giving different types of waste to collectors on different days? Why?

The results of this data collection process revealed there were four different concerns that affect perceptions of waste separation—financial, convenience, development, and self-image. The people who practiced waste separation had different beliefs than those who did not. In some cases, people who did not separate waste had the misperception that sorting occurred after the garbage was combined in one waste container. For them it was a dirty, grimy, time-consuming task. Those who sorted waste believed it was a fast process made easier because wet and dry wastes were deposited in separate containers. They also believed waste separation gave them a positive image with neighbors and family members, whereas those who did not separate garbage believed it was a demeaning task (Hernández 2000).

In some cases community members may collect this information themselves, while in others you will present the results to the advisory committee. In either case, the community must have access to the information and should be involved in directing the program to the next step.

In some circumstances, the data collected to design the program can also be used to establish a baseline of public knowledge, attitudes, and behaviors. Several years later, a repeat of the same survey can help indicate the success of the program by measuring change in key factors. In this way, the planning phase can contribute to the evaluation phase. The six-year Rouge Project to build public awareness, knowledge, and behaviors in the Rouge River watershed documented that, prior to their efforts, nearly half the respondents thought business and industry were the source of poor water quality. A post-program survey revealed more accurate responses. Those blaming business and industry were reduced to one-third, while another third believed water pollution was the result of combined sewer overflows, and the remaining third believed stormwater runoff was the major cause of water quality problems (Powell and Bails 2000). This increase in knowledge about causes and solutions of water pollution was attributed to the program.

Some of the most commonly used tools for collecting information about an audience include interviews or focus groups which yield rich detailed explanations of experiences from a limited number of people. Questions asked in a survey format usually provide a set option of responses, making it easier to tally and interpret a large number of respondents. Direct observation is a useful method for documenting changes in the target behavior. More information about these data collection tools can be found in Chapter 1.

9.1.2 Understanding barriers and opportunities

The audience assessment is used to develop a plan for communicating information and reducing barriers. Once the results of the assessment have been presented to the community advisory group, they can help identify the information that will help people adopt the new behavior by considering questions such as:

- What information about the problem or solution will change perceptions?
- What information about how to implement the solution is needed?
- What information about the consequences of the action is not well understood?
- What do people care about? What do they want to keep? What do they want to change?
- What will motivate them to support change?
- What changes in location, opportunity, and service will reduce barriers?

Sometimes a comparison of people who are already doing the behavior to people who are not yet doing it will help identify the information and opportunities needed.

It is not surprising that many people care about the same things. These common motives can be used to attract a large audience to a program or workshop that can help them achieve their goals as well as yours. In general, people are concerned about:

- spending quality time with their family
- health, for themselves and their family
- status and public appearance
- participating in community improvement
- saving money or other resources for an uncertain future
- freedom, equality, or justice
- clean air and water
- connections with nature

In addition to motivations, the advisory group can identify barriers that should be reduced, new alternatives, more convenient solutions, and the community benefits that should be highlighted in the communication campaign. Messages involving respected people and those who already practice the target behavior may be useful.

The identification of these basic elements comes directly from the community assessment. A good understanding of the perceived barriers and benefits will suggest which social marketing tools should be used in the program. Although tools can be combined, it is more expensive to use multiple strategies. Since each technique has somewhat unique characteristics, it is important to carefully consider which tool will help achieve the program objectives. The information on each technique provided in this chapter gives suggestions for how it is most effectively used.

9.1.3 Creating and selecting messages

The next step is where some might think "magic happens." Actually, it is the creative work of converting the information to messages that will be understood easily. You might want to say: "Stop eating endangered birds," but you know from experience this will not help people find other sources of food. So you might present the message more subtly: "Will this bird live long enough to raise its own chick?" Complementary messages might reinforce the community's sense of responsibility to protect the birds and the importance of a more modest harvest to sustain the population. You can use the information from the audience assessment to create your message:

- Recall what people care about. How can you use the motives and values they have to make your information more acceptable? Because people care about wildlife, messages about forest fires may remind people not to play with matches because some animals might lose their home. Because people care about spending time with their family, a message about recycling might emphasize the role children can play in preparing and sorting materials.
- Recognize that most people do not want to be the first one to try something new. We are social animals and like to be with the pack. Use respected community

leaders to convey the message that this is a good change, because they do it too. Movie stars and athletes are often courted to promote products and causes, as they help define what society deems is appropriate.

- Choose one or more basic elements for the foundation of your message: (1) rational elements that provide factual information; (2) emotional elements that elicit a positive or negative response, such as pride or fear; (3) moral elements that refer to the audiences' sense of right and wrong; (4) nonverbal elements that use visual cues and symbols to make their point (Kotler *et al.* 2002). Any element can be effective, and a message only needs one, although using more than one may help the message appeal to a wider audience. Adding either a lighthearted or serious tone to the element also can its appeal.
- Fear appeals should be used only when followed by reasonable information about what people can do to reduce the threat. If the information is complicated or difficult to use, people could feel paralyzed and helpless by their fear. This does not lead to change.
- Humor is a good strategy and often makes the message more memorable. It is not useful, however, for a complicated message.
- If you know where your audience is in the evolution toward a new behavior, you can better target a message. For example, someone who is still thinking about the issue should get a message that promotes the positive outcomes and expectations. Someone who has already decided to change their behavior would be better served by a message that details how to conduct the behavior and how to know they are doing it correctly.

You may choose to hire creative people to generate possible messages (such as an advertising agency) or ask a group of staff and advisors to brainstorm suggestions over snacks. Either way, a variety of potential messages should be generated. Some may be more appropriate from certain social marketing techniques (such as prompts or commitment) than others. Ask the advisory group to review, comment, and choose the messages that resonate best. Use their information to help decide which are the most persuasive and memorable messages, and the materials that best fit the community. Which ones enable viewers to say, "Yes, this would work for me"? Then take the selection to a sample group of the audience to confirm that these materials are useful, appropriate, attractive, and speak to them. The advisory group may be able to help you with this.

It is critical to test the messages on the audience, although too often campaigns rely on only staff and experts to determine the best message. After one expensive failure, it is easier to remember to test the translation, ask local people to review the images, or make sure teachers will use the suggested materials. This pilot testing process is essential in the development of an effective campaign. Mistakes caught at this stage are much easier to fix than after the program is implemented. Revised tools should be pilot tested again, to make sure the new version is acceptable.

It is generally easier to change behavior if there are many different reasons why someone would find this a good thing to do. Try to use all of those reasons in a campaign, since your favorite reason may not attract many people. Additionally, using a range of benefits will help promote the behavior to more people. Do not worry about which is the "right" reason, either. It does not matter if people ride bicycles to work because they do not want to pay for parking, because they want the physical health benefits, or because they want to improve air quality—what matters is that they do it.

9.1.4 Program implementation

When you are pleased with the results of your pilot test, you are ready to produce the materials and launch your outreach campaign. Implementing social marketing tools involves identifying what will be done, when, where, and how. Tasks and responsibilities should be outlined in an action plan with a timeline. Constraints, such as time and money, may restrict the activities that are conducted.

Complicated projects may evolve over time in phases that first target a geographic area and then spread to others, or that communicate one introductory message and then progress to more detailed ones.

Continue to involve your advisory group and community members in the program delivery process. Do not forget to establish strategies to monitor the results of the campaign. Once again, consult with community members to decide how to measure changes.

Programs will be implemented differently based on the social marketing techniques that are chosen. Research studies give us an idea of the effectiveness of techniques when used separately, but most effective social marketing programs use them in combination. Even then, however, techniques vary. In some cases, personal contact and communication is an attractive alternative to public outreach campaigns that use newspaper ads and billboards to emphasize information (McKenzie-Mohr Associates and Lura Consulting 2001).

9.1.5 Program evaluation

Communities and agencies will want to know if their social marketing efforts are successful, making program evaluation imperative. Since it is typical for social marketing programs to use several tools in a campaign to bring about behavior change, an evaluation should assess the entire program. Unlike some conservation education programs that only measure participant satisfaction or knowledge, social marketing programs are evaluated by measuring saturation (how many people saw and remembered the message) and actual behavior change.

This is done most often with observation (counting the number of single-passenger vehicles, the lawns being irrigated, or the homes with native plant gardens) or using data-collecting systems that measure behavior, such as bar-code readers at grocery check-outs, electric and water meters, and traffic counters. Many studies have demonstrated the unreliability of asking people if they changed their behavior; respondents are often generous with their memory of good intentions and tend to over-estimate desired behavior. It could be expensive to train and outfit observers to watch and record behavior, but in some cases this will be the best way to achieve an unbiased record.

Another strategy is to design the program to collect evaluation data internally. If your brochure includes a coupon or ticket that can be collected when used, you will be able to know how many people picked up the brochure and acted on the message. Coupons of different colors could be distributed in different venues, such as newspaper flyers, school newsletters, or door-to-door. When they are turned in you have a record of which technique was the most effective at reaching and convincing the intended audience.

Local universities may be able to help provide student assistants to collect and analyze data. It may be sufficient to measure behavior before and after your campaign with the population that you are addressing, unless you need to know which tool was most effective. A small pilot program, for example, may be used to compare several tools to decide how to expand the program in the next phase.

A comprehensive social marketing program

Rare, an international conservation organization based in Arlington, VA (US) uses social marketing techniques in partnerships with local grassroots conservationists to develop innovative programs that help conserve biological diversity in more than 40 countries around the world. One of Rare's signature programs, Rare Pride, raises awareness, influences attitudes, and changes behavior to increase public support for conservation of threatened species and ecosystems. Pride campaigns are based on a charismatic species that becomes a campaign mascot and symbol of local pride. When coupled with effective marketing vehicles—such as billboards, television and radio advertisements, church sermons, buttons, posters, leaflets, beer bottles, water delivery trucks, and more—campaigns can rally local support for environmental protection. Grounded in social marketing principles and managed by a local conservationist, Rare Pride is tailored to fit the needs and beliefs of the audience and makes conservation fun and engaging. The key to their success is involving and engaging every segment of the community, from elected leaders to teachers and average citizens (Rare 2003).

A Rare Pride program in Guatemala's Tikal National Park uses jaguars (*Panthera onca*), an endangered big cat, as a mascot to discourage the poaching in the surrounding villages. Miguel Vásquez, the campaign coordinator, has become a local celebrity by dressing as a jaguar and speaking to school children and adults to raise awareness of the effects of wildlife poaching (Figure 9.1). According to Sharon Price, the coordinator of Rare Pride Programs, after 20 months, the campaign has recorded measurable results. "Nearly 85% of the primary school children know the values for conserving animals in the park and more than 85% of the adults recognize Tikal as a World Heritage Site. In addition, restaurant owners have expressed interest in using 'farmed' wild game for meat and area youth are actively involved in two environmental clubs."

Although the community-based education programs (e.g. puppet shows, costumed critters, school visits, and religious sermons) and mass-marketing techniques (e.g. billboards, popular songs, bumper stickers, television announcements, and music videos) are the visible products of a Pride campaign, they are not the only elements. These techniques are chosen and designed after a lengthy process of

Fig. 9.1 At community events this jaguar draws attention to Rare's conservation message (photo by M. Vásquez).

training the campaign manager, creating a project plan; building a group of stakeholder supporters; conducting a survey of the public's knowledge, attitudes, and behavior; and developing a measurable conservation objective for each. A survey is conducted at the end of the campaign to measure changes in knowledge, attitudes, and behavior.

In many cases the results of the campaign can be seen without a survey. Campaigns have facilitated the creation of new reserves in Indonesia, Costa Rica, and the Philippines and the passage of new natural resource management legislation in Montserrat and Yap. After a campaign in Mexico's Sierra de Manatlán Biosphere Reserve promoting best practices for slash and burn techniques, forest fires were reduced by 50%. A Pride campaign helped the Palau Conservation Society launch itself as Palau's first homegrown environmental NGO and one of Micronesia's leading voices for conservation.

9.2 Social marketing tools

Every social marketing program needs to inform the audience about the targeted issue and the desired behavior. Most of the techniques mentioned in this book can be used to do that. What might make a brochure or a workshop more likely to change behavior, however, is the use of social marketing techniques (Box 9.1)

Box 9.1 Social marketing tools at glance

Tool	Value	Tip
Sign, poster, billboard	Raise awareness of issue	Keep the message short, simple, and positive.
Prompt	Provide a reminder to do an easy-to-forget action	Use for habitual behaviors, after knowledge and intent are in place
Feedback	Inform people that they are making a difference	Provide both personal and collective feedback
Commitment	Obtain the promise of future action	Usually requires personal interaction.
Model	Creates a new social norm and provides procedural information	Use to suggest how, when, and where after people already know what and why.
Incentives and disincentives	Provides a different reason to stop or begin a new behavior	Can be difficult to continue providing the incentive or sustaining the change but useful to kick-start a new behavior.
Press interviews	Provide information and views to the public	Provide background for the reporter but give simple, clear soundbites for the public
Advertisement	Raise awareness and provide testimonials	Can reach a large number of people with very simple message

explained below. A brochure that persuades by highlighting local leaders uses the technique of modeling; a workshop that asks for participants to sign an agreement form uses commitment. These techniques could make the brochure and workshop more effective at changing behavior than those that just provide information.

9.3 Signs, billboards, and posters

A variety of written communication tools can be used in a social marketing context if they are designed to inform and persuade. Brochures, posters, door hangers, stickers, news releases, exhibits, editorials, stories, and billboards are but a few of these. The basics of graphic design and writing skills to construct many of these tools are explained in Chapters 10 and 12, including how to achieve readability, direct eye movement, and develop and use a theme.

Signs, billboards, and posters are posted messages printed on flat surfaces that can be quickly understood. Because they can utilize images, they can communicate across language and literacy barriers. Some signs and billboards have moving parts and flashing lights, though some communities and nations actively discourage roadside messages that might distract drivers. They are less likely to be used by conservation outreach efforts because of expense.

Billboards are large, permanent structures usually read from moving cars. Signs are typically site-specific and smaller permanent structures. Posters are printed on paper for large-scale distribution.

9.3.1 Planning

You may wish to develop a summary of what you know from your audience assessment as you begin to plan your sign, billboard, or poster. Describe your audience and your ultimate goal. Then describe what you want them to know, believe, and do as a result of your campaign. Also list the benefits of and reasons why they may want to do this.

A billboard is a mass media tool usually used to raise awareness (Figure 9.2). Signs should be posted at the site where they can inform and change very specific behaviors. The sign shown in Figure 9.3 is mounted on a platform for easy use and effectively tells hikers what they should do and why before walking in this area. Signs that provide too much information may be much less effective. With your advisory group, decide on a tone and basic element for your message. Do you want to be stern and moralistic, or lighthearted and emotional? Will you use a testimonial from someone who has already done it?

Because posters are generally distributed for others to post it is difficult to know where they will be placed. Therefore, it is most important to design an attractive image that the audience will want to display. Posters distributed to Washington (US) high school principals were so effective at attracting high school students that many were stolen, prompting the agency to duplicate smaller flyers for distribution (Kotler *et al.* 2002).

Try not to threaten or demand a behavior because people could feel coerced. Messages that are negative or threatening (e.g. Don't litter) are more likely to be ignored or generate the opposite reaction than messages that reinforce the socially acceptable behavior or offer choices (e.g. Keep your park clean) (Reich and Robertson 1979).

In addition to your message, it is essential to have a good design for a sign, billboard, or poster because they stand alone as a communication technique. There is no accompanying presentation or explanation. The layout of the sign, billboard, or poster will include several elements, often a headline or phrase and an image. These elements should be attractive and balanced so the eye moves through the whole design. Large areas of unused space are important to prevent the design from looking cluttered or tight; 30–40% open space is recommended (Carey 1996).

When you have several ideas for your message and design, circulate drafts among your advisory group and collect their feedback. Ask them to describe the main message and the purpose of the sign, and how it makes them feel. Since your advisory group may be more knowledgeable about your goals than your audience,

Fig. 9.2 Highway billboards can be eye-catching places for a short, memorable message designed to raise awareness (photo by M. Monroe).

make sure you also ask the audience the same questions. With this feedback, apply necessary changes to images, colors, layout, and wording to improve the design.

9.3.2 Implementation

Posters usually require a distribution plan, whereas signs and billboards are erected by the campaign. A campaign may have one sign at a park entrance warning people of the consequences of feeding animals, or could have multiple copies distributed to every community in the vicinity. Consider how long you want the signs to be posted and how you will distribute them. A statewide campaign to remind travelers of lands that are managed with prescribed fire in Florida produced and distributed 150 1.3 m^2 painted signs, made from 2 cm thick outdoor-grade plywood. Unfortunately the signs were so heavy that the organizers had to rent a large truck to carry them to workshops. The workshop participants were also informed of the need to arrive in a large vehicle so they could transport signs back to their work sites. Aluminum signs would have been equally sturdy, very light-weight, but more expensive.

Fig. 9.3 To protect natural areas from the spread of Phytophthora Root Fungus, this sign effectively explains the risk, the consequences, and the action hikers can take. It contains more information than a road sign and is still graphically pleasing (photo by M. Monroe).

Consider how fast people will be traveling when they read your sign or billboard. Highway travelers have 6–10 s to understand a billboard, so use less than 10 words and a large font (Figure 9.2). If your audience is walking or the sign is placed at an intersection where they stop momentarily, your font can be smaller and have more words (Figure 9.3). Brief messages, however, are generally easier to read and remember. Images and symbols can be understood very quickly and should be used on large signs and billboards. Remember to pilot test any symbols to make sure people comprehend the intended message.

Post signs where they can be read easily. Merely moving a notice about a water conserving behavior to the middle of the public shower room increased compliance from 6% to 19% (Aronson and O'Leary 1982–83). Unfortunately, because some users did not believe the message was important or simply do not like to be told what to do, they knocked the sign over. Psychological reactance is a response that is opposite of the intent of the message; people go out of their way to engage in the very thing that the message wants them to avoid (Brehm and Brehm 1981).

Pilot testing the sign prior to implementation could detect this reaction and give you time to revise the message.

9.3.3 Evaluation

You can determine if your signs, posters, and billboards are effective if a survey of the audience remembers seeing your signs and can state your message and if perception of the problem has increased from your baseline level.

9.4 Prompts

Sometimes people know what to do and intend to do it, but just forget. Many of the habitual activities that we do everyday are hard to change, even if we know we should. Conservation educators can use prompts to help people remember to do something they already know is an appropriate behavior. Prompts are brief reminders that tell people what to do and when to do it. The location of the prompt often tells them where or how to do it. Stickers, magnets, posters, billboards, and signs are some of the opportunities educators have for prompts. All of these techniques share several common features—they contain a short phrase that is easily seen, quickly read, and strategically placed.

The popular wallet-sized Seafood Watch pocket guide (Figure 9.4) is produced at the Monterey Bay Aquarium in California, US to remind consumers which seafood is harvested relatively sustainably and which should be avoided. The goal of the program is to shift consumption patterns toward species sustainably harvested. Several criteria are used to rank species, including bycatch, population size, and farming practices. Over a million copies of the region-specific guide have been

Fig. 9.4 The Monterey Bay Aquarium Seafood Wallet Card enables consumers to eat from more sustainable fisheries.

distributed to organizations, agencies, restaurants, and consumers. The program also includes workshops for chefs and information packets for restaurants. Because it can be carried to restaurants in a wallet, it makes a perfect prompt for customers.

9.4.1 Planning

Since location is one identifying characteristic of a prompt, a number of advertising opportunities are excluded. Ball caps, T-shirts, pencils, and bus advertisements are excellent choices for creative and memorable messages, but the program developer cannot control where these messages are seen. Therefore, these are best used in situations where the goal is to advertise to a large audience anywhere and everywhere. Because they are not likely to be associated with the ability to immediately perform the behavior, they are not prompts.

When planning your prompt, consider placing it strategically where a person would need a reminder. For example, you could place a "No Compost" sticker on a garbage can to remind residents to take their organic waste to the compost pile. If the compostable waste were dumped into the garbage bin in the kitchen, however, this prompt is useless. An attractive pail on the kitchen counter might be a better prompt.

"Be bright, turn out the light" is a frequently used slogan for light switches. "Recycle this container" is a reminder printed on some beverage cans and bottles. Shelftalkers, or signs on grocery shelves that inform consumers about the recycled content of the product, have been effective in several communities. Prompts are usually visual reminders but can also be an auditory aid, as in the spoken message in airports to "mind your step" at the end of a moving sidewalk. Prompts do not convey enough information to convince someone to conduct the behavior, and therefore are only effective when people already intend to do it (McKenzie-Mohr and Smith 1999; McKenzie-Mohr 2000).

Use your advisory group to give you ideas about where a prompt would effectively remind them to participate in a certain behavior. You might even ask each of them to give a prototype to a neighbor and see if it works.

9.4.2 Implementation

For prompts to be effective they must (McKenzie-Mohr and Smith 1999):

- Deliver the reminder close to the behavior—in time and space. Consider where you could put a prompt that would be appropriate, and how it would be affixed. A key chain message could remind people to measure tire pressure to improve gas mileage, magnets can go on refrigerators, and stickers must be affixed to something that people will not mind covering.
- Be attractive and interesting so that people notice them. A clever message or logo is a good way to attract attention. A butterfly label on native plants in the nursery could remind people to consider wildlife when they landscape.
- Not be so noticeable that people tire of seeing them. Prompts only work if people notice them. If they are so ubiquitous or obnoxious that people avoid them, they are not working.

Fig. 9.5 This prompt reminds people not to use storm sewer drains for waste material. The campus mascot, an alligator, makes a good spokesperson for local water quality in Gainesville, FL (photo by M. Monroe).

- Be self-explanatory and understandable. Prompts are reminders, not explainers. The message needs to be specific, clear, appealing, and simple (Figure 9.5).

To achieve these goals, prompts should follow appropriate graphic design techniques (Chapters 10 and 12) for color, white space, images, and fonts.

9.4.3 Evaluation

Studies show that prompts can be effective at increasing compliance. You can measure the effectiveness of your prompts by observing the behavior of people approaching public prompts (on litter receptacles perhaps) or by asking a random sample of the public where they saw the prompt and what it means. That will help you know if the prompt is conveying the appropriate message. To know if it is achieving the behavioral goal, it may be necessary to use surveys or observations. These tools can compare the population that received the prompt with a population that did not. Since prompts usually are used in conjunction with other tools such as persuasive information, it is often difficult to know how valuable the prompt is alone.

9.5 Feedback

People need to know that their attempts at a new behavior are (1) correct and (2) worthwhile. They also like to know they are not the only people doing this new activity. Thus, providing feedback to participants is an important element of a social marketing program.

Writer Bill McKibben's experience with feedback indicates that when provided while the behavior is being conducted, feedback can be interesting, and even addictive. His new hybrid vehicle has a gauge that tells him how many miles he gets to a gallon of fuel (McKibben 2003). In this case, the feedback mechanism is so sensitive that drivers see a difference when going downhill or stopping at an intersection. McKibben observes that his driving became a competitive game to maintain high mileage. Like an itch, he notes, you cannot help but pay attention to the feedback and try to better your last result. McKibben suggests that the ability to measure results on a host of environmental actions will help change behaviors.

While it may be difficult to give each person an accurate summary of the consequences of their own actions, it can be possible to paint a collective picture of how a neighborhood or city is performing. During the 1979 gasoline shortage in the United States, a clever study in Texas used the televised evening news to provide feedback to viewers regarding the number of gallons of gasoline purchased from a selection of local gas stations. Every weekday an announcer provided a conservation tip while showing a graph displaying that day's gasoline consumption. When the graph was shown, consumption decreased by 25%, 27%, and 38% in the first three weeks compared to a baseline figure taken during the previous 2 weeks. During the following 2 weeks when the graph was not shown, consumption remained at 25% below the baseline figure. Showing the graph again for three weeks dropped consumption to 32%, 34%, and 34%. Three months after the last graph was shown, consumption was 15% below baseline (Rothstein 1980).

A program designed to reduce energy consumption in apartment complexes in Voorschoten, Netherlands compared three techniques: feedback, reinforcement, and information (Midden *et al.* 1983). Every household received information on energy conservation and one group received only this information. A second group received weekly feedback on the energy they saved compared to their own consumption during a previous period. Another group received feedback on their energy savings in comparison to other residents. The fourth group received the feedback and a financial incentive. The results indicate that individual feedback and financial reinforcement with comparative feedback are effective at reducing energy use and not significantly different. Comparative feedback can be effective if the comparison group is perceived as similar to those receiving the information. Information alone is not effective.

Another experiment used feedback with information on individual behavior as well as group actions (Schultz 1999). Households were asked to set out recyclables on the curb for observation and collection; feedback was provided on the day of collection for 4 weeks. Observations continued 4 weeks after the treatment period ended. Both individual and group feedback conditions resulted in an increase in recycling participation and an increase in the amount of material recycled. Neither the initial plea nor the information-only condition led to a significant change in behavior (Schultz 1999).

9.5.1 Planning

Collective and individual feedback are effective at providing participants with information they need to continue their actions. Thus, feedback tends to be used only for actions that are repeated frequently. In a campaign to promote a one-time behavior, such as replacing a toilet with a water-conserving model, there is little opportunity to provide feedback to the participant. The number of water-conserving toilets installed can be used to help convey the social norm (Chapter 3) and attract interest among those who have not yet converted.

It may be that feedback, such as prompts, is most useful when associated with the opportunity to change that behavior. Feedback on gasoline consumption provided at the gas pump would not help the driver who is already out of gas, but if provided the night before may help a family plan their upcoming travels.

In addition, feedback may be most useful if the audience believes the information is relevant to their situation. If data are provided from a previous year, another village, or a different season, it may be harder for people to believe that the information is helpful to them.

Testing your ideas for feedback with your advisory group will help you choose the message and the location that will work the best for your community.

9.5.2 Implementation

Since the information provided will change daily or weekly, from individual to individual, or from home to home, it may not be practical to use printed communication devices to share feedback. Brochures and posters may be out of date too quickly to justify the expense. The exception, of course, is personalized utility statements. A campaign could use these bills along with informational tips to further reduce energy or water use and a comparison chart for average home and family usage. Electronic forms of communication may be feasible for providing feedback. Web sites, radio announcements, electronic signs, and television news programs could be considered for community feedback. For example, an electronic billboard tells the residents of Melbourne, Australia the current level of their water reservoir and which category of water conservation rules applies.

To use feedback effectively, consider these tips:

- Provide feedback for behaviors that are done repeatedly.
- If using personal feedback, try to include comparative information that is relevant to the situation so people have a realistic sense of how they are doing.
- If using collective feedback (e.g. how the town is doing), consider comparing data from other similar groups outside the target area.
- Add an incentive to encourage participation, such as a competition (see Incentives).
- Provide feedback through communication tools that are easily changed and updated.
- Provide feedback through communication tools that are easily noticed and understood.
- For complex or unfamiliar tasks, provide feedback in conjunction with procedural information so people remember what they are supposed to do.

- Consider how to display feedback so it is understandable. If announced on a highway sign, it must be read at high speeds. If announced over the evening news, a chart or graph might be useful. If provided on the Web, detailed pages could be included for those who wish to know more.

9.5.3 Evaluation

To understand how feedback affects people's willingness to change behavior, a survey is necessary. Using a written or phone survey will help you know how many people changed their practices and if the feedback they received had any impact on their behavior. You could also conduct an experiment that compares several treatments in different geographical areas to determine which techniques are most effective for this behavior and audience.

9.6 Models

Models and other similar tools like cases, examples, or demonstrations are important components of a social marketing program. Models typically involve the use of a few leaders or early adopters to model a behavior for others or use a physical example to act as a model. A demonstration (Chapter 12) is a type of model used at an interpretive site, but other types of demonstrations are possible, such as one that demonstrates a certain process or action developed for a frequently traveled route. Models and demonstrations encourage changes in behavior by physically showing people how it is done or what the end result will be. A case or case study is an educational tool to explore a problem and models a solution in a classroom setting (Chapter 5). A model paints a picture of the possible by providing an example. Models are effective because people often like knowing that they are not the first person to try something and can learn by watching others (Chapter 3). Therefore, if you are persuading people to do something different, and that action or result can be witnessed, consider the power of models.

9.6.1 Planning

Research on models indicates that when someone is doing the 'right' behavior, other people are more likely to do it too, even if they do not specifically talk about it. For research purposes, these models are 'plants' trained to perform. In the real world, models could be regular people who have already made the switch to the new behavior. Seeing someone model the activity is an indication that new behaviors are possible and feasible. If the activity is observable, it is also a chance to find out how the activity is done, what one needs to wear or do, or how long it takes. Polling place stickers proclaiming, 'I voted' or home-window water conservation stickers announce who in the community has participated in this behavior and let people know how widespread the behavior might be. Not only might these people answer questions about their experience, but the frequency with which one sees these stickers helps change the notion of what is 'normal.'

Fig. 9.6 This sign helps make obvious the landscaping techniques that reduce the risk of fire around these structures (photo by M. Monroe).

To use models in your conservation program, think about the barriers to participation in this activity. If the behavior is so new that people are not sure how to do it, consider establishing a demonstration at a convenient location in the community (Figure 9.6). Include brochures or signs so newcomers know what is being demonstrated and who to contact for more information.

If people do not know that community members have already begun to practice the new behavior, consider strategies that advertise who has already taken the plunge. Putting a large box of recycled materials on the curb helps tell everyone in the neighborhood "I am recycling" and could make non-participants feel guilty. It is more difficult to devise similar proclamations for behaviors that are personal or hidden. Consider simple but effective strategies to advertise participation. Some organizations use car license plate holders, T-shirts, ball caps, and bumper stickers to advertise membership (e.g. World Wildlife Fund member), behaviors (e.g. "I brake for snakes"), or beliefs (e.g. "Growing trees grows jobs").

Another strategy for using models is to organize an event or festival that engages people in conducting the behavior in public for fun. Making bird feeders, planting dune grass, or removing invasive vines are activities that can be done with a group of people in a festive mood. The act of participating creates an opportunity to learn how to do the activity from others. Additionally, the opportunity to have fun while starting a new adventure may help insure the success of the behavior change program.

Finally, examples (a type of model) can be used easily in nearly every communication technique in this book. They portray details of what and how and suggest that others are doing it. Testimonials from community leaders can be powerful stories that model by example. They communicate not only that someone is doing the activity, but are happy enough with the results to be quoted. Newspaper articles can feature demonstrations of organic gardens, brochures can provide quotes from people who returned from an ecotourism vacation, and a public service announcement (PSA) can feature bicycle commuters. The more relevant, concrete, and vivid the example, the more memorable and useful the information will be (Gonzales *et al.* 1988).

Your advisory group will be helpful at determining the examples that are best for communicating the appropriate message in their community and how models might be most effectively used. They might even become the models!

9.6.2 Implementation

Since models show concrete actions and behaviors, they are most effective when someone already has the background information and is interested in considering the new behavior. Thus, models should come after basic persuasion activities and provide the details people need to move from "I ought" to "I will." Models are most meaningful when personally experienced, so a model in each neighborhood may be more effective than one for a whole town. They are most useful when seen in person.

Testimonials and examples, which are a type of written model, can be used to make persuasive materials more effective in brochures, posters, stickers, PSAs, and newspaper advertisements.

Characteristics that make models effective are

- *Visibility.* Models should be located so they can be seen, such as a model garden at the entrance to a neighborhood, or at an intersection where people stop for traffic. If the action is not visible, make obvious some type of evidence that people are participating (McKenzie-Mohr and Smith 1999).
- *Information.* Models should have brochures or signs telling people what is going on, how it was created, and what they can expect will happen.
- *Vivid.* Models and examples are good communication tools because they are specific, concrete, and near by.
- *Simple.* If the action is complex, a series of models may be needed to break the action into achievable steps.
- *Locally applicable.* Models demonstrate that the action is locally feasible. The essence of a model is communicating to people that this can happen here.

9.6.3 Evaluation

You can evaluate the success of your model by comparing compliance rates in areas with models to areas (or times) without models. Conversely, you can observe the behavior prior to and after the implementation of a model. Finally, asking people if they have seen the model and understand it is helpful to know how well the model conveys your message.

Modeling can work when people do not know each other and do not even acknowledge each other. In an effort to conserve water, the athletic complex at the University of California-Santa Cruz posted a sign that encouraged showering athletes to wet down, turn off the water while they soap up, then turn on the water to rinse off. Only 6% of the users complied with the instructions. Making the sign more obvious (posted on a tripod in the middle of the shower room) increased the compliance rate to 19%. But when an accomplice modeled the water-conserving behavior, a full 49% of the athletes did the same. If two models were showering, compliance rose to 67%. None of the models spoke to or even looked at the athletes (Aronson and O'Leary 1982–83). Making the act of conservation obvious could have a significant impact on the audience.

Models are effective tools for bringing about behavior change. In rural communities, the success of a modeling technique may depend on the closeness of the community and the level of comfort people have with each other. Extension workers often use this technique by including visits to models as part of their group tours and in workshops.

9.7 Commitment

Some interesting aspects of human behavior have been harnessed by social psychologists to help predict behavior. For example, if people say they will do something they are more likely to do it. Voicing the commitment creates a self-image that people are likely to honor. In fact, asking someone if they will do a behavior might be all it takes to obtain greater compliance.

In one study, registered voters were called the night before an election and asked "Do you expect you will vote or not?" Those who answered that they would vote were asked to provide the most important reason for voting. Since people tend to overpredict a socially desirable behavior, it was not surprising that more respondents thought they would vote than actually did. But there were a significantly greater number of voters among those who said they would vote compared to those who were not asked (Greenwald *et al.* 1987).

Commitment increased bus ridership in Portland, Oregon US (Bachman and Katzev 1982). Everyone was given information about bus routes and a card identifying them as a study participant. Those who were asked and agreed to participate in the study were more likely to ride the bus for the specified two trips per week compared to those who were not asked. Interestingly, those in the commitment group still chose to ride the bus even after the conclusion of the study.

When home energy auditors were trained to include several different techniques while visiting homeowners, residents were much more likely to retrofit their homes. The simple techniques included using vivid language and meaningful examples (all the cracks in your windows amount to a football-sized hole in your wall), describing the problem in terms of resources lost (money) instead of acting for resource gain (energy), and asking homeowners to make a verbal commitment to make the recommended changes (Gonzales *et al.* 1988).

Another aspect of human behavior is that people are more likely to agree to a request if they have previously been asked to comply with a larger or smaller request. Initially asked a large but still legitimate request, people are more likely to agree to a second more reasonable request than if they are only asked the reasonable request (Schwarzwald *et al.* 1979). If the first request is outlandish, however, they are not likely to agree to a second request. A small initial request (called the foot-in-the-door request) is also likely to increase compliance with a second, more onerous request (Zuckerman *et al.* 1979). In this case, people perceive themselves as someone who supports this issue. Agreeing to the second request enables them to act consistently with their perceptions of themselves.

9.7.1 Planning

Most cultures value and promote the characteristics of honesty, trustworthiness, reliability, and consistency. In those cultures, asking for commitment may be an important tool in a social marketing program.

Getting verbal or written commitment is almost always done in face-to-face conversations. This means your social marketing program must have trained helpers to knock on doors, to staff a booth at an event, or to intercept visitors at a public area. The conversation should be carefully scripted to include persuasive information about the problem and exact wording of key questions to ask for commitment. Many telephone solicitors know how to do this very well by asking, "Can I put you down for a $50 contribution?" For a topic that is not something they care about, most people do not hesitate to say "no." Consider what you know about the audience and what kind of commitment is possible. Carefully word the kind of question people might agree to, and consider how much information about the resource problem they need in order to make a commitment. Would distributing a brochure or other tool help? Can you give them a prompt to remind them of their commitment?

Obtaining commitment was an important element of a social marketing campaign, Turn it Off, that asked motorists to turn off their engines when idling. In Toronto, Canada, idling cars are an air quality concern at locations where people wait to pick up family members—at schools and at public transportation stops. It is also a behavior that is more easily changed than commuting, for example. Local partners and agencies worked together on the steering committee to direct and guide the project (McKenzie-Mohr Associates and LURA Consulting 2001). Recognizing that freezing temperatures are one reason that people wait in an idling car, the campaign focused on idling behavior in warmer months. The program helped establish a community norm for turning off an engine when waiting for more than 10 s.

Fig. 9.7 This window sticker reminds drivers to turn off their engine while idling and lets others know of their good intentions. The graphics from this campaign are available free of charge for use in anti-idling campaigns from Natural Resources Canada (http://oee.nrcan.gc.ca/idling/home.cfm).

To help remind people to turn off their car, signs were posted at idling locations. To help establish a community norm, volunteers provided motorists with an information card, explained the importance of turning their engine off, and asked them to commit to doing so. Those who agreed received a window sticker (Figure 9.7). Over 80% of the drivers who were asked to make a commitment put the sticker on their windshield. These stickers served to demonstrate their commitment to others and to remind drivers to turn off their engine. In a pilot test, the signs alone did not reduce idling incidence or duration. When coupled with the commitment strategies, however, idling incidence dropped by 32% and duration by 73%. This combination of personal contact, commitment, and prompts could significant reduce engine idling in locations like Toronto, where over 50% of the motorists wait for family members with their engines idling.

9.7.2 Implementation

To make commitment an effective tool in your program, follow these tips (McKenzie-Mohr and Smith 1999):

- The commitment must be voluntary. People must never feel like they were pressured or coerced to commit.
- Commitment should be requested only after people express an interest in the activity. Therefore, plan to provide information and ask several questions to ascertain their interest before moving toward the commitment questions.
- Written commitment is more effective than verbal commitment, as in signing a pledge card, a petition, or a form.
- Verbal commitment is more effective than no commitment at all.

- Publicly displayed commitments are effective, such as a poster or newspaper article about those companies that agree to a new behavior.
- Since commitments are usually obtained through personal presentations or conversations, it may be easiest to use opportunities that already exist to speak to citizens. Therefore, try intercepting visitors at check-out counters, entrance gates, or registration areas.
- Actively involving the participant in practicing the behavior increases their confidence, their ability to give a commitment, and their ability to carry through.

9.7.3 Evaluation

Since using this technique involves contacting individuals, it is easy to develop a form to record each contact. If you are knocking on doors, your form might include date, address, whether or not a conversation happened, whether or not the person expressed interest, and whether or not the person made a commitment. You could contact those who made a commitment in three months to ascertain whether they carried out the behavior.

9.8 Incentives and disincentives

There are many different opinions about the use of incentives and disincentives (e.g. fines and punishment) to encourage or discourage certain behaviors. Some argue that if the ultimate goal is for citizens to maintain the behavior on their own, a program should not rely on an incentive to reward good behavior—such a practice cannot be sustainable (De Young 1993). In addition, the use of financial incentives can undermine the intrinsic motives that may help people continue the behavior (Zuckerman *et al.* 1979). The popularity of social incentives like awards programs, however, indicate that there may be cases where these public incentives work to generate a new awareness of a problem, increase the motivation for those who take some risk to change behavior, and help build a new norm. In some cases realizing the positive benefits of conservation can become an incentive. Residents of small fishing villages in northern Quebec, for example, may be interested in conserving seabird populations as long as the birds stimulate economic development through jobs for youth, ecotourism programs, homestays for bird watchers, and media programs (Blanchard and Monroe 1990).

As these examples indicate, incentives and disincentives can complement other tools and be part of a behavior change program. They may be particularly useful when the motivation to change behavior is low (McKenzie-Mohr and Smith 1999) or when a significant financial investment is required.

9.8.1 Planning

Awards programs are an example of an incentive-based program that culminate an educational effort, generate media attention, and can work to change a social norm. In some cases the award is a designation for achieving a new standard. Florida's Clean Marina Program (2004), for example, designates marinas and boatyards that incorporate best management practices to protect water quality. A clear connection between clean marinas and economic profitability establishes a strong incentive for marina owners to consider voluntarily improving their environmental quality. Federal grants for sewage pumpout facilities prompted the Department of Environmental Protection to design a holistic program that addresses water, air quality, waste, and land concerns. The marina business community participated in a statewide process to identify best practices and design the program. A workshop informs marina owners what they must do to achieve the Clean Marina designation and benefit from the grant program; participants sign a pledge card committing them to reducing pollutants from local waterways. They conduct an assessment of their facility with provided worksheets and reference materials and have access to supporting mentors if they have questions. A third party visit will confirm that the marina has adopted the practices leading to designation, and recognition as a Clean Marina is given in an awards ceremony with a press release. The marina is then authorized to use the Clean Marina logo on letterhead and advertisements, fly the flag, and hang the plaque (Figure 9.8). This combination of training, incentives, commitment, and prompts has worked well.

In other cases the awards are prizes for winners of a contest. In Tanzania and The Gambia (Allen 2000) contest-based award schemes have been highly successful at increasing awareness and generating community projects to enhance the local environment. Women's clubs, schools, civic organizations, and small businesses were recognized with awards for developing drought-tolerant school landscapes, performing songs and skits about community sanitation, launching agroforestry initiatives, and cleaning-up local beaches. Increased awareness throughout the community is often the most important outcome, but in the case of Tanzania, the awards scheme also helped mobilize public opinion in support of improved coastal resources. The district-level awards ceremonies drew thousands of people, enough

Fig. 9.8 The Clean Marina flag acts as a prompt to boaters and an incentive to marina owners to participate in the program and attract more business.

to attract the attention of policy-makers. Suddenly there was a political constituency for the new coastal policy that could not be ignored (Day 2001).

Planning a large awards program can be a significant process. Carefully consider what behavior you want to encourage, and whether a contest will encourage people to do so. Decide what rules and guidelines will support your goal. Is group participation necessary, or can one person on a bulldozer be eligible for a cleanup award? Create categories so that different types of projects can win. Categories could be by entrant (e.g. school group, youth project, women's club, business) but also could be by topic (e.g. water conservation, gardens, litter). Provide criteria for a 'good' project. This may be difficult for your first contest, but it will define your program and give participants an idea of what they should strive for. Provide criteria for submitting the project. Will you require photos, testimonials, or evidence of accomplishment? Identify what the awards will be, and choose incentives that are relevant and desirable. Organize the judging process, identifying reputable and unbiased people to judge each category of submissions by standard criteria. A point system is often developed for each criterion. And finally, plan the awards ceremony where all of the submissions can be celebrated, the winners announced, and the awards distributed.

Most non-awards incentive programs are monetary. People who reduce their quantity of weekly trash by recycling, reusing, or reducing consumption are rewarded with a smaller trash collection bill. In communities that charge for the trash bag, the fewer bags used, the more money that is saved. People who invest in energy-saving appliances and insulation are rewarded with a reduced energy bill. Free bus tickets can be used to encourage increased ridership, or coupons can be distributed to encourage people to buy native plants for their flower garden.

On occasion, an incentive program can provide something people want so much that they will even pay for it! To reduce the heaps of waste from the annual German Fest celebrations, one community festival committee created a reusable, clear boot-shaped mug for "Boot Night." The reusable mugs generate less trash than disposable paper mugs and allow security staff to see the liquid, controlling underage drinking. Boot mugs sold for $1 or $2 more than the cost of a beer in a disposable cup; refills were available at the regular price. T-shirts and sponsors helped promote the boot mugs. Changing the shape of the boot every year motivated people to collect the mugs as souvenirs. Sales of the boot mugs have been very successful, selling out each year and reducing waste on Boot Night (Giving Trash the Boot! 2004).

When changes in behavior require changes in business productivity or economic investments, a financial barrier can be overcome with monetary incentives. In the Sacramento Valley in California (US) agricultural runoff contributed significant amounts of nutrients, pesticides, and sediment to the water supply. Established in 1998, the Feather River Water and Air Quality Project was charged with improving irrigation systems to reduce water use and runoff and reducing pesticide application. Farmers were given financial incentives to improve and maintain their irrigation systems and $500,000 in federal cost-sharing grants to cover the expenses of using pheromones instead of pesticides. Significant water savings have been realized as well as water quality improvement (Feather River Water and Air Quality Project 2004).

9.8.2 Implementation

An effective incentive program has a few key characteristics (McKenzie-Mohr and Smith 1999):

- The incentive must be desirable. It must be something people want, or they will not bother with the behavior change.
- The incentive must be the right size. It cannot be overwhelmingly large, or people will be suspicious. It cannot be miniscule either, or people will not pay attention. The best incentives are reasonable rewards that get people to try something new, with the intent that maintenance of the new behavior will occur when people experience the benefits firsthand.
- The incentive should be accompanied by a media campaign that (1) announces the incentive, (2) helps people acknowledge the benefits of the behavior, and (3) provides an example of someone who has been pleased with the behavior and the incentive.
- The incentive should be closely associated with the behavior. Completely irrelevant rewards are not likely to help maintain the behavior because people may not link the action with the incentive.
- The incentive should actually reward the behavior you are trying to encourage. This may require research to identify how people will react to the program. Bottle bills enable people who return beverage containers to receive a small monetary reward. The "rebate" is actually a deposit included in the increased cost of the beverage. The buyer who does not return the container is punished by paying an unclaimed deposit in the increased price of the beverage. Although this does not work to encourage recycling, it does work to encourage someone to pick up discarded bottles, resulting in less litter.

9.8.3 Evaluation

Incentive programs often have a built-in monitoring tool because you can count submissions for a contest, fines collected, or coupons distributed. This evaluation function can be quite useful when deciding whether to reduce, continue, or expand the incentive program. What is harder to determine, however, is the extent to which the incentive program prompted people to continue practicing the behavior without the incentive. Pairing incentives with other tools, like commitment or modeling, helps to make the program continue long after the incentives are gone.

9.9 Press interviews

Many conservation educators are nervous about the possibility that someone from the press might interview them for a story. It is an honor to be interviewed, but it is also a risk since the journalist writes the story, not you. You can provide information along with a fact sheet (Chapter 10), but exactly what is conveyed is not up to you. Therefore, you need to be prepared. It helps to have several brief, clear, easy to understand messages that you repeat several times to the journalist who is writing the story. The journalist can help distribute your message at virtually no cost to you. In some cases an agency message may gain in credibility if it is carried by the news media. It helps to be prepared and to cultivate a good relationship with the local press.

One strategy for informing many journalists about an important story and providing material is to hold a press event. In an effort to halt the declining Russian sturgeon population, the US Fish and Wildlife Service launched a campaign to let caviar eaters know the new limits on international trade and to let illegal smugglers know that DNA testing is used to determine the origin of eggs in a can of caviar. A press conference in New York City, US was held featuring a huge fish tank encasing a 7-foot Atlantic sturgeon. The Director of the Service said, "These magnificent fish survived the catastrophe that wiped out the dinosaurs and most of the other species existing 65 million years ago. But in the space of one century, we may do what nature could not—drive this fish to extinction."

The event took about 6 months to organize and included press kits, posters, stickers, public service announcements. Journalists from television, newspaper, gourmet magazines, and a cigar aficionado publication ran the story. Information was included in theatre playbills. Only one mistake marred the event: midtown Manhattan traffic was so bad that some media personnel could not reach the press conference that morning (NCTC 2004).

9.9.1 Planning

You get a message that a reporter called. The first thing to do is call back and find out what they want. It could be a simple question to clarify a term, or it could be a request for an interview about your work. If the latter, ask to schedule the interview in an hour and take the time to collect your thoughts. You should not start an interview unprepared.

If using mass media is one element of your communication campaign, you have a sense of your audience and you know your message. Choose the message(s) that this journalist should convey to his or her audience and devise how to present them. Frame the messages in complete sentences, not phrases, to make it easier for you to use them in conversation. Instead of saying, "Cats: Indoors" try, "Protect small birds and mammals and keep your pet safe from cars by keeping your cat indoors."

Although your message needs to communicate with the ultimate audience—the public listening to or reading the news—you first need to communicate with the journalist. Provide some basic background information about the importance, the consequences, the actions that can be taken, and the opportunities people have to be involved. Most communities do not have a journalist that specializes in science or environmental issues, so you should assume that the reporter needs this background (Nelkin 1995). The words you use to explain the problem may be used verbatim in the story. In this way, you can have a major influence on the article.

Write out a few sentences that are most important to your story and practice them aloud. Use action verbs and do not ramble. Ask a colleague to review your statements with you; think about what the journalist is likely to ask and be ready with good, tight, answers. Journalists are trained to start their stories with who, what, why, where, when and how. Make sure you have answers for each that connect back to your goals. Specific, concrete examples are usually helpful. Rather than saying, "Lots of plants are endangered," it would be better to state, "Here in the park we have six endangered plants and 30 others which are likely to become endangered."

The reason resource managers complain about the media is that complicated stories get condensed into soundbites. That is particularly true for television. A TV reporter will be looking for about 15 s of your interview. So keep your sentences short and clear and give them a soundbite they can use. Since they may not use the interviewer's question on air, try to answer in a complete sentence.

If you have advance warning, try to test the key messages on your audience. This is an important role for your advisory group. They can help collect information about how the various messages work and make sure that the examples are culturally appropriate.

9.9.2 Implementation

When you are being interviewed, try to relax. Sometimes it helps to pretend you are talking to a neighbor over a cup of tea. If you are on camera, do not look at the camera and keep your eyes focused on the interviewer. Comb your hair beforehand and refrain from fidgeting. Do not wear bright or busy fabrics; instead, stick to neutral colors and traditional styles.

Remember to use your assessment data to frame your story. You have an understanding about what the audience knows and understands. You can tell the journalist what is important to stress, based on your findings.

Use these additional tips for a successful media interviews (Rodekohr 1996):

- Do not wait for them to ask the right questions. You can volunteer the information they need.
- Do not offer your opinion as fact. Let the audience come to their own conclusions.
- Make sure the reporter understands what you have said. Use the questions he or she asks to determine if comprehension occurred.
- There is no such thing as "off the record." It implies you know something that others should not know.
- Never lie to a reporter. If you do not know the answer, just say so. Making up data is not helpful.

9.9.3 Evaluation

The results of your press interview will be broadcast that evening or within a few days. You can review the story and decide if the main points you intended to make were apparent. You can also ask the journalist to give you some tips about your performance. It might help develop a professional relationship with the media if you appear to be interested in doing a better job and helping them do their job. Calls to your agency or organization and hits on your Web site are additional strategies for learning about the effectiveness of your interview.

9.10 Paid and public advertisements

Paid and public advertisements use print (e.g. newspaper and magazine) and electronic media (e.g. radio, television, and Internet) to communicate brief messages. Fifteen or thirty seconds is quite a bit more time than a billboard gets, but it still is not enough time to explore the complexities of natural resource issues. Paid advertisements can be very expensive, depending on the cost of the professionals who design it, the size of the ad, and the market coverage you want, but this is one way to get your message out to large numbers of people. If your budget cannot support a paid advertisement, you may be eligible for a PSA. In some countries, newspapers, radio, and television stations air free announcements as a public service. These announcements usually promote activities of government agencies or non-profit organizations, or are seen to be in the community interest. In other countries it may be more common for a business to support the work of conservation education in exchange for mentioning their name as a sponsor.

PSA's and paid advertisements are generally considered to be the main tool of public media campaigns to convince or encourage new behavior. Research on media campaigns indicates that people are more likely to benefit from the message if they care about the information or are motivated to think about it. Higher levels of education are often associated with groups of people who might care more about environmental messages, but this is not to say communities with less education cannot be made more receptive to environmental campaigns (Martinez and Scicchatano 1998).

Advertisements were an important element of a multifaceted campaign organized by WWF-UK and TRAFFIC to deter trade in endangered species by improving enforcement and penalties in the United Kingdom (Traffic International 2003; WWF-UK 2004). A secondary goal was to raise awareness of the impact of illegal trading among the public. The 2002 campaign used a variety of tools—press releases, media events, advertisements, television personalities, petitions, direct lobbying, and the release of key reports and findings to engage the public and convince the Parlimentarians. Several elements and strategies helped the effectiveness of their campaign:

- The issue was timely; legislation was due for a review.

- The petition and legislation were carefully worded so that no one but uncaring, anti-animal individuals could be opposed.
- Credible scientific reports provided graphic details on illegal activity, the organized crime connection, and frequently used trade routes. Press releases and photos accompanied the release of these reports, orchestrated to gradually build momentum in the campaign.
- On-line advocates from among their membership delivered messages to representatives overnight in response to urgent email notices.
- They did their homework: the law enforcement officials and other relevant agencies wanted the proposed changes.
- They relentlessly applied pressure through press coverage until the support for the changes was obvious.

Most campaigns strategically schedule the release of certain messages and the use of different tools, moving from general awareness to specific instructions of what one can do. In the case of the UK legislation, members of parliament were the target audience, but a public campaign was vital to show ministers the nature of support for the proposed legislation. In obtaining that support, the campaign was able to promote a large number of reasons why people should care about trade in endangered species and how wildlife products can be used sustainably, minimizing the impact of illegal trade.

The campaign was ultimately successful, with the UK government authorizing legislation to make illegal trade in endangered species an arrestable offence in the United Kingdom and increasing the length of a prison sentence from 2 to 5 years. In July 2004, the Institute of Public Relations awarded WWF-UK and TRAFFIC with the highly prestigious Public Affairs Award for this Wildlife Trade Campaign. TRAFFIC Coordinator Crawford Allen commented, "The award was won because the campaign's approach and substance were dynamic and engaging, building the required support in Parliament to the great benefit of law enforcement and wildlife conservation alike." Explained coordinator David Cowdrey, "we organized a campaign to win and offered good support for key changes; we didn't organize a campaign to protest the current state of illegal trade."

9.10.1 Planning

Think about the message you want to convey and your audience. How can you best reach that audience? What motives and values will attract their attention and move them toward a change? Your advertisement should have

- one clear and memorable message
- a focal point—either an image or a headline. Action images and clever headlines are best
- relevance to the audience—they should know why and how this message will help them
- a final directive of what they should do—call, write, volunteer, contribute, etc.

- brevity—for a 30 s message, limit your advertisement to about 75 words or 150 syllables

Consider which media channel(s) to use for your advertisement. Television, while popular and familiar, is expensive. Nevertheless, to reach a large audience with a basic, simple awareness message, a 30 s television spot might be helpful. Invest the time to make a creative, memorable message. Learn from both good and bad examples of product marketing advertisements.

A radio advertisement allows a campaign to target a more limited audience with a specific message. Who listens to classical music? What about the morning traffic report? Radio messages can be targeted by geographic area, program time, language, and sometimes profession and at a rather economical cost. Unfortunately, many radio listeners tune out the advertisements, or listen to a station without advertisements. In nations with multiple language groups, this means translating and recording the same message for everyone.

A magazine advertisement offers another opportunity to narrowly define the audience—people who subscribe to or read that magazine. Many conservation organizations sell advertising space in their newsletter or magazine. Other relevant magazines may be commercially available, such as Backpacker or Organic Gardening. If you want to reach a larger segment of the public, try a news or sports magazine. Large magazines will be able to tell you whom they reach. You can also use your audience assessment to find out what magazines and other media they use. Knowing about the audience, you can design an advertisement to inform, motivate, and appeal to them. A drawback to magazine advertisements, however, is that they are less helpful during crisis campaigns as there is a long lead time for printing.

Local organizations have newsletters that may be useful for your advertisement. Some employers will allow relevant notices to be placed in newsletters that go to all staff.

Newspapers reach national or local audiences on a daily or weekly basis. Simple ads, in black and white, with limited text and graphics, can be placed quickly. Color ads or ads in national papers may reach a larger audience but can be much more expensive.

9.10.2 Implementation

Advertisements can play a strategic role in a campaign. Because they are expensive, they tend to be used sparingly, just at the moment when action is needed. A plethora of advertisements precede elections, for example. In Wooster, Ohio (US) a failed campaign for a sales tax to fund a Purchase of Development Rights program offers insights into the importance of the right message in advertisements. Although the campaign used a variety of tools (letters to the editor, yard signs, and a public forum), the newspaper was a key communication medium for this semi-rural population. Images that looked like the city of Wooster dominated the ads and mailers instead of photographs taken across the county. Similarly, although research indicated that the $4.00 increase in taxes was not a burden, campaign organizers thought mentioning the tax increase was too risky. Instead, they promoted farmland preservation. The opponents were able to exploit the fear of paying more taxes in their attack advertisements (Bostdorff and Woods 2003).

There is a great deal of competition in this Information Age. For your advertisement or PSA to be noticed and remembered it should be interesting, professional, important, relevant, and appealing. Use the planning process and your advisory group to help you invest in the best advertisement possible.

9.10.3 Evaluation

Try to link your advertisement to something else you can monitor, such as calls to a telephone number or hits on a Web site. By matching the dates that you ran the ads to the records from the Web site, you can infer the impact of the ad.

9.11 Summary

You can increase the likelihood of designing a successful social marketing program by carefully combining the techniques described in this chapter. Prompts with commitment, incentives with demonstration, persuasive documents with feedback, or press interviews with an awards ceremony are a few of the possible combinations of techniques that can meet specific goals. Communication tools that convey messages through personal contact help provide the audience with information, opportunities to learn how to perform the desired behavior, and a chance to commit to continuing the behavior. Mass media tools (e.g. PSA's, press events, advertisements) are excellent ways to spread the word, increase awareness, and set the stage for problem solving. Group meetings, demonstrations, and workshops (Chapter 8) are usually good strategies to answer questions, provide details, offer concrete information, reduce the barriers to the behavior change, and obtain commitment. Testimonials from local opinion leaders and respected officials can help sway public opinion and encourage more participation.

Group activities and family festivals (Chapter 8) are other strategies that are helpful to communicate information and increase the attractiveness of a new behavior. A neighborhood work day to maintain local community hiking trails can be an enjoyable way to accomplish a needed goal. The Bear Festival in Umatilla Florida, for example, is an opportunity to tell residents how to avoid attracting bears to their birdfeeders and garbage cans.

Using multiple strategies to promote a behavior will increase the likelihood that a greater variety of people will participate. Typically, there are several benefits associated with a conservation behavior. Use all of them in your campaign as it does not matter which good reason people use for practicing the behavior. Family time together, community participation, health benefits, and the satisfaction of helping may be used to motivate people to participate in a new behavior (Kaplan 2000, De Young 2000). The more people who participate, the more quickly a new social norm will form favoring the conservation activity.

A good planning process that engages community leadership in the design and implementation of a social marketing program is an important investment in producing a behavior change. Local expertise is critical to select the appropriate behavior and identify barriers. Working with the community also can strengthen their skills and motivate them to work on marketing other conservation behaviors. The activity of working with neighbors to create a more desirable future is a powerful motive.

Further reading

Fostering Sustainable Behavior: An Introduction to Community-Based Social Marketing by Doug McKenzie-Mohr and William Smith (1999). New Society Publishers of Gabriola Island, BC, Canada (accompanying Web site at www.cbsm.com).

Getting out your message with the written word

Conservation educators commonly use print media to inform distant audiences, to provide information to visitors at preserves, or to build awareness about conservation issues among target groups, ranging from legislators to farmers. Editorials in newspapers, news stories released via mass media, and brochures, flyers, guidebooks, and other printed material can reach your target audience with information, ideas, and illustrations of concepts.

Publications are easy to disseminate, and can be used how and when the audience desires. Printed materials also are relatively easy to produce and revise. The PIE framework, Planning–Implementation–Evaluation, for creating effective print materials follows similar guidelines for any effective education and outreach. This chapter explores the design of print materials for a variety of audiences. Although the goals and audiences may differ, the basic concepts for producing effective print communication are similar. For any print material, following the A, B, C, D's of print communication—making your material attractive, brief, clear, and dynamic—will ensure your success (Ham 1992; Jacobson 1999).

The design and format of publications largely dictates whether they will be read at all. Editorials and news releases have standard formats that must be followed to help ensure publication. Similarly, following guidelines for graphic design and easy-to-read text is critical to attract readers to your publications. From a letter to the editor about snakehead fish to a guidebook on tropical plants, this chapter explores ways to develop successful print material to achieve your conservation objective.

10.1 Opinion articles

Opinion articles are published regularly in newspapers. Also called op-eds because they often appear *op*posite the *ed*itorial page, they are an attractive technique for building awareness about a conservation issue among informed citizens and policy-makers. Op-eds can be up to 700 words in length, but succinct writing resulting in fewer words will help ensure your ideas get published and read. This means you will need to stay focused on one issue and make only three or four strong supporting arguments. If possible, talk to the op-ed editor before submitting your piece and see what interests and angles might be published. Cite your credentials for writing the piece, and why it is newsworthy. Remember all four functions of mass media—to inform, entertain, persuade, and educate—as you work on your piece. Facts are key, yet entertainment value will help ensure that editors print your piece and that readers finish it. Your first sentence must grab people's attention and entice them to keep reading.

Starting with a concrete example or image that illustrates your topic can arouse curiosity. Next state your opinion clearly and succinctly. If you can not summarize your point in one sentence, think about it some more before writing. Once you've made your point, briefly give any background needed. Then make your argument and back it up with a mix of facts, quotations from authorities, and specific examples. When possible, use anecdotes that support your opinion. People are more likely to read and remember stories than a list of facts. The ending of your op-ed also should grab the reader. Restate what action should be taken and make your last sentence, like your first, strong and memorable (Meadows 2000).

Op-eds should focus on local or regional angles for local newspapers and on broader topics for national papers. They should be tied to current news or trends when possible to appeal to editors. Relating your editorial to an upcoming event, a historical anniversary, or a holiday provides an immediate hook to catch readers. Ecologist Gary Nabhan, at the Arizona-Sonora Desert Museum, published an op-ed timed for release at Thanksgiving. It launched the museum's campaign to conserve pollinator species in order to protect rare plants and economically important food crops. Nabhan reminded readers that their Thanksgiving dinner could be devoid of traditional foods if we do not protect the habitat of 'forgotten pollinators' that pollinate everything from pumpkins to cranberries (Jacobson 1999).

10.1.1 Planning

Planning an op-ed article involves identifying your objectives, understanding the media, and targeting your audience from policy-makers to a broad base of readers. Writing the op-ed involves deciding on a hook or angle for the piece and the key messages to be delivered, and delivering the message in a clear manner. This is not the only challenge. Contacting editors and maintaining contacts with the media are an important aspect of getting the printed word out.

An editorial written by Victoria Tschinkel, Florida director of The Nature Conservancy was published in the Miami Herald (Box 10.1). The timing coincided with warm summer months, and trips to the beach. Her editorial helped launch a campaign to promote legislation in the United States to cap greenhouse gas emissions under a bill known as the Climate Stewardship Act. The goal of the op-ed piece was to raise awareness about the issues of global warming among Florida residents, a coastal state in the United States that would be greatly affected by sea-level rise. The Nature Conservancy's media relations staff contacted the editorial page editor at the Miami Herald newspaper before submitting the article.

Box 10.1 Op-ed example written by the director of the Florida Chapter of the Nature Conservancy.

Headline: *It will affect Florida, too: Global Warming*

Byline: By Victoria J. Tschinkel; www.nature.org/florida

As we head to the beach this summer, to favorite family-reunion spots and to places we share with our children and grandchildren, few of us will wonder whether these cherished places will be here for generations of Americans not yet born. That arc of blue-green water and sand dunes where we gathered shells as a child will always be there, right?

Wrong, say scientists who study global warming.

Decades of research have brought about a fundamental scientific consensus that global warming is real. Average global temperatures are rising rapidly, at a pace never before experienced by humans. If left unchecked, global warming will have very serious consequences for our health, economy and environment.

Sea-level rise is accelerating worldwide, and the most expected estimates are that, in Florida, the rise will be 15 to 20 inches over the next century. In the next 25 years, the state sea level is likely to rise 5 inches. Along with rising sea level comes beach erosion. The number of miles of eroding beaches has increased over the last 100 years in Florida for various reasons, and this trend is expected to continue. Beaches with shallow orientation, especially those on the Gulf Coast, will erode inland one foot to 10 feet with each one-foot rise in sea level. With sea-level rise comes increased impacts from storm surges.

Coastal wetlands and the fish that depend on them will likely be the most harmed by sea-level rise in Florida. At the upper estimate of 20 inches, a rise in sea level would affect approximately 800 square miles of freshwater wetlands in South Florida alone.

But we can choose a different course, one in which our country's natural assets are handed to future generations intact.

This summer, an important piece of legislation—the Climate Stewardship Act (S 139)—is likely to come for a vote before the Senate. Introduced last year by Senator John McCain, R-Arizona., and Senator Joseph I. Lieberman, D-Connecticut., this proposal would encourage more innovation and automotive and industrial efficiency to curb global warming.

The Climate Stewardship Act is a moderate bill calling for capping U.S. greenhouse gases at 2000 levels by 2010 and using market incentives and a flexible trading system to spur emissions reductions. It can succeed. After all, the bill is modeled after the successful federal acid-rain program enacted in 1990, which achieved more total reductions in sulfuric acid than required and at lower costs than predicted.

Moreover, BP a leading oil and energy company, recently proved that it is possible to reduce emissions while remaining profitable. A few years ago, BP voluntarily pledged to limit its 2010 heat-trapping gas emissions to below its

1990 emissions levels. Given flexibility and internal incentives, the company met its target eight years ahead of schedule, saving BP about $600 million.

The European Union has passed mandatory controls on emissions beginning in 2005. In ratifying the Climate Stewardship Act, the United States, the single largest emitter of heat-trapping gases, would be showing appropriate and significant leadership around the world. Our actions would send a much-needed signal to developing countries that they, too, should and must act to reduce emissions.

Victoria J. Tschinkel is Florida director of The Nature Conservancy.

With an expression of interest from the editor, TNC submitted the 500-word article and a photograph to the Herald and 2 other major newspapers in Florida.

10.1.2 Implementation

Tschinkel's editorial was submitted in the heat of July, providing a timely hook to snare editors and readers. Note how Tschinkel appeals to the reader's *self-interest* in presenting reasons for caring about climate change issues—the disappearance of our favorite beaches as sea-level rises. Tschinkel presents both the threat of loss from beach erosion and impacts of storm surges, and an alternative bright future if we support appropriate legislation. Tschinkel also dispels a main objection to regulation of emissions. She offers an example of good stewardship coupled with strong economic return. Note how Tschinkel uses an active voice and tells a story to keep readers interested.

When submitting an op-ed piece, it is helpful to call or email the newspaper and communicate with the editor of the op-ed page. If you are able to make contact, you can discuss the piece with the editor to see what would be of interest. If you have relevant credentials you can send a copy of your resume with your op-ed piece, and mention them again in a brief cover letter (Salzman 1998).

Another strategy is to convince a columnist to write about your issue. Read editorials by columnists in your newspaper to identify someone who covers the kind of issues you are involved with and might have similar opinions. A phone call to introduce the topic, and a written explanation to fax or e-mail to the columnist if interested is all that is required.

10.1.3 Evaluation

Before sending in your op-ed, have several people read it. If they have difficulty understanding any part of it, newspaper editors will too. Questions to ask include:

1. Does the lead address an audience need, concern, or interest?
2. Is the solution presented in a clear, concise manner?
3. Are the consequences of leaving the problem unresolved clearly presented?
4. Does the op-ed help the audience think through or mentally rehearse the action you want them to take?

Read the piece out loud to eliminate awkward wording. Keep revising until it is clear and organized. Avoid stirring up what editorial page editor of the Chicago Tribune terms a three bowler. The phrase refers to the possibility a reader will be so bored by the unrelenting dullness of a newspaper article as he sits at breakfast that his face flops into his cereal bowl once, twice, or, if the article is especially boring, three times (Woods 2003). Ultimately the goal of an op-ed is to serve as a catalyst to increase readers' awareness, foster positive attitudes, or stimulate pro-environmental behaviors. Some indicators of success would be the response to the op-ed, such as letters written to comment on your topic, direct comments to you from other members of the community, or increased interest by policy-makers.

Tschinkel's editorial appeared in the Miami Herald two days after she e-mailed it to the editor. The other two papers did not run the story. The climate-change legislation stalled in Congress; however, the TNC media staff felt the piece contributed to their large-scale effort to draw attention to the ongoing issue.

10.2 Letters to the editor

Research on newspaper readership shows that letters from readers are widely read by citizens and policy-makers (Smith, J. 1995). Letters to the Editor are an excellent device for commenting on conservation issues in the news and making a point about your program or activities. You can urge readers to support your fund drive, clean-up campaign, or special event. Letters can also thank community organizations and particular individuals for help while reminding readers about the goal of your organization. Letters to the editor provide a useful forum for getting your message out quickly and succinctly.

10.2.1 Planning

When composing your letter, assume that readers know little about your topic. You can increase your chance of being published if your letter addresses a recent story in the paper or a current event. Addressed to the Editor, your letter should be brief and to the point. Your letter should state your opinion about the conservation issue or problem, give supporting facts and evidence, then conclude with an action you wish the reader to take.

10.2.2 Implementation

A successful format for a letter to the Editor begins with an attention-getter. The first sentence should grab the attention of readers so that they will continue to read. What gets attention? A startling fact, an unusual example of the subject of your letter, an interesting piece of information, a provocative quote, a catchy or humorous phrase, an analogy, or a story will hook a reader. If the letter is designed to be persuasive, start with an argument with which most people will agree.

Once you have the readers' attention, present the main idea or argument of your letter. Readers' seldom have patience to keep reading if they do not know what the primary focus of the letter will be. After presenting your main idea, provide supporting information. To persuade the reader, you should demonstrate your knowledge of the area by providing facts and any relevant expert opinion. The letter should indicate that you have a good reason for writing. If you have the support of other people or groups that support your cause, indicate this along with your own credentials.

Finally, request the action you desire (e.g. a positive vote on a land-planning initiative, or compliance with new water conservation measures). Ask readers to consider your arguments and take appropriate action.

Many newspapers have a word limit of a couple of hundred words or less, so you must deliver your message concisely. Make your point and stop. A short letter—one page—has a better chance of being printed in its entirety than a longer one.

Consider also including a photograph with your letter. A compelling photograph can illustrate your point and catch the attention of editors and readers. A photograph will help focus the reader's eye on your letter before any others on the page.

Letters should be clear and well written. They should be factual and rational. Tirades, personal attacks, or petty-sounding concerns will not be published. Letters should follow newspaper style, with terse writing and short paragraphs of only two to three sentences. It is helpful to read the letters page regularly to identify compelling styles. Notice how humor or catchy phrases can provide a change of pace and help attract attention.

A letter to the editor about snakeheads (*Channa argus*), an exotic fish from China that appeared in Maryland and Virginia waters, emphasizes the devastation snakeheads could cause to freshwater ecosystems in the United States (Box 10.2). The authors provide clear guidance on the action they wish readers to take to circumvent this potential disaster.

10.2.3 Evaluation

Before mailing your letter, have colleagues or friends critique it for you. Was the message clear? Did it catch the reader's attention and awaken a concern? Did it clearly state what actions the reader should take?

Of course the first stage of feedback you will get is from the editor. Is your letter published? Do you receive comments or responses to the letter? The longer-term outcome, such as votes for a land preservation bill, attendance at an event, or changes in local policy, will help provide the ultimate evaluation.

10.3 News releases

News releases, also called press releases, are a device for sending your message to the public via the news media. News releases are short stories describing a newsworthy activity or event of interest to newspaper, radio, television, and internet audiences. Depending on the scope of your news and the objectives of your outreach program, you can send news releases to local, state, national, or international media outlets.

For conservation organizations, having a story covered in the press translates into "free advertising" for the organization and the ability to reach an enormous

Box 10.2 Letter-to-the-Editor (adapted from Audubon take Action, National Audubon Society Web site (2005).

Letter to the Editor—Federal Legislation Can Help Stop the Snakehead

Dear Letters Editor:

The snakehead fish that now swim freely in the Potomac River represent not only a potential ecological disaster, but also a policy failure. Snakeheads can devastate freshwater ecosystems in the US because they are a top-level predator with no natural enemies. They can breathe air and survive outside water for several days.

When the snakehead was first found in a pond in Maryland in 2002, the response was entirely local: a special Maryland task force, and an eradication effort in the Maryland pond.

Now, the problem is much worse, but this time it has struck the neighboring state of Virginia, and resource agencies have no idea how far the problem could spread.

Local solutions are needed, but will never be enough to control these species that spread quietly across local and state boundaries and then expand rapidly. The Congress should move quickly to pass the National Aquatic Invasive Species Act, which would create a nationwide approach to shielding our water from invading species. The law will come too late to protect the pond in Maryland and may even be too late to prevent lasting damage to the Potomac River, but we can protect the rest of our lakes, rivers, and waterways from ecological harm at the hand of these invaders. Our children and grandchildren will thank us as they swim. Contact your representatives about this critical threat.

Sincerely,
Name
City

audience, depending on the amount of coverage. For example a story picked up by major US newspapers might reach 2 million readers. One picked up by the British Broadcasting Corporation can reach 185 million homes worldwide. Businesses pay millions for that kind of coverage. An added value is that the story is in the form of hard news, which is regarded by the public as a more legitimate information channel than advertising (Jacobson 1999).

The catch is that your news release will compete with releases from other organizations and industries stacked on the news editor's desk. You will have to work hard to get your message out. It is important that your release is structurally correct and newsworthy. It must be brief and clear. You must attract the reader to your news by appealing to their interests and concerns, with topics ranging from unusual events and people to saving money and providing clean water.

In the past 5 years, The Nature Conservancy has produced over a thousand press releases in the United States. They have appeared in papers in every state and all are available on their Web site (The Nature Conservancy 2005). In 2005, their big story was the rediscovery of the Ivory-billed Woodpecker (*Campephilus principalis*) in an Arkansas forest that The Nature Conservancy had helped to preserve. Well-documented sightings of such a large bird, thought to have gone extinct 60 years ago, made front page news. Of course, discoveries of this magnitude are not made every day, but many activities of conservation organizations and agencies are newsworthy. The Nature Conservancy ensures new protected lands, partnerships, people, programs, and activities make the news through their press releases (Box 10.3). They also look for ways to tie in current events to their conservation goals. For example, they took advantage of the release of a

---

Box 10.3 Sample topics of press releases by The Nature Conservancy published in US newspapers.

Long Though Extinct, Ivory-billed Woodpecker Rediscovered in Big Woods of Arkansas

The Nature Conservancy Partners with State, County, Town to Preserve 308-Acre Rare Forest

Conservancy Announces New Leaders of Salmon Program

Celebrate Earth Day at One of The Nature Conservancy's Preserves

The Nature Conservancy to hold Earth & Ocean Gala

Neotropical Migratory Birds Can Count on Critical Habitat

Volunteers Needed to Plant Trees

The Nature Conservancy to Give Away 10,000 Educational DVDs

Prescribed Burns Restoring Muleshoe, Arizona

Guided Nature Walks at Arizona Prairie Preserve Begin June 1

Fire Management Grant Awarded

2003—Year of the Hawaiian Forest

Largest Land Conservation Purchase in Hawaii History

The Nature Conservancy Establishes Hotline To Assist With Battle Against Non-Native Invasive Species

Get Your Feet Wet and Learn About Freshwater Ecology

U.S. Forest Service Recognizes Nature Conservancy Employee

Don't Let Bat Myths Scare You This Halloween

New Study Highlights Economic Benefits of Nature Tourism in Arizona

Beware: There May be Bad Plants in Your Tucson Backyard

Nature Conservancy Offers Opportunities to View Seasonal Wildflowers

Harry Potter film depicting "messenger" owls, to promote saving threatened owl habitat. Their press release caught the attention of editors: "Nature Conservancy Scientists Hope New Harry Potter Film Will Increase Attention to Saving Threatened Owl Habitat."

10.3.1 News release structure and content

The following guidelines focus on news releases for print media, which do not require expensive equipment or technical expertise. Technical help for broadcast media releases can be hired or sometimes obtained as donations from advertising agencies.

The format of a news release for print media helps ensure that it is easy for reporters to use. Contacts, dates, and the five W's—who, what, where, when, and why—must be carefully composed and structured for easy reading. The structure of a news release places the most important information in the first or "lead" paragraph. The five W's that may be included in the lead paragraph cogently answer: Who is the story about? What happened? Where did it happen? When did it happen? Why did it happen? And, finally an H—how did it happen?

The lead paragraph must capture the essence of your story and tell the reader something of value. It also needs to hook the editor so that your story will be printed. The remaining paragraphs flesh out the story by providing necessary details. The body of the story often uses quotes from the people involved to give the story a human element and to add interest. The release should stay focused on your communications objective. If there is an action or behavior that you want from the reader, make sure your message states it clearly. Make an outline of the major points of your story and sequence them from most important to least important.

The structure of a news release can be diagrammed as an inverted pyramid (Figure 10.1). Information that is least important is placed at the end. Editors will cut the article from the bottom up to fit the space available. By placing key material first, you minimize the risk of losing it. The closing paragraph often describes the mission of your organization or agency. It often, but not always, will

Fig. 10.1 A news release format resembles an inverted pyramid.

be cut. Mention your organization's name and location near the beginning of the story to ensure that it is included.

The headline of your news release should get the editor's attention while highlighting your message. Headlines must be accurate, brief, and tell who did what. They may contain a humorous or dramatic statement to catch the editor's eye. The headline that you write for your news release is unlikely to be used by the paper; its main purpose is to catch the editor's eye.

The story should be of interest to a wide range of people. Emphasize the readers' needs, concerns, and desires and make your message fill the need or satisfy the desire. Self-interest compels attention. When possible, tell your message as a story and put people in it. Editors of local papers in particular must emphasize the local angle of a story. Readers want to know the relevance of a news story to their lives.

As you will notice when reading a newspaper, brevity is key. Sentences are short. Paragraphs also are short, made up of only a few sentences. News releases are seldom longer than one page. The trend toward lower reading abilities or lowered appetites for reading among the public means your press release must use simple words and sentences to reach the reader. Additional information can be included with the release in the form of fact sheets and background material for use by the media.

The professional appearance of your news release, with a double-spaced type, ample margins, and simple-to-read style can help predispose the editor to react positively to its content. Send your news release to a specific person at the media outlet. Use their name, not just "editor" and begin to establish a rapport with the people who may cover your activities. Check with colleagues at other organizations that get good press coverage for a recommendation of relevant reporters or editors.

The example in Box 10.4 demonstrates the format, with the release date, contacts, headline, and body of the text. If the news release is more than one page,

---

Box 10.4 Example of a news release announcing a new national park (http://www.nature.org).

Letterhead: The Nature Conservancy—saving the last great places on earth

For Immediate Release

Contact Information:
Christine Broda-Bahm
Phone
Email

The Nature Conservancy and Partners Complete Acquisition of Baca Ranch:
Secretary of the Interior Designates New National Park in Colorado

Alamosa, Colorado—September 13, 2004—After more than a decade of work to conserve the 151-square mile Baca Ranch, The Nature Conservancy today announced it had completed the last of a complex set of real estate transactions, clearing the way for the protection of the ranch and the designation of the nation's newest national park, the Great Sand Dunes National Park.

At a ceremony today at the Great Sand Dunes National Monument and Preserve, Secretary of the Interior Gale Norton officially re-designated the monument as the Great Sand Dunes National Park, the nation's 58th national park.

"This is an emotional day for the hundreds of people who poured their heart and soul into this 10-year conservation odyssey," said Steve McCormick, the Conservancy's president and CEO. "The success of this effort lies in the dedication and commitment of the unprecedented coalition of individuals and organizations that came together, determined to protect the San Luis Valley's water and conserve one of the nation's most spectacular and ecologically-important places."

Upon closing the Baca Ranch transaction, the Conservancy transferred management responsibility for 27,000 acres of land within the designated national park boundaries to the National Park Service. The Conservancy will continue to manage the remainder of the Baca lands in partnership with the US. Department of the Interior pending a final $3.4 million federal appropriation. Once the remaining monies are appropriated, the full ownership will be transferred to the National Park Service, the US Fish and Wildlife Service and USDA Forest Service to create the Baca National Wildlife Refuge and expand the existing Rio Grande National Forest.

The 97,000-acre Baca Ranch is one of the largest, unfragmented landscapes in the state. Its awe-inspiring terrain feature grasslands, alpine lakes, sand dunes and two 14,000 foot peaks, Kit Carson Peak, and Challenger Point. The ranch harbors more than 70 species of rare plants and animals, including several species found nowhere else in the world. The landscape provides vital habitat for a wealth of wildlife species, including a large elk herd, fox, mountain lion, bighorn sheep and numerous migratory bird species, including Sandhill Cranes.

"The completion of this transaction is the realization of a vision put forth by the people of the San Luis Valley," said Charles Bedford, the Conservancy's Colorado associate state director. "The local community's desire to preserve the valley's agricultural and natural heritage is ultimately what made conservation of the Baca Ranch possible."

The Nature Conservancy is a leading international, nonprofit organization that preserves plants, animals and natural communities representing the diversity of life on Earth by protecting the lands and waters they need to survive. To date, the Conservancy and its more than one million members have been responsible for the protection of more than 15 million acres in the United States and have helped preserve more than 101 million acres in Latin America, the Caribbean, Asia and the Pacific. In Colorado, working with local communities and partners, close to 600,000 acres have been protected.

write "more" at the bottom of the page. Indicate the end of the piece by typing "####" or "-30-" below the last line.

Whenever possible, a photograph or graphic should be included with the release to illustrate the story. As the saying goes, one picture really is worth a thousand words. An intriguing, high quality photo can help get your release published and catch the reader's eye. For example, charming photographs of monk seals and fishermen were available to accompany a press release produced by the Mediterranean office of the World Wildlife Fund to promote their conservation project in the Aegean Sea.

News releases are often a key part of an entire press packet that attracts and facilitates the media to cover your conservation program. The World Wide Fund for Nature (WWF) organized a press trip to publicize the story of a fishing community that turned into guardians of critically endangered monk seals (*Monachus monachus*) on the Turkish coast. The steps for conducting the press trip and providing a news release for a feature story and other materials for the media are recounted by Sampreethi Aipanjiguly, Communications Officer for the WWF-Mediterranean Program.

The story of their work with a local non-governmental organization (SAD-AFAG), fishermen and monk seals in Aydincik, Turkey, began 15 years ago. The monk seals were critically endangered from hunting and habitat loss, and more recently from declining fisheries. The few remaining monk seals in Turkey live alone or in groups of two or three on islands and remote coastlines. The local fishermen also were suffering from the declining fish populations, in part caused by commercial trawlers fishing illegally close to shore. SAD-AFAG worked with the fishermen to tackle this problem. They were able to obtain a patrol boat to monitor and reduce illegal fishing practices. With help from World Wide Fund for Nature, they next established a fishery cooperative to improve the fishery and the livelihoods of the fishermen. Aipanjiguly's task was to publicize the achievements of the project using the opening of a fishery cooperative in Aydincik as a hook to catch media interest.

10.3.2 Planning

The objectives for attracting media attention were to use the press to help build local and international support for the WWF-funded project in Turkey, says Aipanjiguly. At the local level, this would help validate the beliefs of the fishermen in their sustainable fisheries work, and increase local government support as a result of international attention. Internationally, it could help build recognition and support for WWF programs. Aipanjiguly targeted international—French, German, Italian, and British—media that reached audiences in places where many people take summer vacations to the Turkish coast and other Mediterranean destinations. The magazine, Environment and Development, which covers the Middle East, was also a target.

Aipanjiguly wrote two press releases for the Turkish media about the project the year before the cooperative opened. She realized that the story was so compelling it could garner international media interest.

Aipanjiguly planned a feature article and a press trip to Aydincik in the summertime to publicize the latest achievement of the project—the establishment of the cooperative. The project would make a good story article—a general interest piece that was not tied to hard news. The feature story would not only provide information about the event, but could include more depth and color to make it entertaining as well as edifying. The timing would make it good holiday reading as people headed to the coast. About a month in advance, Aipanjiguly contacted international press correspondents to invite them to participate in the press trip. Media contacts developed over the years by WWF and new contacts gleaned from the International Press Office in Istanbul, Turkey, helped recruit 10 journalists for the 1-day trip to Aydincik. Several others could not make the trip but asked for written materials, or called in and conducted interviews by phone.

Aipanjiguly worked with the WWF-Turkey communications staff, the coordinator of SAD-AFAG, and the head of the marine program for WWF-Mediterranean to make the trip a success. Aipanjiguly stressed that each WWF communications person could only handle two to three journalists, particularly because there was a mix of media involved. In addition to the press release and written background materials that the print journalists needed, television media required moving visuals, interesting images, and interviews with people. Radio journalists needed 30-s soundbites, recordings of noisy monk seals and people being interviewed. Because her program had only a small travel budget, Aipanjiguly was unable to do a reconnaissance trip to Aydincik to arrange the logistics for the press trip, and instead relied on local partners to arrange boat trips for the journalists to join fishermen and see the monk seals. Aipanjiguly's team planned the event from start to finish with a detailed checklist. Each activity on the agenda had people assigned to complete specific tasks before and during the press trip.

The opening of the fisheries cooperative provided an event to make the story newsworthy. Aipanjiguly knew that people find news about other people interesting, and emphasized the human interest of the struggling fishermen solving not only their problem, but protecting an endangered species at the same time. Aipanjiguly made sure that the press material sounded like news rather than WWF propaganda, by emphasizing the concrete impacts of the project. WWF invited the mayor and governor of the region to officiate. Their attendance would provide political support for the project. Because the international press was invited, it helped ensure that politicians would want to participate and take advantage of the public relations opportunity.

10.3.3 Implementation

The journalists received a press kit before the trip. This included the press release telling the story of the establishment of the fishery cooperative; an agenda for the field trip; a five-page background paper explaining the problem, solution, impact of the program, and future activities; a map showing the location of Aydincik and the seals; and an institutional brochure from WWF. For the broadcast media, a compilation of video images of the fishermen and seals was provided. The press package prominently displayed the WWF logo and contact cards.

In order to make the press release interesting to readers, Aipanjiguly wrote it as a feature story (Box 10.5). She (1) put specific people into the piece, (2) told a story, and (3) let the readers see and hear the situation for themselves through quotes and descriptions. Of course, as for any news release, Aipanjiguly followed the rules to keep it short and clear. She wrote a good headline that captured attention and included essential information. The press release provided the backbone of the story the journalists would experience on their trip.

The agenda for the press trip was an ambitious one:

0600—Join fishermen in their morning catch
1000—Press briefing (with Ministry of Environment and Forestry, Governor and Mayor of Aydincik, Head of Aydincik Fisheries Co-operative, Head of Conservation for SAD-AFAG, and WWF Marine Officer) as follows:
    1000–1015: Presentation of project framework by WWF Marine officer
    1030–1045: Presentation of SAD-AFAG activities at the field site
    1045–1100: The story of the monk seals and fishermen
1130—Opening ceremony of Aydincik Fisheries Cooperative and the new guard boat
1200—Interview opportunities with local fishermen and staff of the guard boat
1300–1600: Lunch and rest
1600–1800: Boat trip to no-fishing zones in Aydincik and monk seal caves
1900—Dinner
2030—Departure

Although the informal meals seemed like a relaxed setting, Aipanjiguly reminded her team to remain alert throughout the trip. "There is no such thing as 'off the record'."

At the end of the field trip, Aipanjiguly checked with the journalists to see if they needed any additional material. Some contacted her later to fill in missing data or obtain extra photographs.

10.3.4 Evaluation

To monitor the coverage of specific press releases or issues, clipping services are available that scan and "clip" specific news coverage for a business or organization. This costly expense was not within the WWF budget. Instead, Aipanjiguly and the communications staff monitored their results informally by contacting the journalists that covered the stories to see how widely they were disseminated. The WWF network in Europe and the Mediterranean also notified their office about press coverage. The story appeared in the media throughout Europe and the

Box 10.5 Feature news release from the World Wide Fund for Nature announcing their project with fishermen and monk seals on the Turkish coast.

WWF-Mediterranean Letterhead Stationary

For Immediate Release
Date
Contact: Sampreethi Aipanjiguly, Communications Officer, WWF-Mediterranean
Yeim Aslan, Communications Contact, SAD-AFAG
(address, phone, and e-mail listed)

Guardians of the Monk Seals: Fishermen are helping endangered seals through a WWF project on the Turkish Coast

It's still cool at five in the morning. Ahmet "Charlie" Orhan steers his small boat across the glassy Mediterranean waters, glowing pink in the early morning light. He's heading for his fishing nets, laid the night before just off the rocky shoreline, and now hopefully full of fish.

All around the Mediterranean, thousands more fishermen are similarly fetching the night's catch. But although the scene is timeless, today is a special day for Charlie and the other fishermen from the small town of Aydincik, Turkey. Today the governor will open the new shop of their fishery cooperative—the latest achievement in ongoing work that's giving the fishermen, and the endangered Mediterranean monk seal, a better chance for survival.

Like some other artisanal fishermen around the world, he's concerned about declining numbers of fish.

"In the 1980s, there were so many fish," he says. But now we don't see the same number or variety.

The culprits are trawlers. These boats, which come from bigger cities and whose huge nets indiscriminately catch everything that crosses their path, are not supposed to fish closer than 3 miles from shore. The inshore area is reserved for artisanal fishermen like Charlie—locals who set a small number of nets to supply their town with fish.

But as they have fished out the deeper water, trawlers have—illegally—come closer and closer to shore over the past 20 years.

Fewer fish is bad news for fishermen. It's also bad news for the Mediterranean monk seal.

These shy animals once lived in colonies along the coasts of the Black Sea, the Mediterranean, and the Atlantic Ocean from Portugal to Senegal. But today they are one of the rarest mammals in the world.

"Only 500 remain," says Yalcin Savas, Head of Conservation at SAD-AFAG, a Turkish group that works with WWF on monk seal conservation issues.

Mediterranean. It even appeared in a Chinese airline magazine. "The timing of the press release, during the summer when people are thinking about the coast, and the charming subjects—adorable monk seals and local fishermen—helped 'sell' the story," says Aipanjiguly.

At the local level, the international attention helped build regional political support for the project. The mayor increased his support for the fishermen using the patrol boat to halt illegal fishing activities. WWF's local partner organization received more attention for their model program and plans to build on this experience in other areas. The fishermen reported they appreciate the attention the publicity created for their new cooperative fish store.

A lesson learned from the media trip included the importance of doing a reconnaissance to prepare for a press event, says Aipanjiguly. Some journalists had to travel for many hours to reach Aydincik, and the 0600 start time the next day may have caused less intrepid reporters to miss the activity. Although a remote location is generally an impediment to getting the media to cover a story, WWF's long-term media contacts helped facilitate the participation of journalists. This emphasizes the importance of building a relationship with the press. Over time, WWF has come to be viewed as a reputable and authoritative source. Success in having a story published depends on the media's belief that your release is factual and that your institution is trustworthy. Getting journalists to travel to a remote site to report on an event is a complex undertaking. The success of the Aydincik press trip in generating newsworthy stories helped bolster WWF's reputation for the next media action.

10.4 Brochures

Brochures are typically produced to introduce an organization's goals and activities for dissemination to a broad audience. Brochures can describe environmental management objectives, special facts about an agency or site, or details about an important conservation campaign. Text is combined with graphics or photographs to create an appealing appearance. Brochures often are useful for increasing membership or raising funds. They can fit in an envelope for easy mailing, or the back cover can include an address so an envelope is not needed. They can be grouped in a dispenser to target different audiences. A brochure is often the front-line information for an organization—the first thing a prospective member or supporter sees about your cause or learns about your site.

Brochures are the most commonly used written format for interpretation at parks and refuges. They are distributed as hand-outs at sites, exhibits, and trails, mailed to groups planning a site visit, and used for membership drives. Brochures also are dispensed at nearby hotels and at other public parks or agencies.

People can pick up brochures and read them at their leisure. This allows you to provide detailed information about a conservation activity—from steps to attract backyard birds or remove exotic plants, to measures to protect groundwater or maps of regional parks. People also can save brochures for a souvenir or a reference for future use.

10.4.1 Planning

Your communication goal and specific audience will dictate your design and content for a brochure. To attract most audiences, a brochure should have a catchy title, bright colors, and an inviting layout. Collect and study brochures that catch your eye from other organizations before deciding on a design for your own brochure. For the lay public, if the cover of a brochure does not look inviting, few will make the effort to delve into the text, no matter how scintillating the writing. Beyond the cover, you need relevant and interesting content to keep their attention and accomplish your communication goals.

Careful planning will ensure that all the elements of a brochure—paper size, color, illustrations, layout, and text all work together (e.g. Brigham 1991; Monroe 2005). Once you have identified the main goal of the brochure, consider the interests and concerns of the audience as you frame the message and design the brochure to attract them. Important considerations about the target audience also will dictate the format—from inclusion of more than one language and a reliance on illustrations for multilingual audiences, to large-print for older audiences. For the general public, text written at an eighth-grade reading level is appropriate, although draft versions of your brochure should be tested with the audience before proceeding to publication.

Your budget will influence your choice of colors and papers for your brochure. Full color photographs on glossy paper may be easiest to attract attention, but it is also the most expensive. Careful use of space and catchy titles and graphic design can make even a single color brochure practically irresistible. Many word processing programs include a function for creating different types of brochures. With these easy-to-use programs you can experiment with different layouts and get feedback before printing.

The Earth Sanctuaries Foundation of Australia produced a brochure to publicize their program to residents and tourists. The Earth Sanctuaries Foundation provides a refuge for wildlife by restoring native habitat and excluding alien predators, such as foxes and cats. The brochure features an attractive montage of photographs of unusual-looking, endangered Australian wildlife including platypus and numbats to catch their audiences' attention. The cover solicits interest with their theme: "Saving Our Wildlife from Extinction." Like most organizations' brochures it describes their mission, their objectives, concrete examples of achievements, and areas of funding needs. After briefly describing the problems and solutions for Australia's threatened animals, the brochure suggests the action readers should take. Brief text follows each of the following headings:

- The sad history
- Now—a life raft for our precious species
- Making a real difference
- Providing a long-term solution
- You can help us

A convenient membership form can be detached from the brochure and mailed in with a contribution—"Yes I want to help save Australia's threatened wildlife and their habitats!"

10.4.2 Implementation

Brochure cover

Keep the title brief and thematic. Short titles of less than 10 words get more readership. For example, *"Farm Birds: Nature's Pest Controllers" or "Water–Lifeblood of the Everglades"* give a snappy overview of the theme of the brochure. Keep your target audience in mind; if the brochure is for recreational users, mention hikers, boaters, or other audiences, in the title. For example: *"Birders' Guide to Seeing and Protecting Shorebirds"* is tantalizing to bird-watchers. *"Native Bees are Valuable Crop Pollinators"* targets a key concern of farmers. Draw readers in by emphasizing their personal interest.

A single photograph or illustration on the cover is usually more effective than multiple visuals (Figure 10.2). Make the visual interesting by showing something happening. Action shots of a bear eating berries or people hiking through a forest have more appeal than an inactive bear or a forest setting. Use bright colors or a high contrast design to attract people. Select colors that help emphasize your theme, such as red for a brochure on prescribed fire or blue for aquatic ecology.

Fig. 10.2 Brochure covers with exciting graphics and clear messages from the NSW National Parks and Wildlife Service, the Wildlife Conservation Society, and the Bahamas National Trust.

Body

The body of the brochure should make use of subtitles, photographs, and other graphics to break up the writing. Many people will only read the headlines, so you need to make it obvious why the reader should continue. Some tips for increasing readability include:

- Use wide margins and extra white space (empty areas) around the headlines and between sections to make the brochure look easy to read. Empty areas attract the reader's attention.
- Use a simple font to make the page look inviting.
- Try using direct quotes or question and answer approaches. These often entice people to continue reading.
- Use bullets or check-boxes to add interest and organize the text.
- Use simple graphics, maps, and charts.
- Put captions (in a different size or font from the text) under photos and graphics. People often read captions, second only to headlines.
- Follow the tips for effective writing (Box 10.6).

Box 10.6 Tips for writing well to ensure your story will be read (Consult writing guides, such as Zinsser 1985 or Strunk and White 1979 for more tips.).

- Write with nouns and verbs. Adjectives and adverbs seldom add vigor to a story. For example, "The wolf howled," is more compelling than, "The wolf called loudly."
- Use action verbs to keep a story moving and interest the reader, rather than "to be" verbs that show little action.
- Use active voice:—"The biologist darted the tiger;" not passive voice—"The tiger was darted by the biologist."
- Use simple, ordinary language. Avoid jargon and elaborate words, for example, write "use", not "utilize", and "now", not "at the present time".
- Avoid using qualifiers, such as very, rather, and little. They sap strength from your statements.
- Be clear. As Mark Twain admonished: Say what you propose to say, don't merely come near it.
- Choose the right word, not its second cousin.
- Know your reader: relay your message using elements of interest to the reader.
- Use personal words, where appropriate, like "you," "we," a person's name, or direct quote to appeal to the reader.
- Be specific and provide details; do not be vague.
- Be concise. Short sentences and paragraphs are easier to read.
- Revise, revise, revise. The waste basket or delete button is your best friend.

10.4.3 Evaluation

All education and outreach materials benefit from feedback. Have some members of your target audience evaluate your brochure before you invest in publication. They can help you avoid the common mistakes in brochure design that reduce readership. The problem ABC's of the brochure reader include:

Aggravation. The page is too jammed with information. The type is hard to read or too close together. Organization is poor or nonexistent. The page lacks white spaces and does not provide breathing space for the poor reader.

Boredom. Nothing in the brochure stands out; it looks like a page of gray. Headlines are too small. Columns are too wide. Text is too long. Paragraphs and sentences are too long or complex.

Confusion. The reader cannot follow the flow of the text, visuals interrupt reading, or the headings and subheadings are indistinguishable.

Pilot testing a draft of the brochure with the target audience can help ensure its success. After publication, tracking the numbers of brochures taken or disseminated suggests how attractive the audience may have found them. Surveys of the audience, through a tear-out response sheet, or formal sampling, can provide data on the specific impacts of the brochure. Some organizations include a specific phone number for readers to call for further information. This allows an organization to track the number of inquiries related to a brochure.

10.5 Fact sheets and flyers

Fact sheets and flyers serve a variety of purposes. Many large conservation organizations have fact sheets available for the press and public that can be mailed to specific audiences or perused on their Web sites. Fact sheets or flyers can be sent to constituents or reporters with information about new study results, activities, and updates on projects. They can be used to answer questions from the general public. They can be posted on a bulletin board or pulled from an envelope.

As a result of their many functions, fact sheets and flyers come in a variety of styles and forms. Yet, good ones have several things in common. They are short (1–2 pages), written messages that grab someone's attention. They are well-organized and easy to read. They generally focus on a specific topic and provide enough details about an activity or issue that the reader can use to understand the subject. Longer fact sheets can be designed to cover a complex topic, yet, breaking the information into several fact sheets is often preferable. People are more inclined to read brief materials.

Fact sheets and flyers are often used to give the press and the public a better understanding of a subject area in which they are already interested. The Wildlife Conservation Society has a series of fact sheets on their Web site describing their programs and projects in countries across Asia, Africa, and Latin America. They distribute the publications at press events and refer them to people that request information about the organization.

The design of a fact sheet or flyer has two jobs to do. First, it has to get the attention of the reader for your main message. Second, it has to help the reader absorb the information that you present. It will need to include information and data that readers will find interesting and engaging.

Visual displays of information, unlike plain text, encourage a diversity of individual viewer styles and rates of reasoning and understanding (Tufte 1990). Colors, graphic design, illustrations, and photographs can attract and invite the reader to engage in the information in an entertaining and thoughtful manner. Flyers that are posted and passed out at functions are characterized by a large, catchy title to convey the essence of the flyer, large type that is easy to read, a lot of white space to make it inviting, and a professional quality photograph (Chapter 6) or illustrations to enliven the page and emphasize the message.

Fact sheets and flyers are an economical and time-saving technique to communicate with multiple audiences. Good fact sheets help replace time-consuming chores of writing detailed letters to interested individuals or long explanations on the phone. The information provided in fact sheets is clear and consistent. Also, it is easy to add new information to an organization's collection of information materials.

10.5.1 Planning

First, imagine the typical reader you need to reach. Like all communication techniques keep this image foremost in your mind as you design a fact sheet or flyer with a message that is relevant to them. Similar to planning a brochure, you will need to consider how many copies you will need, as well as other printing decisions about ink, paper, and color, dictated by your budget and audience. If you are designing a fact sheet to support a press release, think like a journalist and include historical, factual, and anecdotal information that could be used in an in-depth story.

The Xerces Society for Invertebrate Conservation, produced a series of fact sheets targeting farmers in California (Xerces Society 2005). A fact sheet targeting watermelon farmers, "Native Bee Pollination of Watermelon," provides information about watermelon pollination, native bee ecology, and conservation. Because of the very specific audience, detailed information could be presented. The first sentence in the fact sheet catches at the heartstrings of any farmer: "Pollination by bees is critical for a successful watermelon crop." The fact sheet continues with information about the effectiveness of native bees for pollination, how to recognize and protect native bees, and how to enhance pollinator habitat (Figure 10.3).

To develop a fact sheet or flyer, make a list of key ideas that are crucial for you and your audience. Prioritize them and select the one basic theme, or idea for your print material. Develop three to five main points or examples that explain your theme. People have trouble remembering more than five main points (Jacobson 1999). Once you have focused on your message, you can consider

INTERNATIONAL CONSERVATION

WCS staff meeting   © WCS Tiger being captured   © Bart Schlevel

Cambodia

China

India

Indonesia

Iran

Kamchatka

Lao PDR

Mongolia

Myanmar

Pakistan

Papua New Guinea

Russia

South Pacific

Thailand

ALLEVIATING TIGER-HUMAN CONFLICTS

Wherever people and large carnivores coexist, conflicts between the two are usually inevitable. This lesson applies to Amur tigers in the Russian Far East. Livestock depredations and attacks on people, although exceedingly rare, impede conservation efforts. Traditionally, such conflicts were usually resolved by simply killing the tiger. Cumulatively, such deaths represent a significant mortality factor for small populations of endangered species. We have been working closely with a special branch of the Russian government to address such situations and resolve them to protect both the tiger as well as human life and livelihood.

The Human Aspect

If Amur tigers are to survive in the wild, they must coexist with people on multiple-use forestlands. Human-tiger conflicts in the Russian Far East generally fall into two categories: attacks on people and predation on domestic animals. Tigers that attack people are considered problem tigers unless they are defending cubs or themselves. In most instances, tigers that attack people are wounded animals (often wounded by poacher bullets) that cannot effectively hunt wild prey. Unfortunately, such animals remain a serious threat to unsuspecting and innocent people. The second scenario, in which tigers prey on domestic animals, is considered a problem if they do so very close to human habitation. That is, a tiger that kills a cow or a dog that wanders a kilometer into the woods unattended is not considered a problem, but one that kills cows in barnyards does require management.

Threats

Before the creation of the special Tiger Response Team, most conflict situations were either ignored or resolved with a bullet. This situation resulted in unnecessary tiger mortalities or in wounded cats that became even more dangerous. Such conflicts also fostered dangerously negative attitudes towards tigers and "retaliation killing" by local residents.

Highlights

Total Area
Russian Far East:
• Primorski Krai
• Khabarovski Krai
• Total area: 185,000 km^2

Habitat Types
• Temperate / Boreal Forest

Wildlife Present
Siberian (Amur, or Northeast) tiger*, far eastern leopard*, eurasian lynx, far eastern forest cat, brown bear, asiatic black bear*, wolf, wolverine, sika deer*, roe deer, wild boar, goral*, snow sheep, caribou, stellar sea eagle yellow-throated marten, sable

Amur Tiger   © Russia/ WCS

Fig. 10.3 The Wildlife Conservation Society provides fact sheets on their regional programs, such as their initiative to reduce tiger-human conflicts (http://www.wcs.org).

pictures, illustrations, graphics, and layout to attract your audience. Make sure your call-to-action or objective of the flyer is clear and that the reader can see these elements without having to search for them. If your organization intends to create more than one fact sheet, you will need to develop a standard format and to identify your organization with a logo or recognizable style. Remember to provide easy to find contact information, including a name, telephone number, e-mail address, and Web site address (if available).

The distribution of your fact sheet or flyer will influence the format. The planning process helps you determine the audiences, how you will reach them, how much money you can spend, what public places are available for distribution of your material, how many people will help with distribution, and how fast the information must get out. You will need to decide if you are planning to hand out the material, send it in the mail, distribute it through literature shelves or bulletin boards, or electronically via e-mail or a Web site.

The format for your fact sheet or flyer will vary according to the distribution method. Flyers destined for bulletin boards must have big, simple headlines, lots of space, little verbiage, and distinctive graphics to make it stand out from the myriad of other flyers posted. A mailed flyer must look attractive when it appears at the door. A graphic or provocative headline should attract attention on the cover and space must be left for addresses and stamps. Web pages demand equal care in their layout to ensure easy reading for the viewer (Chapter 11).

10.5.2 Implementation

Your publication will need to attract the reader's attention, hold their interest, awaken their desire, and lead them to action. To grab the reader's attention, the title should tell what the fact sheet or flyer is about. It should highlight the theme and attract attention to the page. An effective title is short, less than eight words long. If a longer title is needed, a smaller subheading can be used. A good photograph or graphic image should help attract attention and highlight your theme.

Once you have the reader's attention, you need to hold their interest with important relevant details about your activity, issue, or subject of concern. Short, simple sentences, and everyday language help keep a reader's interest.

Excite the reader about your topic. Awaken a desire to learn more about the subject. A beautiful picture or descriptive detail showing why the reader should care about the topic can help them focus on how it might affect or improve their lives.

Once you have people's interest and concern, you will want to lead them to action. Depending on the objective of your fact sheet or flyer, provide a compelling reason for the reader to act. Be clear and direct in exactly how they may take action.

The visual design of your fact sheet or flyer can help ensure you move the reader from attention to action. Pictures are important for telling a story. Even newspapers that once were composed mostly of text have increased their use of pictures and graphics to attract readership.

Now they can be viewed as well as read. The weekly news section of the New York Times newspaper used to be mostly text. Now pictures, drawings, big catchy headlines, and bold layouts with a lot of white space fill the pages. For most print materials, devote two-thirds of your publication to pictures, white space, and headings (Beamish 1995).

Good graphic design can help ensure your print material is read. A number of factors should be considered as you are designing your fact sheet or flyer, or almost any printed material.

- Have a dominant focus on the page to attract the eye. This can be a headline or a picture. Too many bold features can cause confusion. If you emphasize everything, you emphasize nothing. Creating a hierarchy of design elements emphasizing a bold title and photograph or illustration and de-emphasizing less important information will help lead the reader through the brochure.
- Pictures do more than decorate your text. Choose the right photograph or graphic that will help make your point and hook the casual reader. Other pictures should illustrate key points, so readers can tell the general message at a glance. Remember, people like to look at people in your pictures (Chapter 6), just as they like to read about them in your text.
- Clarity and simplicity are key to an effective page. Respect the readers' time and attention by providing information as concisely and attractively as possible. If readers have to search for clues about the purpose of the flyer, or the contact information for the organization, they probably will not bother.
- Stick to one or two varieties of typeface in a flyer. Although your computer might offer a hundred fonts, refrain from using more than a couple. Designers use a large, bold type for headlines and key captions, and a smaller, easy-to-read typeface for text. Dark letters on a light background provide a stark, legible contrast. White letters on a dark background can highlight a headline, but are more difficult to read.
- Make it easy for people to read the text. If the type is too small, the columns too wide, or the paragraphs too long, few readers will want to work hard enough to read it. Blocks of text can be broken up by interspersing short paragraphs with longer ones, indenting paragraphs, and using bold subheadings and bullet points. Even better, rewrite your text to be more concise and use more photos, graphics, and white space.
- Use illustrations that are relevant to your theme and objectives. Drawings or photographs should draw attention to or dramatize your message.
- Use charts and pictures to help explain complex ideas. Make sure your audience recognizes any symbols, maps, or diagrams you use by pilot testing your visual aids with the target audience.
- Use color creatively. Color enlivens and decorates your printed material. Color can serve as a label, such as green for forest, blue for water; or as a measure of quantity on a chart, such as showing more and less whale sightings by varying the value and saturation of the color.
- The first rule of color is that pure, bright, or very strong colors have loud, unbearable effects when they stand unrelieved over large areas adjacent to each other, but extraordinary effects can be achieved when they are used sparingly on or between dull background tones (Tufte 1990). Small, bright areas will stand out vividly against gray or neutral backgrounds.

- To maintain a sense of unity, use the same colors throughout the printed material, repeating color combinations. The text, illustrations, graphics, and white space in-between should feel balanced visually.
- Allow the eye to move in one general direction when reading the flyer, instead of a serpentine type pattern. Make sure there are proper column breaks; with pleasing, but not unnecessary spaces.
- Have a clear, visible logo or organization name so that readers will know whom to contact and how.

10.5.3 Evaluation

Feedback from the target audience suggests if your message, content, and design are doing the job. Organizations can track the number of "hits" on fact sheets appearing on their Web site, as well as requests from the target audience for further information about the environmental issue. Assessing whether your information is used by journalists in subsequent media articles will reveal if your material was appealing and useful. The ultimate sign that the fact sheets or flyers have been successful is by monitoring whether your specific audience takes the desired action. For the Xerces Society, an increasing number of watermelon farmers leaving habitat for native bees will indicate success. If there is little response to your material, your method of dissemination, content, and graphic design all need to be re-tested with the audience to determine how to make improvements. Expert review of your materials by public relations specialists or graphic designers can help you evaluate the quality of your fact sheet or flyer. Continual review of materials produced for similar audiences by other organizations and businesses can provide ideas for making your material more effective.

10.6 Guidebooks

What's that bird? Where are the hiking trails? How do I monitor my management program? Guidebooks serve many purposes, from identification of plants and animals to park guides for visitors. They have helped resource managers improve their management activities and landowners to protect threatened species. Guidebooks can target a variety of audiences using text and illustrations to disseminate ideas from basic to technical.

Guidebooks can help people become aware of conservation targets and learn new information and skills. A study of the development of field guides to tropical forest plants was initiated by the Forestry Research Program at Oxford University, UK. They found that guidebooks could promote global exchange of information about plants from international to local grassroots levels. Information about tropical plant species allowed better use of plants by collectors of non-timber forest products and by tour guides. Plant guides also could help influence farmers not to burn a valuable tree on their farm. In Bolivia, older community members felt that young people would be an ideal target audience for guidebooks to help retain indigenous knowledge of plants and animals that was being lost (Hawthorne 2004).

The sale of guidebooks also can generate income for local organizations. A plant guide, Trees of Brazil—*Arvores Brasileiras*, was produced in Brazil by author Harri Lorenzi. Researchers use the book for reference. The high quality photographs and large format also make the books popular among middle-class people with an interest in nature. Lorenzi directs a private research institute funded through the sale of the guides. The proceeds of book sales fund employment of four scientists and an airplane for botanical research in the Amazon (Lawrence and Hawthorne 2005).

10.6.1 Planning

Clearly identifying the purpose and target audiences for your guidebook is a critical first step. No matter how good your writing and graphics may be, the guide will fail unless it serves the needs of your organization and the audience. Audiences for guidebooks have varying technical knowledge, ways of communicating, and levels of formal education. Creating a scientifically accurate guidebook that is useable by readers with different literacy levels or languages is a challenge.

The content, format, length, design, and size should be geared to achieving your objectives with the audience in mind. Who will use your guidebook and how will they use it? Examining guidebooks that you like can provide models for layout, type, graphics, and writing style. Getting feedback from members of your target audience at this beginning stage can help steer you in the right direction. Take some time to research what the audience might be looking for—what styles, formats, and approaches resonate with them. If your audience intends to use the guide book in the field, a portable size or laminated cover might be most useful.

Consulting the users of your guidebook during the planning process will help ensure its usefulness. The authors of a tree guide for agroforestry production in Bolivia selected the species to be included in consultation with local community members and botanists (Lawrence and Hawthorne 2005). The authors visited communities in the four regions to be included in their guidebook. In each community they asked men and women to rank the top 10 species they felt should be included in a guide. Botanists added to this prioritized list to include species which they considered to have potential for agroforestry, but which were not yet well known to farmers. Thirty of the lowest priority species were cut from the list of 100 species to produce a guide that was within the available budget (Lawrence and Hawthorne 2005).

Once you have some ideas about the ideal guidebook design, talk with a printer to discuss the costs of various layouts, paper types, colors, and length. You may need to compromise on some choices to stay within your budget. Will the guidebook be for sale or provided free of charge? Or both? The Bay Islands Conservation Association in Honduras provided free copies of a full color guidebook to the natural history of the islands to all school teachers through a grant to produce 1000 copies from a development agency; revenues from the sale of the guidebooks to tourists subsidized additional printings (Jacobson 1992) (Figure 10.4).

Fig. 10.4 The cover of a guidebook about marine conservation shows people enjoying the marine habitat.

Budgets often force compromises in size, quality, and format of guidebooks. Be sure that your skills are a match for the requirements of producing a guidebook. You or your organization's staff must have good writing skills, good photographic or illustration skills, and good layout and design skills. Keep in mind that you also must know what will appeal to the target audience, as well as the people writing the guidebook.

If particular expertise is missing in your organization, additional staff or volunteers may be available to supply the needed skills. Oftentimes businesses and community members are willing to provide support to conservation organizations for a worthy cause, such as developing an educational guidebook. Photographers may be willing to donate pictures for the venture. Professional writers and artists also can be hired if your budget is adequate, or photographs purchased from stock photo agencies. Some businesses may underwrite your costs in exchange for the opportunity for free advertising on your publication, or the creation of good public relations.

Once you are clear about the type of guidebook you are producing, establish a realistic schedule for the dates to complete research, preparation, audience feedback, and finally publication.

10.6.2 Implementation

To maximize your resources, it is important to have a plan and follow a structured approach. Begin by planning the details of your writing and thinking through the information required. It may be beneficial to draft a working document that includes special equipment you may need, expertise, or illustrations you may need to obtain, a timetable for fieldwork, an itinerary, and time for writing. It is important to focus your attention on the goal of your guidebook and exclude material that does not contribute to the goal.

When it comes to the time to write the guidebook, always keep the reader in mind. The objective is to be clear and easily understood. The material should attract the attention and interest of the reader, and elicit the desired reaction. The aim may be the successful identification of a butterfly, understanding of a camping technique, or a new skill in native plant gardening.

In order to achieve these results, the writing must be concise, direct, and jargon-free. Follow the rules of good writing reviewed in Box 10.6 . Define or illustrate new terms, when needed. Glossaries, keys, charts, and illustrations can help simplify complex terms or concepts. Easy-to-read typeface, an uncluttered design, and attractive photographs, drawings, or graphics, all help ensure that your guidebook will be used. Similar to other print materials, a thematic title for the guidebook should state its content and target audience. "Valuable Tree Species for Agroforestry on Small Farms" or "Guide to Aquatic Insect Ecology for Fly Fishermen," should pique the curiosity of target farmers or anglers.

Keep in mind the audience for the guidebook. The Education for Nature Organization in Viet Nam produces a guidebook about ecosystems in a comic book format (Figure 10.5.). They incorporate student activities, games, puzzles, and stories with abundant illustrations to appeal to their young audience.

Fig. 10.5 A student activity guidebook about ecosystems uses a comic book format to attract the young audience (Education for Nature—Viet Nam).

The guidebook, Flora of the Ducke Reserve (Ribeiro *et al.* 1999), required a team approach to produce this useful guide. The audience for the guide was tropical biologists who needed to identify plants in the field. Thirty-seven specialists and several local tree-identification experts helped in the identification of plant material for the book. A group of Brazilian botanists and students helped integrate the knowledge from the specialists to design a user-friendly guide to their reserve in the Amazon Basin where little information is available. Their decision to include photographs of the leaves, bark, and venation of each species allows users in the field to identify trees (Lawrence and Hawthorne 2005).

Once you have completed a draft of your guidebook, have colleagues and members of the target audience review it for clarity and relevance, as well as interest. To be successful, guidebooks should provide accurate information that improves the capabilities of the reader. Some factors to test are listed in Box 10.7. Use audience and expert feedback to revise the guidebook. Test it again with the audience if many revisions are made.

You also need to decide on a strategy to market the book. The intended audience must be aware of the book, interested to read the content, and motivated to obtain it. Dissemination of the guidebook at a park headquarters, through an organization membership list, or in conjunction with other guidebook markets, may successfully reach your audience.

10.6.3 Evaluation

Audience feedback and expert review during the production of a guidebook is a key component to ensuring its ultimate success. Review by audience members of draft versions of the guidebook can identify unclear text, graphics, and illustrations. Ambiguous language, inappropriate material, or overly complex directions can be weeded out before you go to the expense of final production.

The World Conservation Union and World Wide Fund for Nature developed a guidebook for managers of marine protected areas (MPA), entitled "How is your

---

Box 10.7 Design considerations for developing guidebooks for specific audiences.

- Purpose and content of guidebook are clear.
- Format is easy to understand.
- Information is accurate.
- Content is relevant for target audience.
- Language, amount of text, and themes are appropriate for audience.
- Keys, indices, maps, captions, color codes, and other systems to guide readers are simple and easy to use.
- Layout and graphic design is attractive.
- Illustrations or photographs aid in identification or elucidation of subject.
- Number, size, scale, and quality of illustrations are appropriate for audience.
- Guidebook is easy and enjoyable to use.
- Size and durability of guidebook are suitable for purpose.
- If guidebook is for sale, audience is willing and able to pay for guidebook.
- Dissemination plans will ensure availability of the guidebook for the target audience.

MPA doing? A Guidebook of Natural and Social Indicators for Evaluating Marine Protected Area Management Effectiveness" (Pomeroy *et al.* 2004). The goal of the guidebook was to help managers assess the success of their management actions. The guidebook was developed, tested, evaluated, and revised over a 3-year period. Initially, the authors hosted a workshop to collect feedback from 30 recognized MPA experts from around the world on the MPA management indicators. A draft guidebook with proposed indicators was reviewed at a workshop for 22 volunteer MPA "pilot" sites. The revised indicators were tested by the pilot sites and reviewed again by experts. Box 10.8 shows some questions from the expert review forms and pilot MPA evaluation forms for the guidebook. The comments from peer reviewers and the pilot sites resulted in the production of a third, "working" draft of the guidebook. This draft was released by the IUCN at the Fifth World Parks Congress in 2003, along with an invitation for review comments from interested individuals. This feedback helped create the fourth and final version of the guidebook.

It is often difficult for authors to receive feedback about their guidebooks after publication. Monitoring distribution and use of guidebooks, demonstrations of changes in audience knowledge or skills, and informal polling of readers can help

---

Box 10.8 Sample questions from expert review forms and pilot MPA evaluation forms used for developing a guidebook to evaluate management of marine protected areas (courtesy of John Parks, coauthor of How is Your MPA Doing? (Pomeroy *et al.* 2004).

A. Expert reviewers considered the following questions in their review:

1. Is the indicator described accurately and the layout a valid method for measurement?
2. Is the description of the indicator written clearly and is it easy to understand?
3. Is the indicator and related information user-friendly?
4. Is there any additional information on how to use the indicator?
5. Are there additional reference materials?

B. Pilot MPS managers evaluated items such as:

1. List the most relevant strengths and weaknesses of your selected indicators in measuring management effectiveness.

Indicator	Number	Strengths	Weaknesses/Problems
Biophysical			
Socioeconomic			
Governance/Administrative			

2. Please comment on problems that occurred in your site while measuring some of the indicators (specify the indicator and methods/instruments used).

indicate success. User evaluation forms can be used to elicit audience feedback if there is a convenient way for the user to return the form, such as at a park or zoo. Feedback is certainly needed before updating and reprinting guidebooks, or creating new ones based on an untested model.

The popularity and use of the MPA guidebook is one indicator of its success. Several thousand English copies of the guidebook are in circulation, and new versions in Spanish and French have been released. Plans to translate it into Mandarin, Portuguese, Vietnamese, Bahasa Indonesia, Korean, and Arabic are in the works. Monitoring of the US National Oceanic and Atmospheric Administration (NOAA) Web site with the free PDF version of the guidebook and information about the MPA pilot sites records over 1200 hits per day. According to John Parks, coauthor of the guidebook from NOAA, the ultimate success of the guidebook will be viewed through two lenses: first, if in 5 years time MPA managers and administrators have integrated the guidebook's management effectiveness evaluations seamlessly into their management actions and budgets; and second, if in 10 years time the marine conservation community can more accurately gauge the degree to which management efforts are resulting in the desired biological and social impacts. Parks hopes that someday the guidebook will become obsolete because MPAs are routinely using the principles outlined in their book.

The feedback from the audience was crucial to developing a useful guidebook. "What really allowed us to produce a useful document was the three rounds of careful peer review and practical feedback from the 22 volunteer MPA sites who tested a draft version of the guidebook," explains Parks. "This cumulative feedback was instrumental in making the publication user-friendly and logical. It helped us think about how to best design the guidebook for ease of use and navigation, particularly because there is a lot of material for the reader to get through, and much of it is somewhat dry. We realized the need to use full color and many photos, in addition to the navigation tools—color-coded tabs for the indicators, consistent indicator outlines and descriptions, and an easy-to-use flow diagram of the four steps in completing an MPA evaluation."

10.7 Summary

Using the printed word for education and outreach is an essential technique for accomplishing conservation objectives of many organizations and agencies. Harnessing the power of mass media through op-ed pieces, letters to the editor, and news releases provides the means to reach vast numbers of people with information in a trusted and reliable format. This type of free advertising is valuable to everyone dealing with critical conservation issues and tight budgets. Knowledge of the structure and format for producing these materials is necessary for success in the competitive mass media arena.

Fact sheets, flyers, brochures, and guidebooks form a backbone of material for use by conservation organizations to build audience awareness, increase knowledge, and foster new conservation skills. Producing effective print materials is hard work. Involving the audience in the planning process can help ensure that the content and format will be appropriate and useful. Studying how words, type, graphics, and space are used in other publications will help you determine why some materials succeed and others fail. The guidelines presented for clear writing and attractive graphic design should help guarantee that your target audience will notice, read, and act on your materials.

Further reading

Getting the Word Out in the Fight to Save the Earth, by Richard Beamish (1995). The John Hopkins University Press, Baltimore, MD.

How to Do Leaflets, Newsletters, and Newspapers, by Nancy Brigham (1991). PEP Publishers, Detroit, MI.

Communication Skills for Conservation Professionals, by Susan Jacobson (1999). Island Press, Washington, DC.

Making the News: A Guide For Nonprofits and Activists, by Jason Salzman (1998). Westview Press, Boulder, CO.

11
Taking advantage of educational technology

The introduction of any type of technology into learning environments usually promises to revolutionize education by making it more efficient, more effective, and more powerful. This was as true for the slide projector as it is for the computer. Some of these innovative technologies, like the digital camera, are genuinely helpful and after training, educators find that these resources become a significant asset to their programs. Others are never widely adopted, like the laser disk. Still others, such as the overhead projector, are used in a variety of educational contexts, but only to enhance the existing educational program. Not much changed.

Classroom teachers, school administrators, and agency administrators have a number of important considerations to make when debating the use of new educational technology. For educators of youth, one goal is helping students become proficient in using technologies that are meaningful and relevant to their interests. In some cases that will be a computer, while in others, a GPS unit for mapping locations or dissolved oxygen probes to measure water quality. Conservation educators can support this goal by creating programs that use the technologies that learners should become familiar with.

Before investing in new technologies, educators will also need to know the direct and indirect costs and how they might change over time. For some technologies, however, the costs change quickly. In the 1990s communities accessed the Internet with dial-up modems, necessitating phone line connections for each computer. In many countries, landlines can provide high speed access (DSL or cable), making larger files more accessible. In developing countries, the advent of wireless technology (e.g. cell phones and wireless Internet hubs) reduces the requirement for landlines. Communities without telephones may connect to the Internet through a satellite. The speed at which technology changes could make prohibitively expensive technologies more accessible in just a few years.

Despite its promises, there is little evidence to suggest that instruction with technology (in this case radio, film, instructional television, or computers) is more effective than traditional instruction (Kent and McNergney 1999). Educators can be effective using any of the strategies and techniques outlined in the other chapters. Films or computers are not likely to replace effective outdoor experiences or community service projects, but these and other technologies can enhance other learning techniques.

Most educational technologies introduced to schools are available to nonformal educators. Rather than preparing learners to become familiar with the technology, however, their interest is more in the efficient and effective communication of information to build knowledge, attitudes, and skills in support of conservation goals. For example, designing an interactive Web site or burning a CD-ROM has become easier than reproducing a slide show, though sending a video may be less user-friendly but more cost-efficient than sending a speaker.

The following questions can help conservation educators consider which technologies might be effective in their programs. Consider how the answers vary when thinking about developing a program with new technology (Rohwedder and Alm 1994; Milone 1996; Kent and McNergney 1999).

- How common is the necessary equipment? How much training is needed to use the tools effectively? Do teacher-training institutions prepare educators to use this technology?
- Do learners interact with the technology? Will the technology enhance the level of learner interaction and challenge, thus improving educational programs? Does this tool teach learners to become more technologically proficient? Would this make educators more likely to use it?
- What kind of expense does this technology entail at present and in the future? Is extensive research necessary to choose which brand to purchase?
- How likely is the equipment to break; how soon can it be repaired or replaced? Is the equipment dedicated to one use, or can it be utilized for several needs? What ancillary materials are required to make the technology educational?
- Does the technology make a complex educational message more consistent? Is the technology an efficient way to distribute your message?
- Can the technology make your message more memorable, more visually appealing, and more engaging? Are learners more motivated to learn the information if it comes via this technology?
- Can the technology enhance learning, create opportunities for using multiple intelligences, allow for different learning styles, and enable users to improve problem-solving skills?
- Is the technology so advanced that it dampens educator professionalism that may be needed to make culturally appropriate adaptations? Does the technology distract learners from your message?
- How many educators and learners can be reached with this technology?

This chapter introduces four techniques and their accompanying technologies that are commonly used to communicate with youth and adult learners in schools and nonformal settings. Because each technique has many possible technologies for production, duplication, or dissemination, we focus primarily on the educational component of the technique. Because technology changes rapidly, we offer generalities that should hold true even as the technology advances.

11.1 Video

In the past, films and movies were produced by exposing and editing film (35 mm for commercial applications, 16 mm for educational institutions, and 8 mm for home movies). Although this may produce the highest quality image for commercial purposes, many educational applications migrated to video formats because of the ease of editing and producing material for television broadcast or personal use. There are several types of video formats available around the world, so videos and their players must be compatible. One disadvantage to videos is that they are played by physically moving the tape forward or backward across a head. Although it is much easier to rewind and replay a video than a film, it is difficult to navigate quickly.

With the evolution of still photography to digital formats, so too has the video business become digital. Digital video images can be accessed randomly, in addition to sequentially, making the "fast forward" command more like "skip." Editing software is cheaper and easier to use than its video counterpart, and enables the editor to work on portions of the product in any order. Today, one can use a digital camcorder to shoot a video, edit it on a computer, store it on a CD-ROM or DVD, and play it on a computer, television with DVD, or Web site. Some editors find the digital format so appealing they convert original video to digital for editing and then back to video when complete. A variety of technologies can be used to produce or distribute videos (Box 11.1).

Box 11.1 Video technology terminology

Videodisc or Laserdisc: a large flat disk that holds vast amounts of digitized information, images, and sounds. It is played on a machine like a VCR and attached to a television for viewing. Because videodiscs require specialized equipment to play they have not been widely accepted.

Video Formats: VHS and 8 mm film are popular for home movies but not appropriate for professional work. S-VHS and Hi-8 can be edited for serious projects. Betacam SP is still recommended for analog video among professionals. Digital formats have surpassed video in quality; digital betacam is preferred.

Moving Picture Experts Group (pronounced m-peg)—(MPEG): a group of digital video compression standards that produce a high quality video. During the editing process, MPEG stores the changes to the original file instead of every frame. MPEG-1 produces a video with a resolution of 352-by-240 pixels at 30 frames per second. This is slightly worse than a taped video. Videos on the Web are usually 160-by-120 pixels. MPEG-2 can handle 720-by-480 and 1280-by-720 at 60 frames per second. It is usually used for television broadcasting and DVDs. MPEG-4 compresses graphics and images into a narrower bandwidth and provides more graphical options.

This section refers to the basic technique of the video itself. Video is an appropriate educational technique when the topic is visually appealing and interesting. It is often the tool of choice when the topic involves movement or when a skill is being demonstrated. Note the variety of ways it can enhance a conservation program:

- *Videos can be produced by learners.* In Nepal, rural communities expressed their experiences and thoughts about community forests in a video letter that was shown to top-level policy-makers in Katmandu (Grieser 2000). Secondary students in San Diego CA participate in the Montgomery Media Institute to produce and edit their own programs, which are broadcast in a school-wide closed circuit network (Milone 1996). Digital photography has made it even easier for learners to produce a video to convey information and meaning.

- *Videos can be produced by professionals interviewing local citizens.* By providing an opportunity for people's voices to be heard, this technology can be a powerful strategy for communication. In Mississippi, US, a video document-ary, "Green," featured disenfranchised communities living in the shadow of petrochemical industries and brought their concerns to the public. In Egypt, the production of a video was a key element to help the government agencies understand the complex issues surrounding the maintenance of local irrigation ditches. The camera became a tool that enabled both men and women to articulate their experiences. In particular, women were able to explain that there was no other place to put household garbage—information that had not been part of the previous understanding (Grieser and Rawlins 2000).

- *Professionally produced videos can be broadcast over wireless, cable, satellite, or television networks to homes and schools.* The popularity of television enables video to reach an enormous number of people. The availability of public television and cable channels makes it feasible for agencies to air their videos. In countries where the public is accustomed to watching television or movies, the expectation for high quality shows may be quite high. Viewers will also expect to be entertained by a broadcast video. It is unclear how much infor-mation is learned while viewing an entertaining show, though positive emotions and attitudes are important in learning (Chapter 2). The prolifera-tion of channels in some countries, however, makes it less likely that viewers will find and watch your show amid the cooking shows, situation comedies, travel shows, talk shows, sports shows, and old movies. Nevertheless, high quality nature films are powerful venues for taking the public places they are not likely to visit and showing sights that few people will see. *Winged Migration*, a French film, is a good example of a powerful movie. Dozens of birds were photographed in flight, from the air, giving viewers a "bird's-eye view" and appreciation of the birds' grace and beauty. Although more expensive, investments in equipment, production, and technological effects make films like *Winged Migration* much more intriguing and visually appealing than educational videos.

- *Professionally produced videos can be used by agencies and organizations to introduce their location, explain concepts, and gain support for their projects and perspectives.* Videos can be distributed to schools or field staff, shown during programs, used in the classroom, or posted on the Web. They can be an effective component of a training program or a visual introduction of a site. Nature centers and parks often use a video to orient visitors to the seasonal changes of their site and the things they are not likely to see on their visit.

Although high quality videos can be extremely powerful, the technology is not used by all educators. Barriers to using videos in classrooms include difficulty in finding the right match between the video and the curriculum objective, cost of purchase, equipment upkeep, time, and potential discipline problems in a darkened classroom (Kent and McNergney 1999). Because the cost of producing a good, quality video is fairly high, conservation educators should consider if they will be able to use the product for several years, or if the video might become out dated before it is finished. As technology advances and the costs come down, producing a digital video will become more feasible.

Once a video is produced, it is a form of one-way communication. The final product cannot be reorganized or regionalized by users. Historically, the cost of production limited the adaptability of video. Digitally produced videos, however, can be regionalized easily in the production phase. Elements can be edited out or added, producing a suite of different versions. When videos accompany a program that engages the audience in discussion or enables a presenter to answer questions, they promote two-way communication. Depending upon the purpose of the program, a stand-alone video may suffice, or a personal presentation with the video might be necessary.

In areas of the world where television is not available, radio often takes this niche for entertainment, news, and information. Radio programs can be translated into different languages and broadcast to rural areas. Radio has been used successfully by Rare, an international conservation education organization, to build support for conservation (Box 11.2).

Regardless of the purpose of a video, the format in which it is produced (digital or film), or the means by which it is distributed (Web, DVD, etc.), there are guidelines to making a high quality product that should be considered.

11.1.1 Planning

The development of a video should begin with a clarification of the objectives for this tool. Agreement on the purpose, the audience that will view it, and the message that they should understand as a result will help writers, planners, and funders move forward in the development of a coherent and meaningful video. It helps to have a one-sentence theme that clearly conveys your message, such as "Mosquitoes spread deadly diseases" or "Mangroves protect the land and create important wetland habitat." It may be helpful to sketch out the development of concepts in a video in preparation for script writing.

Box 11.2 Radio plays a key role in behavior change

In many parts of the developing world, radio is an important medium for mass communication. In Saint Lucia, where a "soap opera culture" makes American dramas very popular, several organizations worked together to create a local radio series, Apwe Plezi (Rare 2003). The scripts were filled with heroes, villains, local storylines, and important messages about resource conservation and family planning.

Rare, Saint Lucia's Forestry Department, and the Planned Parenthood Association used focus groups, research, script reviews, and outreach to create the successful series. Relevant topics, such as unemployment and teenage pregnancy, were addressed so listeners could think about how these issues affected their own health as well as society.

And it worked. By the second season, about 35% of the island's adult population was listening to the program. Planned Parenthood clinics reported a 32% increase in new family planning users. Seventy-one per cent of Apwe Plezi regular listeners reported making a change in their behavior as a result of listening to the program. Most significantly, researchers and family planning specialists agree that Apwe Plezi, which aired for three seasons from 1996 to 1999, played a role in supporting positive changes in Saint Lucia's fertility trends: Teen pregnancy dropped from 21% in 1990 to 16% in 1999. The overall birthrate dropped 13% between 1996 and 1999.

A two-column script is used to describe the progression of the audio and visual portions of the video (Figure 11.1). The left half of the page describes the visual scenes and the right half, the audio content, which includes the script. The columns are synchronized so that a scene and its audio accompaniment are presented together. Onscreen graphics and text are noted on the left side. Directions to the producer are also written into the script, such as "pull back to reveal landscape" or "dog barking."

Producers use their own language when referring to the components of a video. A long shot refers to an image from a distance; a close shot fills the frame with the subject. Cut is a sharp transition to a new scene, whereas fade or dissolve refer to more gradual transitions. Pan directs the camera to move across a scene, zoom slides the camera lens in and out. A narrator speaking the script without being shown in the video is called a voice-over and is often cheaper than hiring someone to act as well as narrate the script.

The video script should be more similar to spoken communication than written, because the audience will be listening, not reading. Short sentences, vivid imagery, and easy-to-understand phrases will help communicate the message. Do not use numbers or dates, unless they are short. Consider using onscreen text to emphasize key words, new vocabulary, or phrases (Telg 2004).

	Wildland-Urban Interface Video Script – page 2	
	Graphics	Script
6	INTERVIEW: RESIDENT NAME LOCATION (About 15 seconds	RESIDENT– Discussing how living in the interface improves quality of life and one or two different concerns or challenges. Something like, "We love that our kids can run and play in the woods, but we do worry about the potential for wildfires."
7	- ATTRACTIVE FOREST LAND - CHARISMATIC WILDLIFE - A RIPPLING BROOK. - TRASH-FILLED RIVER - PEOPLE IN WUI– GARDENING - RESIDENTS AND RESOURCE PROFESSIONAL WALKING IN THE FOREST, TALKING	NARRATOR: Living in the interface has its advantages and challenges, and as these residents have realized, their presence in the interface inevitably affects their surroundings. You can make choices that will help reduce your impact on natural resources and improve conditions that affect forest, wildlife, and human well being. Different issues require different approaches. Let's see how interface residents and natural resource managers across the south are working on these issues together.
8	- FIRE FOOTAGE IN THE WILDAND-URBAN INTERFACE (2–3 shots)	N: Fire is often the main wildland-urban interface issue people think about. Although most wildfires are caused by human carelessness, some happen naturally. If allowed to burn , they help maintain fire-dependent ecosystems. But with more and more houses scattered throughout the interface, almost every fire that occurs threatens human safety and property and becomes a major emergency.
9	FIREFIGHTERS SPRAYING BURNING INTERFACE HOMES	N: When fires are out of control, firefighters are forced to make split-second decisions about what's most important to save. Protecting fragmented interface areas where people's safety and property are top priority can be more difficult than defending large forested areas.

Fig. 11.1 Video scripts are developed in two columns.

The following should be considered while developing the video:

- How long will the video be? Most adult audiences pay attention for 8–10 min, and children for less. Changing the pace and context of the video can help to hold the audience's attention (Telg 2004).
- What will the video include? Interviews can help the audience imagine their own experiences and allow the message to be conveyed by multiple voices. They can also be more expensive and time-consuming. A narrator can be used to link interviews together.
- What is the audience interested in? What do they already know? What specifics do they need to witness to understand your message? The video script should supplement and not duplicate what people see on the screen. It is not helpful to have a play-by-play narration of what people can easily see. Instead, identify what will help support and explain the visual imagery.

- Will you use graphics or music? Both can enhance the message, make the video more interesting and enjoyable to watch, and signal topical changes in the program.

As you develop your ideas for the video, consider who will produce it. If you have access to cameras and editing equipment, learn to use them by attending a class or workshop, reading the manual or tutorial, or practicing. Amateur productions with new digital technology can be just as good as professional productions, and much less expensive. If you have an in-house production unit, get on their schedule. Conversely, if you plan to contract with a professional firm, look at their recent work, and shop around for someone you can afford and enjoy working with.

Once you have a script, ask others to read it to check for appropriate language, unnecessary jargon, simple explanations, and vivid descriptions. Make sure it will sound like a conversation rather than a lecture. It should sound like you are speaking to one person, even though the video may be broadcast to thousands. This is the easiest stage for making changes in the script—you should not expect to reshoot the video to make changes later.

When your script is approved, and before you begin production, contact agencies and organizations that might have the footage you need for scenes, explanations, and transitions. Gathering pre-shot material can significantly reduce the cost of your program. Adding the organizations that contribute footage to your list of partners and collaborators may make it easier for them to donate material at no charge.

11.1.2 Implementation

During production, you or the hired firm should consider equipment (camera, microphone, batteries, tripod, and the tape), site requirements, and lighting. No matter how steady your hands, tripods are essential in video work. Arrive early enough to test the equipment, set the microphones and lights, and organize the props. If recording outdoors, try to eliminate excess noise from cars or machinery, such as air conditioners. Stop recording each time a plane flies overhead. Natural light on an overcast day is often ideal for taping; strong sunlight may require supplemental lights to remove shadows. Check the results of your set-up frequently to make sure the light and sound are appropriate.

Talking heads refers to a basic, newsy style where the interviewee talks to the camera. If relevant, it may be more interesting to ask the person to do something that relates to the message—walk across a field, test water quality, or take a songbird out of a mist net. Ask the people on camera to wear comfortable clothes with solid, light colors, and no flashy accessories or jewelry. Research shows that most of the audience's perception of the people in the video comes from non-verbal body language. Actions such as reclining in a chair may signal laziness to viewers; therefore ask the interviewees to lean forward, keep their eyes focused on the interviewer, and act interested (Gleason and Holian 1996).

During the taping, be sure to shoot a variety of additional footage that could be used to bridge between edited portions of the video. You may wish to convert a 15-min interview to 5-min by linking several discontinuous pieces together. If the interviewee's head is in a different position, the edited version will be jumpy. A "cutaway" to something in the scenery will enable you to continue the interview as a voice over and then return to the interviewee in a new position. Likewise, you may wish to shoot one scene with several possible transitions—zooming out, fading out, and cutting away to something else—so you can effectively transition to a new scene during editing. Your final product should look clean and simple.

Ask questions of your interviewee that facilitate a complete answer or stand-alone statement. Unless you intend for the interviewer to have a role in the final product, the questions are not usually included.

Some mistakes commonly made by those with less experience in video production include: poor visuals (bad focus, poor lighting, poor color balance, jerky camera movement, too many zooms, repeating shots); not enough cutaways or too few shots of people in transition so that people mysteriously and suddenly appear in a new position; too many graphics; too much or poorly selected music; shots that are too short or too long; people talking too soon after the tape begins; two people speaking at the same time; people who appear nervous; and people who sound or look like they are reading. The most frequent problem is trying to include too much information.

The editors should begin the editing process by reviewing the goals for the video program. With a log of the taped footage, editors and producers can begin to select scenes to use. Begin by identifying the major scenes and then work on the transitions between edited pieces. It is typical to make changes to the script when editors see effective shots and comments. Graphics and sound can be added, as well as names and titles of those interviewed. The opening title and closing credits are among the last details added.

When the video is completed, implement a distribution plan. This may include posting a digital version on your Web site (Box 11.3), advertising an order form on your Web site, mailing copies to members or educators, or distributing copies with a training workshop or program.

11.1.3 Evaluation

An important criterion for evaluating a video is technical accuracy. Long before any cameras roll, you should ask experts to review your script or storyboard. They should review the information carefully for what is said, what is implied, and what is missing. It is possible for a video to convey a message that is not carried by words. Therefore, how images are linked together, the type of background music, and the visual scenes should be considered.

Whether the video "works" is dependent on its original purpose. A video intended to raise public awareness and concern should be evaluated differently than one designed to train employees on specific procedures. A site-orientation video to a state park in Texas was evaluated in 2003 with a short pre- and post-questionnaire. Responses after viewing the video were significantly higher than the initial responses for all categories of questions: general information about

Box 11.3 Streaming video enhances outreach

Video files available on the Web are notoriously large and may require a long time to download, even with a high-speed connection. Streaming Video is a technology that allows small packets of the video file to be downloaded and played in a continuous stream, so that the user views the video as it is being sent. Educators at the University of Washington in Seattle, Washington, have successfully used streaming video as a part of their outreach program to deliver conference presentations and other symposia to a distant audience. The Web site for the Rural Technology Innovation program (www.ruraltech.org) displays both the video of the presenter and the presenter's PowerPoint presentation. Microsoft Producer is a free software add-in to PowerPoint, and allows PowerPoint to record transition times as the presenter goes through the presentation. This makes it a relatively simple process to edit and synchronize the video with the presentation slides. In addition to a computer with a video card and the software, a digital camera and a media server are also needed to produce and broadcast a streaming video program. Students at the College of Forest Resources recorded, edited, and uploaded their Graduate Student Symposium within days of the research presentations, despite the fact that none of them had produced a video previously! A tutorial is available at www.ruraltech.org/video/howto/.

conservation and recreation, history of the site and local culture, and regulations governing human impact on the heritage sites and vandalism (Akers *et al.* 2005), indicating that the audience obtained the intended information from the video.

The following elements can help evaluate instructional videos (Beaudin and Quick 1996):

1. Content
   - Is the video accurate and up-to-date?
   - Is the content generally useful?
   - Does it stimulate, motivate, and inform the learner?
   - Is the video free of bias, such as racial or social class stereotyping?

2. Instructional Plan
   - Did the video begin with a motivating introduction?
   - Were the objectives clear in the introduction?
   - Is the content appropriate for promoting information; does it simplify tasks without providing too much detail?
   - Does the video suggest ways for the learner to practice and apply the new knowledge?
   - Does the video allow for learner reflection?
   - Does the video meet the learning objectives and needs of the learner?
   - Are key learning elements repeated in the summary or conclusion?

- Does the video promote active learning?
- Can the video provide a valuable addition to an existing educational program?

3. Technical Production

- Was the video well planned, organized, and structured?
- Did the camera go where the learner cannot without giving a false idea of reality?
- Is the camera looking at the scene from the learner's perspective?
- Do special effects enhance learning by drawing attention to specific attributes?
- Are varying types of camera shots used?
- Is the vocabulary, speed of narration, music, and sound effects appropriate for the visual tract and the audience?
- Are the audio and visual elements combined well, and do they vary enough to attract and hold attention?
- Are there supplemental instructional materials and do they complement the video's purpose and objectives?

11.2 The World Wide Web

Internet users today are familiar with the vast possibilities of information exchange and retrieval while online: corresponding with friends on e-mail, purchasing airplane tickets, checking for directions and visitation hours, previewing movies, finding reference articles, and accessing weather reports through computers connected by phone, cable, or direct lines. Though it is less popular in some countries, use of the Internet is increasing, especially among students and managers. This network of connected computers, voluntarily agreeing to use a common transmission language (TCP/IP), is called the Internet. The most common system that accesses the Internet is the World Wide Web. It uses a graphical presentation and hypertext (linked text that point to additional information); its common language is Hyper-Text Transfer Protocol, or HTTP (DeVries *et al.* 1996) (Box 11.4).

To access the Web, users need a computer, a modem (which may be installed in the computer), a phone line or direct or wireless connection, a browser (such as Netscape, Mozilla, or Internet Explorer), and an Internet Service Provider. Many local providers offer this service to regional customers, and larger providers are also available (such as Earthlink and Yahoo). Countries with national service may provide Internet access with phone service.

A recent survey of environmental educators in the United States indicates that the majority are Web users, and most of their online activity is work-related information retrieval and e-mail (Heimlich 2003). Additionally, most of their workplaces had Web sites as well. The respondents report that the barriers and difficulties to Web use include information overload, trustworthiness of information,

Box 11.4 Web technology terminology

Blog: A frequent, chronological publication of personal thoughts on the Web, also called Web log. Individuals create their own blogs; each entry is followed by a comment button that allows others to write a reaction. A wiki is similar to a blog but allows for anyone to edit the content posted on the Web.

Bulletin Board: Asynchronous communication that allows users to post messages to a common public forum. Users must know the address or phone number to access the bulletin board for viewing or posting.

E-mail: The most common Internet communication is electronic mail. Messages are sent from the account of the sender to the account of the receiver, which can be accessed through any computer that has Internet access, at any time.

Hypertext: Web-based text that allows links between pages. When users click on a colored phrase or button, a new one loads to their screen.

ISP: Internet Service Providers enable users to access the Internet.

Listserv: A communication sent to groups of subscribers by e-mail. Messages are posted to the entire group and members can reply to the entire list. Organizers at WWF-UK credit this technology to the speedy and efficient pressure they were able to put on parliamentarians to support legislation to restrict wildlife trade (Box 9.3).

MUD and MOO: Interactive virtual worlds that are constructed by users. Many are games, but educators are creating them to engage learners in collaboration and problem solving. MUD is a text-based MultiUser Domain. MOO is a graphical, object-oriented MUD.

URL: Uniform Resource Locator refers to the address of a Web page. It begins with http and then identifies the server on which the page is located by domain name and path.

---

time, and search difficulty. This suggests that conservation educators might benefit from training in using the Web more effectively, and that the sites we create should more effectively direct users to information.

Web sites, of course, may contain a number of different types of information. They can passively present information, like an exhibit or display. They can post digital videos, photographs, presentations, maps, charts, and text. Pages can be connected together with hypertext links, enabling users to click on a phrase and jump directly to a related location. Virtual field trips, a popular Web phenomenon, are just as varied. Some field trips are actually tours of Web sites, linked together in a logical and cohesive manner. Others have interactive maps, photographs, and text to explain beaches in Hawaii or castles in England. Virtual field trips at museum or aquarium sites may include a real time Webcam, enabling virtual visitors to see what real visitors see, like otters at the Monterey Bay Aquarium. A virtual field trip also can include a satellite broadcast of a real field trip, merging technologies of distance learning and Web sites (Box 11.5).

Flash technology enables Web sites to present complex animated information within a reasonable bandwidth. It requires that the user have a Flash plug-in player and, depending on design, could take a long time to download over phone lines. Templates are available that allow users to add their own text and photographs. This technology enables Web sites to have sizzle without size. The Southern Center for Wildland Urban Interface Research and Information contracted with a consultant to develop a Flash presentation to describe a recently renovated firewise home (SCWUIRI 2005). The staff and consultant met together to develop the look, layout, and discuss the content. Staff wrote text, assembled photographs, and developed captions while the consultant created the final product. Annie Hermansen-Báez, Technology Transfer Coordinator for the Center recalls that the process required several months of interaction. The completed program is an important feature of the Center's Web site. Annie says, "We've received a lot of

---

Box 11.5 Web site and satellite help students study shorebirds

---

One of the world's greatest migratory spectacles occurs in relative obscurity along Alaska's Copper River Delta. Millions of shorebirds rest and refuel in these intertidal wetlands on their way to northern breeding grounds. Several units within the US Forest Service, which manages the wetland, worked with the US Fish and Wildlife Service, Ducks Unlimited, Western Hemisphere Shorebird Reserve Network, and Prince William Network under the leadership of the Copper River International Migratory Bird Initiative to create a virtual field trip that would allow students and their teachers to learn about shorebird migration.

"Winging Northward—A Shorebird's Journey" evolved to be a comprehensive educational program with a live, satellite-broadcast field trip from the Delta on May 8, 2002. The bilingual Web site includes a 10-chapter story of migration, told from the perspective of a shorebird, classroom activities, and information about the broadcast. A Webcast was also used to reach students who could not access with satellite coordinates and is still available on demand. An evaluation indicated that students' knowledge about shorebirds increased by 20% (Frost 2003).

The program was popular. Over 850 sites in the United States, Canada, Mexico, and Puerto Rico registered to receive the downlink, with approximately 140,000 students. Over 1250 e-mail messages were sent during the broadcast. Program organizer, Sandy Frost, believes the broad partnership, rigorous attention to educational objectives, and an opportunity to celebrate a truly special part of the world made this a successful program. "We brought the experience of hundreds of thousands of migratory birds teeming on mudflats and swirling in the air to children who may never visit an Alaskan wetland," she explains. "They didn't get muddy but they learned everyone plays a role in conservation."

positive comments on the program. It is an effective strategy for attracting attention and showing before and after changes to the home."

As educators know, learning is more likely to occur when learners manipulate the information. Some educators build Internet resources into tried-and-true activities, such as complementing a study of wetlands with data on wetland loss from Northern Prairie Wetlands Research Center (Czerniak 2004). Newer Web sites for distance learning allow for interaction (Ambach *et al.* 1995). Some sites provide data entry sites for citizen scientists (Chapter 7), enable users to retrieve specified data, chat with other users, and upload course assignments. Interactive computer programs can be posted on the Web or made available off-line on CD-ROM's. As with videos, the technology for producing and disseminating information varies, but at the root is an educational medium with specific opportunities. This section will focus on designing Web sites, while the next two sections will focus on distance education and interactive computer programs.

11.2.1 Planning

Conservation educators who are designing their own Web site should think about the learner and what makes it easy for them to navigate a Web site. Basic rules of thumb include:

- Viewers should be able to get the important information in one computer screen without scrolling. If scrolling is necessary, they should only have to scroll vertically not also horizontally.
- Three clicks, or connections, should be the maximum needed to dive into a site to find information, and pages should download quickly.
- By increasing vividness (audio and animation) Web sites can create more positive and enduring attitudes (Coyle and Thorson 2001).
- Involve the Webmaster or staff person who will be maintaining the site during the design process. If he or she has little time to update the site, it should be designed for low-maintenance.

The Web, by design, is a mechanism for anyone to post anything. As a result, it can be difficult to sort fact from fiction and ascertain value. The Stanford Persuasive Technology Lab has conducted a number of studies over the years and assembled a list of criteria that will help improve credibility and trustworthiness of Web sites. Some, like eliminating errors and making it easy to contact the source, are true for any communication tool. Other tips are more essential for Web sites (Fogg 2002):

- Make it easy to verify the accuracy of the information. Use citations, references, and third-party sources to show that you have confidence in your material.
- Show there is a real organization with real people sponsoring your site. Legitimate organizations have legitimate Web sites. Provide a physical address, a photograph of your office, and links to other organizations or directories that

include your listing. Include photographs of staff and post their employment history and achievements.
- Highlight your expertise. Provide the credentials of your experts and the affiliations with respected organizations, if you have them.
- Include a number of ways for people to contact you—phone number, physical address, and e-mail to each relevant staff person.
- Design the site so it is attractive, professional, appropriate, and consistent. Web sites are judged quickly by the design, so make yours good, easy-to-use, and useful. Do not emphasize the dazzle at the expense of the content.
- Include the date when the site was last updated, and review and update it frequently. Old dates are less credible than current ones.

Like other learning tools and techniques, the prior knowledge and understanding that a learner brings (including familiarity and the density of their cognitive map, Chapter 2), greatly affects how people explore and learn from Web sites. A comparison of novice and experts found that greater topic understanding led to more goals and more complex searches for information (Corredor 2004). Given the same amount of time, novices set fewer goals and thus found less information than the experts. Since many Web sites are designed for the general public, it is critical to think about the users as novices. Sites that provide navigational information, describe interesting features in common language, and make attempts to provide details relevant to users could help novices conduct a more rewarding search.

Like other visual communication techniques, good Web sites are attractive and well designed. Color schemes should be pleasing, information should be easy to locate, and the overall site should be clean and simple to use. Unique to the Web, consider the speed at which your page loads on a home machine and the size of the screen. A page should not take longer than 30 s to load. Photographs may be stored as small thumbnail photos, additional graphics can be put on a separate page, and low-resolution images (72 ppi) can be compressed. Each page should be completely visible on any computer screen. Frames are commonly used on Web sites to provide stationary information like a table of contents or organizational banner. Frames, however, limit the size of the screen that is available for subsequent pages and could make it difficult for search engines to locate your page. They may not be used as often in the future.

The differences between designing for a print medium and the Web, suggest the following principles (Williams and Tollett 1998):

- The minimal cost of posting on the Web enables designers to use a larger number of pages than one would use in print. Step-by-step photos of a process and ample white space should be used. The minimal cost, of course, means everyone can publish anything.
- Web sites can be upgraded and revised frequently. They are an appropriate medium for fast-changing news and current information. The ease of changing a site implies there is no excuse for mistakes and errors.

- Users can respond to Web sites through e-mail forms that are submitted directly to the Webmaster. You can collect suggestions and add to your mailing list.

By linking to other relevant sites you can help create a network of similar organizations and save yourself the work of re-creating their information. People with specific questions often access more than one Web site as they compare information (Eysenbach and Kohler 2002); you can make it easy for them to find what they might need.

11.2.2 Implementation

Once a Web site is designed and developed, it can be loaded on a computer that will act as a Web server (a computer that stores all of your Web site files and makes them available to anyone on the Internet). To keep your site useful, it must be reviewed and updated frequently. Consider rotating feature stories that promote new information, exhibits, or timely happenings at your site. If you want users to contact you, make sure your contact information is easy to find and easy to use. If one address receives all emails from the site, make sure that address is checked frequently and responses are sent quickly. If a response requires significant work, let the requester know that you are working on it. And to make sure users find you, your Webmaster should consider how search engines find Web pages and design your site accordingly. The search engine Google, for example, uses a variety of variables to identify sites, including the number of links to that page, the structure of the site, the use of key words, and the quality of the linking pages. It is possible to register your site with a number of search engines, but marketing your site to your target audience is probably the most effective strategy for increasing traffic to your site.

Having access to information on the Web does not mean people learn more. A recent study in California indicates that having access to the Internet does not impact achievement test scores (Trotter 2002). Like other learning technologies, educators must use this tool to motivate and engage learners in constructing their own knowledge. One such strategy is a WebQuest—a popular activity that uses information from resources on the Internet and typically assigns roles to students to work in cooperative groups to complete a task (Dodge 2005). By incorporating relevant and meaningful questions and tasks, a WebQuest is highly motivating. By assigning roles in cooperative groups, students learn to work together and take advantages of their skills and interests. By providing links and a concrete task, students are able to use the Internet for information and transform that information to meet their needs. Many high quality examples of WebQuests are available . . . on the Web of course! They include links to Internet resources and rubrics for evaluating products on topics such as endangered species, water quality in Chesapeake Bay, acid rain, and climate change.

Little research has explored WebQuest-based learning, and what has been done indicates this is a difficult environment in which to guide learners. Even well-designed WebQuests result in confused students who require much more assistance than the teachers expected (MacGregor and Lou 2004–2005). In addition to learning information, a WebQuest requires that students find needed resources, evaluate them for appropriateness, and select the needed information. We know this is easier to do when students have content familiarity. In addition, these skills are more akin to the higher-order thinking and critical thinking skills associated with cognitive maturity (Chapter 2). In a study of fourth-grade learners, the problem-solving approach to WebQuest learning was compared to a modified WebQuest that included a didactic instructional style that explicitly states what and how to learn (MacGregor and Lou 2004–2005). Both groups received the same WebQuest task, but half of the students also received instructions for where the information could be found and how it should be organized in their brochure. Students with more explicit instructions gained more knowledge about the topic and produced higher quality products than the control. They still had difficulty in using Web resources, however. A follow-up study indicated that providing students with a concept map and instructions for how to organize and synthesize information enabled students to complete the task with greater satisfaction and knowledge (MacGregor and Lou 2004–2005).

11.2.3 Evaluation

The medical community has been active in assessing how people use the Web and whether users are able to accurately assess the credibility of information posted on Web sites. They have generated a number of rubrics and assessments that could be useful to conservation educators (Jones 1999; Price and Hersh 1999). Valuable guidelines, however, can also be obtained from users. A group of fifth-graders compiled the following list when asked to describe the characteristics of good Web sites (MacGregor and Lou 2004–2005):

- Text in concise sentences, simple vocabulary, and headings.
- Information is relevant and broad in scope.
- Additional pages are one click away.
- Design includes colored text boxes, white space, and borders.
- Audio and video elements.
- Photographs that provide clarifying information.

The Web offers a number of sites that provide criteria for users and designers of Web sites. One list (Richmond 2003) includes considerations such as comparable print datasets, connections that multiple users may use to assess the site, visible copyright information, clear authorship and date, and long-term maintenance.

Awards are given to good Web sites. You might consider using the award criteria to help develop or redesign your site, or simply use award-winning sites as a model for developing your site. ARKive is a Web site that has won a number of prestigious

Fig. 11.2 An award-winning Web site is Ark on the Web: Images of life on Earth (ARKive home page: http:// arkive.org).

awards such as Best Environmental Web site Award from the British Environment and Media Awards run by WWF. It archives photographs and information about endangered and extinct animals to raise public awareness for conservation. Naturalists and educators can find a treasure chest of information on animals, such as keas and quaggas. Parallel sites provide information for children, teachers, and parents (Figure 11.2).

Digital photographs and film footage of endangered species are being collected, archived, and made available to scientists, educators, and the pubic through the Internet by the non-profit British organization Wildscreen. Low-resolution images and information are freely available. To obtain high-resolution images, users are directed to copyright holders who may charge a fee.

The project has the cooperation and backing of the dominant conservation organizations like IUCN and film makers, such as BBC and National Geographic, who see this as a public service and effective library for their material. As extinction rates increase, ARKive will become an important repository of valuable and irreplaceable imagery. Harriet Nimmo, the Chief Executive Officer of Wildscreen notes, "ARKive is consistently rated as an outstanding site because it caters to different audiences, is well designed, effective, and impactful. The images are extremely powerful, and making them more available to the world is a win–win–win for photographers, the public, and the wildlife."

Online surveys provide an avenue for extracting some information about users of Web sites. Such forms typically ask for demographic information, the purpose of accessing the site, and level of satisfaction. Because questions with pull-down buttons and predetermined checklists are easier to complete than open-ended questions, these surveys tend to provide only factual, simple data.

If Web site evaluators wish to discover what users think when they navigate a site, how sites compare, or what they cannot find on the site, other evaluation tools are needed. Observation of users searching for information can help identify what information is obtained by quickly scanning sites and how long it takes to locate the information (Eysenbach and Kohler 2002). Focus groups can provide important insights about navigation, content, and aesthetics. An evaluation of a Web site for The Center for Natural Resources at the University of Florida used a focus group format in a computer lab to answer the following questions (Bartels *et al.* 2003):

Navigation

- Is the site easy to navigate? Do you have difficulty finding information?
- Do off-site links that open new windows confuse you?

Content

- Should the scope of information provided be more specific?
- Is the information evenly distributed across various departments?
- Is the information accurate and useful?
- How might you use the information provided on the site?

Aesthetics

- Is the color combination appropriate? What does it remind you of?
- Would you want to change the colors, and if so, to what?
- Are there enough good quality pictures? If not, what types of pictures would you prefer?
- Are the graphics downloading quickly enough?
- Do the graphics complement the text?
- Is the text font difficult to read? Should the text be a different color, size, or font?

11.3 Distance education

In North Carolina, forest landowners take a seat in a community college classroom, look to a television screen, and listen to an instructor on the other side of the state tell them about forest management. During the discussion period, a moderator asks if there are any questions from each of the sites, allowing for personal interaction with the instructor. This video tele-conference is a form of distance education that allows groups to interact at the same time, but in different places (Box 11.6).

Box 11.6 Video tele-conference for natural resource professionals

The North Carolina State University Extension Forestry program developed the Forestry Issues Series to provide interactive continuing education programs to natural resource professionals. The series is delivered by multi-point video tele-conferencing that allows for face-to-face interaction and simultaneous video and audio communication between instructors and learners. Over a dozen sites are part of the state's electronic communication network and use a microwave standard system. Six to 10 programs are produced each year, at a substantial cost savings to participants in travel and lost work time. Participants register in advance, print materials are sent to each site for distribution, and instructors deliver their presentations from a campus conference room. A moderator asks each site for questions and comments, facilitating interaction without pandemonium. Annual evaluations indicate that in addition to time and money savings, participants have increased knowledge, are satisfied with the technology, and are able to apply the information in their work (Bardon and Moore 2005).

Thirty educators scattered across North America sign up to take a course on environmental education program evaluation. They each use a desktop computer to access the course Web site. They download materials and exercises, e-mail fellow students, and upload completed assignments. This is another type of distance education that allows for learners to work independently at different times and places.

Distance education provides educational programs to learners who are not physically in the same room as the instructor. Formerly known as correspondence education, distance education always strives to use new technology (incorporating radio and television, for example) to increase the number of people it reaches as well as improve the quality of the educational program it conveys. With the advent of computers, satellite broadcasts, and the Internet, distance education has incorporated these new dimensions. All definitions of distance education include the following elements: teacher, student, separation of teacher and student by time and/or geography, and technology-mediated communication between teacher and student (Telg 1999b).

There are a number of reasons why institutions promote distance education. In some rural locations, travel time and cost for instructors and learners are prohibitive, making any form of distance education preferable. Specialized instruction that is useful to a dispersed population is another good candidate for distance education, such as a program evaluation course for environmental educators. Finally, institutions striving to build larger programs are finding that the non-traditional student population—people who are raising families or building careers—could be enticed to enroll in degree programs if the courses are offered locally and more conveniently.

Of course there are several challenges to distance education. Almost everything we know about high quality and effective education assumes an interaction between the instructor and the learner. Facilitating experiences, guiding inquiry, developing cooperative groups, and meeting individual needs, for example, are easier when the instructor and learners work together (Chapter 2). Distance education, in comparison, runs the risk of being impersonal, stilted, and based more on the transfer of information than on learning skills or concepts. The challenge in distance education is to use technology to create high-quality, effective instruction that engages learners in experiences, inquiry, cooperative groups, or any other desirable learning theory. Newer technologies, such as those that facilitate synchronous communication, can do a reasonable job of mimicking non-distance education. The technology itself, however, presents another challenge to distance education. Both students and faculty require training to create and use the necessary technology effectively (Rhoades 2005). One study noted that some students' anxiety was reduced as they became more fluent in the technology, but for others, no amount of training increased their interaction with the course (Kelsey 2000, as cited in Rhoades 2005). "Scavenger hunts" can be used to introduce students to the various elements of a distance course, such as reading lists, relevant links, and the discussion board.

Research activities with distance learning opportunities suggest that properly designed Web-based instructional programs can increase critical thinking skills, perhaps because of the opportunity for thoughtful interaction and frequent feedback. The huge diversity in quality of instructional programs, however, makes broad generalizations about the technology difficult. In a comparison of Web-based audio and video lectures with and without organizers for the information (such as outlines and guidelines), undergraduate students preferred the organizers and the audio presentation, perhaps because the quality of Web-based video was poor. When comparing this preferred format against traditional text, information retention and critical thinking skills were equivalent. Thus, distance learning can be as good as traditional instruction (Astleitner 2002).

A number of technologies can be used to create a distance education program, and often several are used in the same program (Telg 1999b; Norton and Sprague 2001).

Correspondence study

This uses postal service or other delivery network to send instructional materials in print, on audiotapes, or on videotapes. It is relatively inexpensive, easily replicated for larger audiences, and very portable. Students can work at their own pace and on their own time. Neither students nor instructors are required to use new technology. Unfortunately the completion rate is low for this type of education, perhaps because the interaction is low and there is little opportunity to meet individual needs. Additionally, the reliance on the postal service may create challenges for timely communication between students and the teacher.

Web-based instruction

Using the Internet, students can work at their own pace on a personal computer. Often this type of distance education is used for asynchronous communication, where users can interact at different times, such as through e-mail messages. Instructional materials can be read on screen or printed on paper. The Web site can include streaming video, PowerPoint presentations, e-mail conversations among students and with the instructor, and if scheduled times are agreeable, live dialogue by e-mail or chat room. Several programs (e.g. Blackboard, WebCT, Vista, Desire 2 Learn) enable instructors to organize all of these functions on the same site. A study comparing two of these software packages suggests that although using this type of learning aid is a benefit, student navigation and uploading procedures are not equivalent; poorer software could require additional training, increase frustration, or compromise learning (Storey *et al*. 2002). Students must have access to a computer and an Internet Service Provider that is usually provided by the educational institution.

Satellite television

This one-way audio and video presentation can be supplemented with telephone communication between students and the teacher. High quality videos of field situations or panel discussions can be created and broadcast (Box 11.5). This technique requires specialized facilities to downlink the satellite broadcast and view it over televisions. It can be expensive to use the satellites, and this tool is not without technical and weather difficulties. Because the program is broadcast only at select times, students have an opportunity to meet and interact with each other at the downlink site, but cannot work at their own pace (i.e. synchronous communication).

Video tele-conference

This form of distance education technology offers synchronous two-way video and audio interaction through remote cameras and televisions or computers that allow all parties to see and hear each other (Figure 11.3, Box 11.6). Like satellite television, specialized equipment is needed at each facility and students must be in the same room at the same time. Instead of a satellite, this technology uses compressed video or a microwave system to convey information. The level of interactivity between students and the instructor and among students is greatly increased with live dialogue. The video, however, tends to be of poor quality, and the costs can be significant.

Video tele-conferencing can also occur on standard phone and Internet lines with a variety of programs and an inexpensive camera with microphone. Because users are talking instead of typing, the conversation can cover more territory in a limited period of time. In addition to audio and video capacity, many tele-conference systems include an electronic whiteboard that allows users to write or draw figures that can be viewed by everyone. Several types of software provide audio and whiteboard capabilities.

11.3.1 Planning

Like every other educational technique, planning for distance education begins with understanding the audience, including their needs, initial knowledge, constraints,

Fig. 11.3 An image of a box turtle is sent by video tele-conferencing software across the Internet where a kindergarten class in Virginia watches the television screen (photo by M. Haddon).

and interests. Unique to distance education, however, is a simultaneous requisite knowledge of the available technology, budget, and access for your audience. Advance planning is needed to match the best technologies to the audience needs. If the downlink sites are not open in the evening, for example, working students may be unable to attend. If students are more likely to participate in online discussions after having a chance to reflect on the material and compose a response, an asynchronous system may be most appropriate.

Planning for distance education should pay special attention to how the instructional program can overcome the challenges of distance. Three key considerations are (1) generating an atmosphere that is most conducive to learning, (2) engaging learners in the learning process, and (3) allowing learners to respond about either the content or the technology (Telg 1999a).

A few characteristics typify most distance learners. They are often older than on-campus students, with jobs and families that keep them from traveling. They may be isolated at a remote site or working away from large numbers of colleagues. They are self-motivated and self-directed. They do not expect a great deal of interaction with the instructor, but are concerned about how they will perform and how they should fix problems with the technology (Telg 1999b). Instructors can use this basic information in the design of their program to make sure they learn about the students' experiences and concerns, offer choices in readings or assignments, provide opportunities for learners to work independently, explicitly teach learners how to use the technology and arrange for local assistance if there is difficulty, and engage learners in opportunities to provide feedback on a regular basis.

Planning for interaction is one of the most challenging aspects of distance education. Choose a learning theory that best fits your teaching style, the learners, and the content, and work to incorporate good teaching into distance technology (Chapter 2). One strategy is forming cooperative groups at each site or among distant learners. Another is asking questions of learners, calling on individuals, and waiting for responses. It usually takes more time to interact with distant learners than in a classroom, because of the lack of nonverbal cues and the technology. Use as many techniques as are feasible to encourage interactivity—fax, e-mail, chat room, surface mail, individual phone call, or conference call.

As you plan your program, consider how to use the technology most effectively and efficiently. If you ask learners to work in small groups at each site on a problem to report to others, for example, you will not want to take satellite time while the groups are working. The groups can finish this assignment before they connect to the program. Encourage participants to state their name and work site when they communicate so everyone can get to know each other. Send printed material to the learners well ahead of the course start date so they have a chance to receive and review it.

Chuck Lennox, a Certified Interpretive Trainer and college instructor offers these tips for developing a distance-learning program (Lennox 2004):

- Design your class or course for online learning; using the technology for what it does best, not what you do best.
- Support your learners with technical assistance and let them know what to expect from the experience.
- Be very specific with your directions for assignments, including exactly where to post them and by when, in whose time zone.
- Use the technology well, such as up-to-date software, high-speed connections, and frequent back-ups of material.
- Check the entire site before going live to reduce confusion and frustration. Make sure the links work and the photographs are not too big.
- Online communication must be handled with care because it does not have the pause, the inflection, the gestures, and the facial expressions that make in-person communication so effective. Make sure students use appropriate netiquette, such as not forwarding junk mail, not sending inflammatory remarks, and ignoring critical comments that are received.

Environmental education evaluation course at a distance

The US Fish and Wildlife Service (FWS) and the Environmental Education Training Partnership (EETAP), a program at the University of Wisconsin-Stevens Point (UWSP), worked together to adapt a 4-day in-person course on program evaluation to distance education. The course was developed and improved over a 10-year period by a team of instructors with the FWS. Originally, two needs assessments were conducted of potential participants (environmental educators working for FWS) and experts in environmental education evaluation to develop the course. In addition to teaching about program evaluation, the course was designed to develop skills in the construction and implementation of evaluation tools through group exercises. An important aspect of the course was working with colleagues to conduct an evaluation of a local environmental program or product with interviews, surveys, and observations. Participants consistently rated the course as highly useful, informative, and confidence-building.

The UWSP and EETAP offered a distance education course on the fundamentals of environmental education, and knew from participant queries that an additional course on program evaluation would be well-received. Rather than develop their own course, they asked FWS to become a partner and provide the framework of an already successful course. Georgia Jeppesen, the FWS course leader was initially concerned that providing their course in a distance format might reduce the number of course participants who came to the face-to-face 4-day version, but now believes having multiple offerings helps to market all the evaluation course options. "The more opportunities we provide for people to learn about program evaluation, the more we are enhancing the field of environmental education. Given the variation in learning styles and the differences in budgets and travel opportunities, we need to provide the similar evaluation courses in multiple formats," she explained.

Instructors of the FWS face-to-face course and EETAP Director and staff converted the course materials and group exercises to online information and individual assignments and reviewed the final product. The online course was designed to be asynchronous: users access the course through the UWSP Web site, read information, respond to questions, download articles, upload assignments, and contribute to discussions on their own time. Since most of the learners are employed full-time and span four time zones, the asynchronous component was essential. The 10-week course begins and ends each semester. Course enrollment is limited to 27 participants per section; a waiting list is maintained for the next course offering. Unlike the 4-day course, the distance course is offered for graduate credit at UWSP and also for no credit. Each semester about four people per section register for graduate credit.

Most of the content in the 4-day course is delivered through instructor presentations with PowerPoint slides. The concepts are reinforced and stressed through group exercises. The content of the distance course is presented through Web pages and readings. Most of the sessions converted to 4–8 screens-worth of text to read and one to three short assignments. About half of the group exercises from the 4-day course were incorporated into the distance format. New exercises were created for the distance course to emphasize topics that were covered in lectures or group discussions in the 4-day version.

Janice Easton, an instructor who helped create the distance version, explains that it was difficult to figure out how much information was enough, what should be included, expanded, or left out. "Entire courses can be taught on many of these subjects, such as surveys and alternative assessment, but the on-line format makes everything a little different," she said. "Finding the balance that would give participants enough information through readings, examples, and assignments so they can apply the information to their setting was the biggest challenge."

11.3.2 Implementation

The UWSP distance course was announced to the environmental education community through a number of different e-mail lists, targeting nonformal educators associated with organizations and agencies.

A course administrator at UWSP handled registration, evaluations, and finances; uploaded the course materials to the Desire To Learn (D2L) platform; and responded to student technical difficulties with Internet access. Participants who use dial-up connections or older software find the materials download very slowly. Instructors were hired to promote interaction among course participants, respond to questions, keep participants on task, and grade assignments.

The first week was designed to orient participants to the technology and each other. The discussion board quickly filled with conversation about similar hobbies and common work experiences. This all-important introductory session helped set a tone that contributed to learning, sharing, and exploring during the course. This perception of the presence of the instructor and other students was found to correlate positively with cognitive and affective outcomes (especially student satisfaction with their own learning) in a study of graduate student attitudes about a synchronous distance education course in genetics (Russo and Benson 2005).

Student–student and student–instructor interaction is as important online as face-to-face. One EE evaluation student commented online, "It *is* great having so many people here to help you out. Still, there are times here where I'd much rather be in a classroom so I could stop someone mid-stream and ask for clarification! Of course, the commute from NJ to WI would kill me . . . so I'll take the online format!"

Although some of these course participants yearn for immediate feedback and group discussions, the majority are not in favor of a synchronous element that would require them to participate at the same time, probably because of the difficulty coordinating activity across several time zones. Neither are they interested in streaming video or slide presentations if it would significantly increase the download time (AEEPE 2005).

As the course proceeded, questions and responses were posted to the discussion board. Although interaction was encouraged by the instructor and the distribution of points, some learners were able to use this feature more than others. Easton explains, "our course evaluation indicates that a number of people had technical trouble using the discussion board. It is frustrating to use a dial-up connection with these large files. As with face-to-face courses, groups develop personalities; some communicate more than others. To encourage communication, we split the sections into smaller groups so learners don't have to read everyone's posting."

Instructional materials for distance education suggest content should be presented in chunks—a 15-min lecture, or three Web pages. After one chunk of content, the teaching style should change to an assignment, conversation, video, or panel presentation (Telg 1999a). The EE evaluation distance course achieved that balance by offering readings, thoughtful questions, examples on the Web, examples from past course participants, and exercises with each unit.

Distance courses that use satellite or video-conferencing rely on technicians and facilitators at each site to organize classes, distribute material, enable the technology to function properly, and help facilitate student interaction. Some distance learning evaluations report that the greatest difficulty that course participants experience is with technology; having experienced personnel at each remote site to reduce technical difficulties is ideal (Londo and Gaddis 2003).

11.3.3 Evaluation

The distance course on program evaluation at UWSP uses a number of tools to determine how it can be improved and how well it works. Any participant who does not complete the course is sent a form to ascertain why. Of particular concern is the amount of time participants perceive the course requires.

Participants who complete the course are asked to submit a final evaluation covering course content, technology, instructors, needed changes, and additional components. One consistent comment is that the course took more time than expected. Busy professionals may find it difficult to set aside 4–8 h a week to devote to the course. If they have not taken a graduate course recently they may have forgotten what the work expectations can be. Those not taking the course for credit were less likely to participate in the discussions or to complete all the assignments. In a typical course section, course participants posted between 15 and 50 messages and read from 100 to 850, though most posted between 30 to 40 messages and read 250–500.

The students are evaluated through their assignments; there is no exam in Applied EE Program Evaluation. Some are scored with a point scale while others use a checklist or scoring rubric. For example, the observation, questionnaire, and interview/focus group units include a checklist to aid participants while developing these tools. The instructors use that same checklist to assign grades. After they have submitted their tools to the drop box, the instructor provides comments on their assignments and the participants can improve their grade by making revisions and resubmitting the assignment.

Participants are also expected to complete a culminating assignment that incorporates what they have learned throughout the course. For this assignment they have three options: (1) create a pre-evaluation report describing their program and their evaluation plan, (2) put together a slide presentation for a conference, or (3) pilot test one of the tools they developed. Very few participants choose to pilot test their tools, unlike the 4-day course where everyone pilots all three of their tools. There may be an important difference between the two formats, with the face-to-face version enabling groups to work together and achieve increased confidence in their evaluation skills.

Fig. 11.4 Participants immerse themselves in learning about program evaluation in the 4-day course at the National Conservation Training Center, WV (photo by T. Harless).

Even though the distance course was developed from an existing 4-day course, instructors believe the courses have evolved different strengths. The distance participants are able to use their own site and program for the assignments, giving them an opportunity to apply the course information immediately to what they are most interested. This is an advantage over the 4-day version, where participants work in groups to evaluate a local program they know very little about, such as a local exhibit or school program.

On the other hand, the 4-day course occurs in a residential training center, giving participants a chance to be totally immersed in evaluation (Figure 11.4). In addition to the 30 h of course contact, interaction between learners and instructors occurs over coffee, during meals, and at consultant times, enabling people to learn a little, think a little, learn a little more, and continue to process new information. The distance participants must add coursework to their workload and home obligations, trying to schedule 40–80 h for their professional development over a 10-week period.

While the distance participants are able to spend more time reading supplemental information and concentrating on program tools, the 4-day participants are able to "think out loud" more effectively. Learning about evaluation and determining what evaluation questions are most effective can be challenging. Some participants ask a question, only to discover that was not really the question they needed answered. Conversations with colleagues and instructors are usually needed to clarify misconceptions and establish more helpful evaluation objectives. Although these conversations could happen in the distance course, for the most part they do not. Jeppesen says, "The discussion board is not used very effectively by some participants. They all are not taking the time to interact with each other, and they do not always respond to the instructor who asks for clarification of their question. The asynchronous nature of the technology might be affecting what people are able to learn, but for others who are not verbally oriented, it might help them to take the time to communicate with other students in writing."

After two semesters of the distance course, instructors meet via a Web conferencing platform that includes an Internet chat room and desktop application sharing to discuss ways to improve the 4-day course, given the experiences and new ideas generated from the distance version. The distance course is improved after each offering, with the majority of improvements aimed at making the assignment directions more explicit. Jenn Dillard, the course administrator, says, "Directions are very key. All of the course interaction is based on e-mail and everything needs to be very explicit so everyone understands what is expected."

11.4 Computer simulation and modeling

Simulations, like role-plays and case studies, can be powerful educational techniques to enable learners to explore new situations without actually being there (Chapter 5). Computer models, a type of simulation, can enable scientists to test assumptions and predict how plants might grow in an environment rich in carbon dioxide or how forest biodiversity responds to fragmentation. Decision support systems, another type of simulation, enable landowners to determine the best species of trees and shrubs for an agroforestry project (SEADSS 2005) or how to control weeds, pests, and diseases in Denmark (Murali *et al.* 1999). With the help of computer simulations, young learners can light their own prescribed fires after setting key parameters, and see if their burn accomplished their objectives. Similarly, adults can see the impact their food shopping might have on economic and environmental variables in Switzerland. Computer simulations are an intriguing educational technique for agencies and organizations to use in exhibits, to post on their Web site, or distribute on a CD to educators or adult learners.

Simulations, sometimes called "toy universes," extract important bits of the real universe, enable learners to control variables, and test learners' responses and skills. Frequently employed in job training programs, these simulations use the power of computers to enable anyone from airplane pilots, doctors, and hazardous waste workers to practice their trade.

For educators, a computer simulation is an educational technology that is repeatable, consistent, and as available as a computer station. Good simulations, like other educational techniques, are engaging enough to hold the learners' attention; enable learners to stop and start from any point; are authentic, realistic, and important, though somewhat random and unpredictable; and allow learners to gain meaningful goals in a reasonable period of time (Lockard *et al.* 1994, as cited in Norton and Sprague 2001).

One recommended product is the Decisions, Decisions series from Tom Snyder Publications. It combines role-playing with computer simulations, allowing students to become characters who must work together to solve a problem. Working in small groups to develop a plan, they enter their decision into the computer and note the consequences. At the end of the program, the simulation offers feedback based on the initial objectives the group created (Norton and Sprague 2001).

11.4.1 Planning

Before creating a computer simulation, review the types of simulations to gain a sense of what the technology can offer, as well as learn about your audiences' level of computer sophistication. If the simulation plays continuously at an exhibit, how long are people likely to spend before they move on? If it is distributed to teachers, what type of computers will it play on and how many are available per classroom?

If you are hiring an outside organization with computer design skills, or using your own staff, you might begin with a planning session that helps the design team focus on your agency or organizational goals, the learning objectives you have for the simulation, and your understanding of audience knowledge, needs, and technological constraints. You could also identify what you will define as success for this project. On what criteria will you evaluate the program—learner engagement, number sold, user knowledge or skill gain, funds expended? Plan to test the product at every possible stage to make sure you are investing wisely.

The development of an outline for the simulation helps the design team talk about what information should be provided, what questions and challenges the user should answer, and what interesting twists and turns the simulation might take. Workshops or focus groups with potential users may be helpful to collect new ideas and meaningful scenarios.

The development of a computer simulation in Brazil, entitled *Carbopolis*, used the thinking of Piaget and Freire to identify their objectives (Eichler *et al.* 2005):

- The main elements of effective learning experiences are activity and discovery.
- Activities are meaningful if linked to students' interests and needs.
- People want to be able to use what they know to evaluate new knowledge and skills.
- A generator theme helps link teaching and learning by providing a topic on which exploration, reflection, and discussions occur.

While these statements are typically reflected in a number of environmental education programs and educational philosophies (Chapters 2 and 3), they are an atypical starting point for computer simulations (Eichler *et al.* 2005). The team combined their pedagogical framework, real experiences, and problem-solving skills to establish their objective: a scenario that enables role-playing and problem-solving activities and provides content-specific information as well as opportunities to research, collect data, hypothesize, and test ideas. Existing computer simulations already incorporate these objectives, but usually are priced too high for educators in developing countries. The Brazilian team added to their goals that their product would be free, in accordance with federal policy on educational software. To accomplish this, they used Java, a free language designed for the Internet. Because of technological limitations, they aimed to design an off-line program that could be downloaded to a classroom computer and, at the end of the session, data generated by students could be uploaded to the Internet. The team created several Java-based tools for the simulation, such as scenario map, data storage, bookmark, notepad, and browser (Eichler *et al.* 2005).

11.4.2 Implementation

Consider how the simulation will be accessed and distributed. If it will be used by classroom teachers, for example, a supplementary printed booklet that describes the objectives and offers discussion questions, student handouts, and assessment ideas may be worth developing. Furthermore, teachers may need training in how to use the simulation before they can be expected to introduce it to their students. Although opportunities to use the CD with assistance is a better training environment, it is not always feasible given computer and time constraints. Susan Marynowski, a trainer who used the Burning Issues computer simulation in teacher workshops on prescribed fire says, "Our 2002 telephone survey reveals that even with only a 20-min demonstration of the CD, over half of the teachers reported using the simulation with their students and 73% of the respondents predict they will use it in the future. Barriers that prevent use are primarily technological, and as more schools obtain computers and LCD projects, use is increasing."

In many cases, computer simulations are most effective when learners have been prepared properly. They must first master the requisite knowledge to be able to apply it with the simulation. In this respect a simulation does not lend itself to the initial experience or inquiry learning, but to the application phase of the learning cycle (Chapter 2).

Similarly, educators and learners who make effective use of debriefing and discussion opportunities tend to learn more from a simulation. It is important for learners to take the time to compare the simulation with the real world and reflect on what they learned that might be transferred to other situations.

Both of these principles point to the benefit of an educator guide to help provide the background and the follow-up discussion questions to make a classroom simulation more effective.

Authoring tools, automated instructional design tools, and expert systems are available to help novices design instructional products (Kasowitz 2000). Contact consultants, universities, or other resource people in your region to discover the latest and most useful tools for your project.

The Brazilian team mentioned above created the computer simulation, *Carbopolis*, to engage learners in environmental problem solving related to air quality and acid rain. The problem, based on a real situation in the State of Rio Grande do Sul, involves decreased farming and cattle production in a region next to a thermo-electrical power plant. After reading declarations from different characters from the power plant and community, sampling and analyzing rainwater and air quality, and using the library research tool for readings and pictures, students form hypotheses about the cause of the problem. When they develop and implement solutions to reduce the severity of the problem, the computer program responds with new levels of key variables. By installing air pollution control systems, for example, students can reduce the level of sulfur dioxide and check to see if water and air quality improve (Eichler *et al.* 2005).

11.4.3 Evaluation

Computer simulations offer some distinct advantages to educators. Because they are not typically used to memorize information, they focus on higher-order thinking skills (Chapter 2). Several studies suggest that the use of computer simulations can enhance problem-solving and critical-thinking skills (Norton and Sprague 2001).

The *Carbopolis* designers used two different formative evaluation techniques while developing the simulation. Thirty teachers were given the program to use. Then they were interviewed to discover how they used it and what their suggestions were for improvements. Those interviewed thought the program would be useful in their classrooms, and complemented the developers on the price (free) and the interdisciplinary features. The only concern related to the amount and complexity of information when used with very young students (Eichler *et al.* 2005).

The second evaluation technique involved classroom observations of teachers preparing for and using *Carbopolis* with students. In one public school, a portion of the software was used, the collection and analysis of air and rainwater samples, to motivate students to think about environmental problems and reinforce what they had learned in chemistry. Students commented with surprise that they had a reason to use what they had learned, "You knew the reason why you were doing that!" (Eichler *et al.* 2005).

In a private school, two teachers created a complex unit that began with 28 h working with *Carbopolis* and 6 field trips to coal mines, a power plant, and a number of environmental offices and agencies on observing the students integrating their Carbopolis experience with their Brazilian experiences, teachers commented, "Carbopolis became real and not virtual anymore" (Eichler *et al.* 2005).

11.5 A suite of environmental technologies

A number of technologies have been developed to measure, record, and analyze various features of our planet. When used by educators, these technologies can greatly enhance conservation education and outreach programs. In many cases the technology accompanies or complements an educational technique mentioned elsewhere in this book: an exhibit, a presentation, or a Web site, for example. This list could be quite lengthy, but for the purpose of illustrating diversity, it is mercifully short! The standard format of technique planning, implementing, and evaluating does not apply to these techniques because they are merely the means, not the educational end.

11.5.1 Geocaching

This high-tech version of orienteering is becoming a popular family activity in natural areas. Geocaching involves finding hidden containers (but not buried) in a park, nature center, or other facility, using GPS coordinates of the container which are posted on a geocaching Web site. Anyone can access the Web site, record the coordinates, and then hunt for the cache with a GPS unit. Caches contain trinkets, a logbook, or small toys, which can be taken by a finder, as long as another item is left in its place. An online form enables anyone to place and record a geocache. The Cleveland Metroparks began placing geocaches in their Ohio parks in 2002 after realizing that private individuals had been stashing caches without permission. Concerned that unmanaged geocaching could damage delicate areas of the park, they instituted a permit system. The 1-year permits enable park managers to control the number and locations of temporary geocaches. The park also prints brochures with directions for finding the geocaches for people who do not have GPS units (Martin 2003).

11.5.2 Environmental monitoring

Educators have encouraged learners to monitor different aspects of the environment for decades. Collecting data on water quality, lichen, or traffic helps people witness changes that may be too small to detect with casual observation. High quality data can be used by resource management agencies to document detrimental activities, such as chemical spills. Water quality monitoring with chemical test kits, electronic probes, and sophisticated testing equipment occurs from Italy to Australia, often sharing their results with others in the same watershed or around the world through interactive Web sites. Citizen science initiatives (Chapter 7) may ask adult volunteers to assess seed preference of birds at their feeder and report data on a Web site. GLOBE, an international monitoring program established protocols for students to collect data on weather, air quality, water quality, ground cover, wildlife, and other aspects of the local environment (Figure 11.5). Students can access the international data and answer their own questions about relationships between variables or similarities among sites. Journey North is a highly acclaimed Web site that engages youth in monitoring and reporting seasonal changes in plants and migrating animals. The Web site allows any user to access the results and witness the biological response to environmental change.

The technology to monitor the environment may be traditional (thermometer) or new (GPS unit); likewise the technology to share the information with others may be traditional (presentation or mail) or new (Internet). Environmental monitoring is a valuable technique and could make a contribution to your program. As with other techniques, consider your primary objectives (learning or data collection) and design the program to achieve them. The conflict between education and science goals can make it challenging to develop monitoring projects, but these examples are good models to explore.

11.5.3 Real-time data

Environmental monitoring can be conducted remotely, automatically uploading data to a satellite, and conveying information to a computer. From there, data can be accessed on a Web site, used in a visitor center display, or observed by

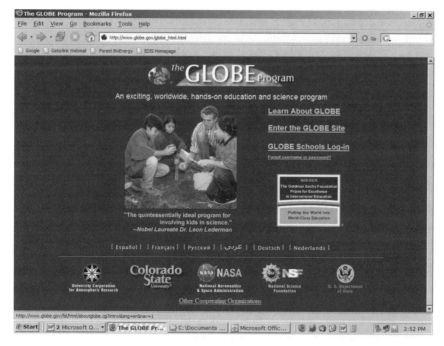

Fig. 11.5 The GLOBE program uses a Web site to provide information and allow participants to share observations (GLOBE home page: http://www.globe.gov/globe_html.html).

students in a classroom. This type of system has been quite useful at the Sabino Canyon Recreation Area near Tucson, Arizona, where visitors need to know the stream depth before venturing on trails with low bridges. Because stream flow is measured by the US Geological Survey remotely and transmitted every 15 min, real-time data (data collected at one location and quickly transmitted to another) are available to the interpreters at the visitor center. They created an exhibit that allows visitors to access the data, answering their own questions about stream flow, flash flood warnings, fire danger, and weather conditions. This year's conditions can be compared with last year's or graphed to better convey information. These technologies are changing the nature of what is interpreted and helping users become more aware and connected to the environment (Blasch and Polley 2003).

11.6 Summary

Educational technology includes a variety of old and new devices that can help conservation educators reach a larger audience or share messages more effectively. Video has been used around the world to bring distant images home, to demonstrate skills, and to lend a realistic element to the message. Similarly, radio can share information in a variety of local languages with people who might not have access to other technologies.

The Internet enables learners and educators to access a myriad of Web sites and other information sources. These interconnected computers can exchange e-mail messages or urgent notices via a list-serve and support bulletin boards or chat rooms that enable a number of users to share information. Distance education can take advantage of video, radio, print materials, satellites, and Web sites to engage place-bound learners. Computer simulations can be created for classroom use, for exhibits, for Internet distribution and download, or for sale.

As technology continues to advance, the newest tools will be marketed to educators, but only the genuinely useful and simple to operate will survive. Those are likely to be technologies that echo and support what we know about learning and education—those that enable learners to grapple with and apply new information in a way that makes a lasting impact on their mental models.

Further reading

http://www.webquest.org for information on Webquests

http://www.blogger.com is the original, free, blog-hosting site

http://www.estme.org/gallery is a collection of NASA and US Department of Education award-winning Web sites.

http://www.geocaching.com is the home of geocaching information and sites.

The Communicator's Handbook: Tools, Techniques and Technology assembled by Agricultural Communication in Education (1996). Maupin House, Gainesville, FL.

The Non-Designer's Web Book by Robin Williams and John Tollett (1998). Peachpit Press, Berkeley, CA.

Environmental Online Communication edited by Arno Scharl (2004). Springer Publications, London, England. Information is available at http://www.ecoresearch.net/springer

Designing on-site activities

First-hand experiences with wildlife or the outdoors are an effective method for increasing people's interest and concern for conservation. Informal settings, such as nature centers, parks, zoos, museums, camps, and farms, have natural advantages over formal school settings. On-site activities at informal settings can nurture curiosity, increase knowledge, improve motivation and attitudes, and engage the audience through participation and social interaction (Brody 2002). Education and outreach techniques are needed to orient, inform, and engage visitors. Signs, guided walks, exhibits, demonstrations, nature studies, and centers help visitors understand their surroundings, explore their personal feelings, and acquire new skills. From an orientation map at a site entrance to a state-of-the-art visitor center, many techniques enhance visitors' experiences.

This chapter provides guidelines for developing on-site activities. An initial planning process at a site paves the way for designing a variety of materials, activities, and areas to achieve the goals of your organization. Components to consider include the total audience experience, from pre-visit materials to follow-up activities to enhance the on-site programs.

12.1 Laying the foundation: initial steps to designing on-site programs

Detailed planning is the best way to insure successful on-site activities as part of your overall conservation program. The design process guides the orderly development of an education and outreach strategy in which you review and select from a variety of alternative actions to achieve your goals.

The design phase starts with an institutional review to determine and articulate why on-site education and outreach activities are needed and which of the site's goals they can address. From reviewing your organizational mission and policies, you can construct the specific objectives for the on-site program.

Once the objectives are established, educational planners describe the resources and potential for the program. An inventory of biological and cultural resources helps determine the resources and sites available for supporting learning activities and for selecting major themes you wish to convey (Veverka 1994; Jacobson 1999). These steps include inventories of:

- Site accessibility, habitat types, unique features, demonstration areas for management or restoration, geological resources, scenic vistas, waterfalls, gardens, and facilities.
- Orientation areas for contacting visitors, such as road intersections, boat launch areas, or campgrounds.
- Regional interactions with nearby sites interpreting related material and regional impacts based on traffic flow.
- Facilities and services needed.
- Actions needed to minimize impact on natural or cultural resources and to protect sensitive areas.

As in all education and outreach programs, you should gather baseline information about the target audiences to design appropriate activities and materials and for later program evaluation. These data paint a portrait of the potential audiences for whom the programs are developed. Key audience information includes:

- Specific target groups (visitors and non-visitors).
- Visitor motivations and perceptions.
- Visitor demographic characteristics.
- Visitor orientation systems (pre-visit, on-site, post-visit).
- Visitor use patterns (time of visit, seasonality).
- Mechanisms for audience participation in the planning process and in an ongoing advisory capacity.

Based on the resources and audiences, the design of education and outreach materials considers questions of how, when, and where the activities will be conducted. During this phase, a thematic concept is developed for each area, including:

- Site-specific objectives, content, and context.
- Recommended media and services.
- Preliminary program, budgets, and justification.
- Draft design for educational activities.
- Flexibility to incorporate new materials and themes in the future.

Once the planning phase is complete, the education and outreach designs are pilot tested with the target audience and the final materials are developed. Contractors or in-house experts are engaged to assemble the program or physical structures for public presentation. The program then is opened to the public. A management plan governs long-term care and maintenance of the materials.

Monitoring and evaluation of the activities will assess whether program goals and objectives are being met. In the classic text, *Interpreting Our Heritage*, park interpreters are instructed to search for the gleam in a visitor's eyes to determine the effectiveness of their program (Tilden 1956). Counting gleams, however, does not provide accountability to managers and decision-makers, nor does it pinpoint problem areas in education and outreach materials in order to modify them accordingly. Methods used for the evaluation of materials and activities commonly include observational measures of visitor behavior, quantitative surveys and questionnaires, focus group studies, and long-term tracking. Staff may track comments from a suggestion box or collect visitor feedback from letters, calls, and suggestion boxes (Chapter 1).

The following types of information are generally collected and evaluated for on-site activities:

- Impact of the activity or materials on visitor knowledge, attitude, or behavior.
- Long-term impact of the activity or materials on visitors after they leave the site, measured through latent effects on schoolwork, homeowner activities, vocational interests, or repeat visits.
- Feedback for designing future activities or materials.
- Cost-effectiveness of activities.
- Time spent on site and money spent at the gift shop.
- Unexpected outcomes (both desirable and undesirable) that were not part of the original objectives.
- Broad impact of the activity or materials on the larger community, as measured through natural area attendance or community trends.
- Restoration and environmental improvement activities in the community.

Evaluation not only improves programs, it also helps meet agency requirements for reporting and cost accounting. It can provide marketing ideas to administrators who make decisions about program continuation and budgets.

Typical on-site education and outreach techniques can be grouped into personal (person-to-person) activities and self-guided activities. Personal services include talks and presentations, information desks, ranger help, guided walks, campfire programs, nature awareness activities, and environmental studies. Self-guided learning activities include exhibits, such as museum displays, signs, and kiosks; demonstration areas; publications, such as maps, brochures, trail guides, and books (Chapter 10); and audiovisual presentations, such as films, tapes, and computer programs (Chapter 11). This chapter describes several examples of personal and self-guided techniques to provide a framework for developing on-site activities.

12.2 Guided walk

On-site guided walks and other tours led by an interpreter offer audiences an opportunity to interact with your facilities and with natural and cultural areas. Guided walks can lead visitors along a forest trail, through a zoo or botanical garden, aboard a tour boat, or almost anywhere you wish to interact with an audience and a particular setting (Figure 12.1). Tours of a demonstration area can show habitat restoration efforts, such as the effects of prescribed burning on a pine forest, the management of rare wildlife species, or integrated pest management on a farm.

Fig. 12.1 A guided boat tour provides participants with a first-hand look at river ecology in Belize (photo by S. Jacobson).

Guided walk activities vary according to the site, educational objectives, and cultures and needs of the audience. In Denmark, two types of guides are available in natural areas (Ulstrup 2001). Nature guides are well versed in biology. They share their knowledge in the field with participants, essentially conducting an outdoor biology tour with a focus on cognition. In contrast, *Friluftsliv* guides offer a different experience. The *friluftsliv* guide's focus is on hiking. The hike may stretch over the course of a day and include an overnight stay. The goal is to teach people to feel comfortable and enjoy nature. The *Friluftsliv* guide provides facilitation in basic areas, such as clothing, food, equipment, and tools. They share basic outdoor skills with their participants (Ulstrup 2001).

Most on-site guided tours have elements in common with effective public talks (Chapter 8). They are entertaining, relevant, meaningful, and organized around a central theme. Making a guided walk entertaining involves the audience in actively observing, searching, thinking, or conducting an activity guided by the educator's narrative and theme.

12.2.1 Planning

Planning a trail for your guided walk should include an understanding of the needs of the visitors, the potential of your site, and your learning objectives. Designing a trail through a forest follows specific guidelines to ensure a unique and refreshing adventure for your audience (Box 12.1).

Box 12.1 Designing forest trails for mystery and variety (adapted from Trapp *et al.* 1994).

- Route trail past large trees, running water, and other unusual or unique sights.
- Introduce or maintain colorful trees and shrubs with diverse textures and patterns.
- Plan vistas to direct visitors to views of lakes, mountain peaks, river valleys, or other vistas.
- Create forest openings or prune vegetation to invite entry and facilitate viewing.
- Provide benches, boardwalks, arbors, or other structures to promote visitor comfort and reflection.
- Screen unpleasant views or artificial structures.
- Use curving trails to draw visitors onward with the mystery of what may come next.

Guided walks have an identifiable introduction, body, and conclusion, like all public presentations. In addition, guided walks have a staging period before the tour starts (Ham 1992).

Staging period

About 15 min before your tour, the audience starts to gather at the departure point. This is your opportunity to greet the members of the audience, learn something about their interests and background, and answer any questions they may have. Your friendly greeting and enthusiasm for the tour helps build rapport with the audience and makes a good first impression. You can mention any physical or safety requirements, availability of food or bathrooms, and suggestions for sun or bug protection at this time.

Introducing the walk

Begin at the designated time and introduce your tour. Your introduction must capture the audience's attention (otherwise, they may disappear!), create interest in the topic of the tour, and orient the audience (Jacobson 1999). For a school or tour group, their specific needs should be incorporated into the walk. Tell them how much time the tour will take and repeat any physical or safety requirements. Most importantly, the introduction will introduce the theme of your tour and the general organization of the trip and your commentary. Set the stage for what they will see along the way by giving a brief overview of the tour. As you walk along, stay in the lead and make sure that you can see the entire group behind you. As you reach each stop, wait if necessary, and make sure that you are talking to the entire group.

Body

The body of the tour is the stops that you make along the way. During this period, develop your theme by describing specific sites, plants, animals, or objects that you pass. Do not talk about "everything," stick to your theme. Remember, your audience will remember only about five main points.

Each stop may take 1–6, depending on the focus and the group. Larger groups take longer to assemble and move. Your narration generally will follow a four-step format in which you:

1. Have the group focus their attention on the specific scene or object of interest. This can be done with a question: "What do you notice about this flower that would make it easy for a bat to pollinate?"
2. Next provide your explanation or description: "The white, fragrant (encourage group to take a sniff later) flower attracts bats flying at night . . ."
3. Now connect the stop with your overall theme, such as: "We depend on pollinators to maintain our landscape." This makes it clear why you stopped there and why the point was important.
4. Lastly, provide a transition sentence to the next stop. A foreshadowing of what will follow or a suggestion of what to look for along the way to the next stop will reorient the audience to the tour. "Have you ever wondered where bats live?"

Experienced guides make their stops more dynamic by involving the audience at each stop in thinking or doing specific tasks. Some tour guides carry a backpack with visual aids, such as animal skins, pressed flowers, bird nests, owl pellets, binoculars, thermometers, photographs, or other objects that might not be seen or used on every tour, but would enhance or illustrate the theme at certain stops. Encouraging the use of all five senses, such as smelling tree leaves for identification or listening to frog calls can make the audience enjoy and remember specific points.

Many guides rely on provocative questions to stimulate the group and encourage interaction. Asking creative questions is an art. Different types of questions solicit a variety of thoughts and answers. Questions may compare differences and similarities between things, such as, "What do bees and people have in common?" Questions can stimulate the group to think about the implications of something, such as: "What will this forest look like in 30 years if timber management remains unchanged?" Follow-up questions can elicit opinions or personal feelings, such as: "What would be a fair solution to protect the forest and forestry jobs?" Questions also may enhance analytical skills, such as: "What needs to happen in order to keep this species from going extinct?"

Like all forms of communication, your questioning techniques and descriptions will depend on your audience. Not everyone will respond to contrived objects from your backpack or wants to be questioned about personal beliefs. Differences in cultural backgrounds also must be considered. Some international tourists visiting US parks find direct questions posed by young rangers to be rude or discomforting based on their own cultures (Machlis and Field 1984). Children in a group may make long stops or explanations impossible. At the same time, children's innate curiosity, lack of inhibition, and size can provoke observations in nature that adults might miss.

The length of the tour also must be geared toward your audience. You may have 20 planned stops along the tour, and anticipate a few unplanned stops to take advantage of serendipitous scenes or sightings that help illustrate your theme. If you have a large audience, you will have to eliminate some stops to stay on schedule. Do not forget that the audience will probably remember only five or fewer main points. Make sure that your stops do not cover much more than can be retained, and that several stops may relate to the same point.

Concluding the walk

The conclusion is given after the last stop of your tour. Like the conclusion to a talk, it should reinforce the theme. The conclusion reminds the audience of the relationship between the stops they had made and the items you discussed during the tour and the significance of the theme. Good conclusions are brief and to the point. They reaffirm the take-home message for the audience and let them know that the tour is definitely over. Thank the group for their participation.

12.2.2 Implementation

Once you have carefully planned all elements of the guided walk, implementation involves a number of components. First, the design and content of your walk should be reviewed by colleagues or outside experts. Second, it should be pilot tested with members of the target audience. After giving a pilot tour, provide a feedback checklist or carefully question the participants about the content, duration, themes, topics, and props of the tour. Also ask about your voice, manner, delivery, language, and interactions with the audience. Make revisions and practice again. You also may want to ask a colleague to accompany you and provide additional feedback.

Successful implementation also is dependent on having a strategy for advertising and scheduling the guided walks. How will the audience learn about them—how will they be marketed? Are staffing and budget needs fulfilled? Finally, consider how you will sustain long-term maintenance of the trail and ongoing operational needs in the way of equipment or materials.

Guided walks should be designed to meet the needs and interests of the target audience. A group of students at the Singapore American School teamed up with the Nature Society of Singapore and the Singapore Association for the Visually Handicapped to implement a sensory nature trail. The goals were to develop a trail on the island of Pulau Ubin, an important area for biodiversity conservation, and to make nature accessible for the visually impaired (Frazier 2002). The design of the trail incorporated advice from biologists for plant identification and ecological references. Students then selected interpretive stations along the trail and created descriptive accounts that visually impaired visitors would find interesting. They incorporated input from their target audience to better understand the range of visual impairments of their visitors.

During the walks, sighted guides helped the visually impaired visitors explore the textures, shapes, and smells of interesting wild plants (many that are used in popular foods and medicines). Shells, rocks, birdcalls, sounds of the sea, and the odor of the mudflats became features of the tour as well. The tours consisted of two student guides for each visually impaired visitor. One student interpreted the natural features of the trail, while the other served as a physical guide. Success of the initial outings led to a commitment by the students to continue to offer their guided walks for the next 5 years. Partnering with the outside groups provided continued advertising, scheduling, and maintenance for the tour and trail.

12.2.3 Evaluation

Evaluation provides feedback to remedy any design or content problems that might be reducing the effectiveness of your guided walk. Direct feedback from visitors provides immediate data to help determine the strengths and weaknesses of the activity. By asking visitors questions or directly observing them, educators can find out whether the visitors used and learned from the interpretation (Ambrose and Paine 1993).

Asking visitors questions about the interpretation determines:

- If the guided walk got its message across.
- Which activities visitors found most interesting or stimulating and why.
- What improvements and changes visitors would suggest.
- What message and information visitors remember.
- If visitors find the guide and walk effective.

Observing visitors at a site determines:

- How many people engage in the guided walk.
- How many visitors stay and interact with the guide or ask for further activities after the walk (often measured by time spent at activity or site).
- How visitors emotionally react to the activity (e.g. smile, frown, talk, or laugh).
- How many visitors talk to each other about the activity (record positive and negative remarks).
- How many visitors ask questions.
- If visitors follow suggested activities or appropriate behaviors.

12.3 Exhibits

An exhibit is generally defined as a strategic presentation of ideas or themes with the intent of educating, informing, or orienting an audience. Exhibits are usually presented in informal settings, such as nature centers, outdoor areas, trails, visitor centers, museums, and building lobbies, where the majority of visitors are exposed to nonpersonal or self-guided interpretation. Visits to informal facilities are self-paced, voluntary, and exploratory, so exhibits reach visitors at their own speed and level of interest. Exhibits engage an audience in nonlinear and creative kinds of learning, as opposed to the more orderly, linear, and verbal learning that goes on in classrooms and seminars.

Types of exhibits include self-guided trails, visitor center displays, educational kiosks, viewpoint markers, natural feature signs, special events displays, campground bulletin boards, regulatory signs, labeled trees or specimens, artistic statements, and 3D models. Exhibits can encompass devices, such as quiz boards, audiovisual programs, continuous radio broadcasts, interactive CD ROMs, talking animal displays, models, and computer simulations. Because exhibits often incorporate real objects, such as petrified wood or collections of shells, visitors can respond to "the real thing." Many exhibits also include take-home materials, such as brochures, posters, fact sheets, or other publications (Chapter 10).

In many cases, exhibits reach more people than personal approaches, making them an important part of an overall education and outreach plan. Because of their importance and durability, care must be taken in designing exhibits to be effective channels for a conservation message. Exhibits can be effectively used in a number of settings and to accomplish a variety of goals. They may inspire viewers to contemplate beautiful vistas, objects, or quotations. They can demonstrate a story or relay an educational message. Or they can simply encourage visitors to make their own discoveries and subtly direct viewers to follow their own interests (Ambrose and Paine 1993).

Exhibits take many physical forms. Indoor exhibits may be free-standing dividers, wall-mounted signs, objects on tables, or displays in cabinets, while outdoor exhibits may be free-standing signs or displays under weather-resistant structures. Exhibits in any situation may be either flat or 3D, such as a model, diorama, object, or outdoor scene (Figure 12.2). Exhibits are visual, usually

Fig. 12.2 Museum artisan, Bob Leavy, creates a life-size diorama of a mangrove swamp for the Florida Museum of Natural History (photo by S. Jacobson).

Fig. 12.3 Live animal exhibits attract a crowd, such as this demonstration of the feeding behavior of an alligator at the St. Augustine Alligator Farm in Florida (photo by S. Jacobson).

including illustrations, maps, graphics, charts, and other explanatory artwork. In addition, effective exhibits strive to excite more than one of the audience's senses, for example, providing an audio explanation and something to touch along with visuals and text. Some exhibits include smell to arouse the interest of an audience, by instructing them to sniff the surrounding air or a particular plant. Other exhibits involve live animals, which are exciting to people, yet require careful attention to animal welfare and safety considerations (Figure 12.3).

Self-guided tours along forest trails, bikeways, automobile drives, demonstration areas, and historic sites follow the same guidelines as other types of exhibits. Self-guided tours lead people sequentially along a series of interpretive stops. The tour flows from an orientation and introduction to the site through 15 to 20 stops that elaborate on the theme. Most self-guided walking tours are about 1 km long and take a half hour to complete. Depending on the specific characteristics of the site and the audience, self-guided trails use signs, brochures, or audiotapes to communicate with visitors.

12.3.1 Planning

The best exhibits attract attention and effectively communicate a message or theme within the attention span of the target audience. Exhibits that use several media are stronger than exhibits that rely on a single medium (such as text), because they reach better the audience members who have varied learning styles and tastes. Researchers have summarized characteristics of effective exhibits into the "ABCD's" of exhibit design (Ham 1992):

- Attractive: attention-getting exhibits that use appropriate colors and interesting graphics and visuals.
- Brief: well-organized and simple exhibits that contain five or fewer main ideas, only enough text to develop the theme, and graphics to help communicate with viewers.
- Clear: the theme of the exhibit is obvious and can be immediately understood by the audience. Additionally, the exhibit is easily visible with adequate lighting and unobstructed viewing.
- Dynamic: the exhibit communicates the message by arousing curiosity, inviting participation, and providing entertainment (Figure 12.4).

Exhibit development process

Planning includes writing text and labels, and developing and testing a mock-up exhibit. It entails working with staff or contractors to have the final exhibit produced and installed. The bulk of the time for developing an exhibit is spent in the planning stage, with less time spent in each subsequent phase in the process. The following describes specific steps that might be included in each of the exhibit development stages (Knudson *et al.* 1995).

During the development of an exhibit, educators set goals, identify and assess audiences, develop objectives, and research and conceptualize the design. Audience research at the San Diego Wild Animal Park, includes the following qualitative methods:

- Listen to visitors' conversations at an animal display.
- Ask visitors what they would like to know.
- Put a tape recorder or comment board at a display.
- Ask animal keepers and other employees what visitors ask about.
- Ask tour guides what visitors ask them.
- Brainstorm questions and topics with external educators and graphic artists (Trapp *et al.* 1994).

Exhibit planning and preparation can take a long time. It can take weeks or months to develop a small exhibit, and over a year for a major exhibit. In addition, the exhibit work plan must allow for adequate time to test a mock-up exhibit and

Fig. 12.4 Dynamic exhibits spark people's curiosity and engage their senses. Visitors are encouraged to tug a toy eel from its rock cranny at the Monterey Bay Aquarium in California (photo by S. Jacobson).

text with audience members, and for a final evaluation to determine whether the exhibit has met the interpretive objectives (Chapter 1).

The exhibit design phase concerns the more practical and applied tasks of determining orientation, panel layouts, lighting, and other design details for the exhibit. Evaluation is used during this phase to test audience and expert opinions of exhibit design ideas. During the planning process your exhibit team must consider a number of design criteria. These describe the "powers" an exhibit should embody (Bitgood and Patterson 1987):

- Attracting power. Does the exhibit get people to stop?
- Holding power. Does the exhibit keep people and for how long?
- Teaching power. Do people learn from the exhibit?
- Motivating power. Are people motivated to find out more or take action?

Box 12.2 provides a summary of how to make your exhibit more powerful.

Box 12.2 Factors leading to effective exhibits that result in longer viewing times (Bitgood and Patterson 1987; Screven 1990; Ham 1992; Jacobson 1999).

- Indicate flow of exhibit by including exhibit title, introductory panel, and conclusion at the end; use lines and angles to lead the eye where viewer should look; provide a clear pathway to follow.
- Identify theme and story line in titles and headings.
- Match exhibit to demographic characteristics of visitors; connect stories to visitor interests.
- Keep text short, concise, and thematic; use short sentences and bullets, use personal pronouns and active voice.
- Feature unusual or rare information, and enhance perceptions of beauty or danger in exhibit to provide variety and freshness.
- Make features easy to view by placing subjects directly behind signs, and main titles at or above eye level.
- Use large size graphics and text; isolate main elements and make sure they stand out from the background.
- Express ideas visually through artifacts, illustrations, and photographs; dominate panels with visual images.
- Create balanced graphic design using edge borders or boundaries, much empty space, harmonious colors and shapes, and consistent and clear typestyles and illustrations.
- Use intrinsically interesting visuals, such as baby animals rather than a data chart.
- Engage senses in addition to visual, such as touch or sound.
- Provide interactive and participatory activities, such as having visitors answer questions, search for something, solve a puzzle, make a prediction, or confront a misconception.
- Ensure that interactive devices can be easily manipulated and provides immediate feedback to the visitor.
- Stimulate interaction among visitors.
- Make exhibit easy to view by providing adequate lighting, and use lighting to focus attention on key themes or objects.
- Place information at visitor's line of sight, close to viewer.
- Demarcate materials for children by special colors or lower placement.
- Eliminate visual interference from neighboring exhibits and auditory barriers from entrance/exit noise.
- Cater to different audiences with audio materials or taped messages.
- Provide comfortable amenities, such as rails, seats, temperature control, and rest rooms.
- Enhance positive social pressures, such as attraction to crowds, walking speeds adjusted to other visitors, and comfortable waiting times.
- Accommodate visitors with disabilities.

Getting an audience to stop and read or listen is an important function of exhibit design. The exhibit design is what initially attracts people, but the text must keep their attention in order for them to learn something. Visitors usually stop less than 1 min at an exhibit—long enough to read only a fraction of the text. Long text discourages many visitors. When preparing text, keep in mind the aim of the exhibit and the characteristics of the audience. These same guidelines apply equally to any nonpersonal or self-guided program—visitor center exhibits, trail signs, computer applications, driving tours, and audio-visual programs. Besides having accurate content, correct spelling, good grammar, and understandable language, exhibit text should convey a theme that the audience can understand, no matter what the medium.

Planning an exhibit also requires consideration of the actual space for the exhibit, from allowing space in the exhibit area for visitors to sit and comfortably contemplate exhibits to providing adequate lighting by windows or artificial light. An inventory of space requirements ensures visitor comfort and effective use of the area. Access and facilities for people with special needs, such as limited mobility, or the very young or elderly may require ramps, wide aisles for wheel chairs, or other accommodations. Safety elements, such as emergency exits, fire extinguishers, and first aid equipment are often dictated by government codes, and need to be followed.

In planning a new monkey exhibit at the San Diego Zoo, the design team included staff from public relations, education, architecture, exhibit, and mainten-ance departments to determine how to make the best use of the limited space available. The team realized they could create access to a vertical exhibit. This would provide more opportunity for viewing the animals; help portray the mon-keys' natural, arboreal habitat; and increase the space available for information signs and displays. Aerial boardwalks would bring visitors into the tree canopy, yet remain accessible to wheelchairs.

Wayside exhibits located outdoors have other design concerns. They must be integrated into the landscape. A landscape designer can help with placement decisions. While exhibits need to interpret the scene, they should not intrude upon it. Exhibits can be designed into pedestrian turnouts on boardwalk trails or blended into railings, hedges, or other unobtrusive borders.

12.3.2 Implementation

Guidelines for text include writing simply, avoiding jargon and technical terms, putting the main point at the beginning, keeping it short, and using vocabulary that visitors use. Similar to the guidelines for good writing (Chapter 10), these are easier said than done. Biologists and technical staff members often have a difficult time conveying their knowledge to the public in understandable language. It is important to write as if you are a friend of the visitor, rather than as a scientist or program administrator. External interpreters, copywriters, or journalists can be hired to assist with exhibit text. An outside person also may provide a fresh and unbiased perspective of your exhibit text and design.

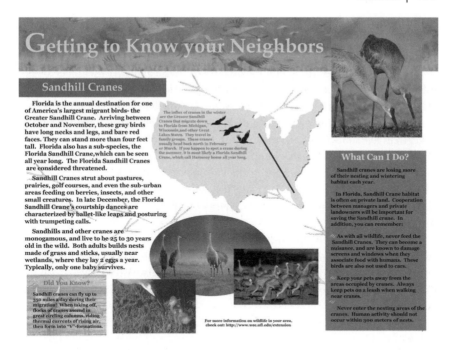

Fig. 12.5 The exhibit panel tells the story of Sandhill Crane conservation using graphics, titles, subtitles, and text (courtesy of M. Hostetler, University of Florida).

The story told by exhibit panels is generally presented in three parts: the title, subtitles, and body text (Figure 12.5). Most exhibits start with an introductory panel, giving the exhibit title and purpose. The audience should be able to grasp what the exhibit is about and why it is worth viewing. It also may be useful to give some information about where the exhibit or program will take the viewer. The direction of the exhibit may be obvious in a small visitor center or museum, but less obvious for a trail, discovery center, or computer simulation.

The introduction is followed by the body of the exhibit, often a series or network of panels. Section panels should follow an obvious order, such as chronology, cause and effect, problem/analysis/solution, or from part to whole or whole to part. Section panels should include thematic headings and give more background information about the exhibit topic.

Museum experts note that since most viewers only read a fraction of the text, the title and subtitles of interpretive panels must communicate the theme quickly. Instead of giving the panel a label that identifies a topic, it is better to use a title that relates the theme, uses an active voice, and is interesting and fun. Thematic, rather than topical titles help get your message across. A topical title might be "Trees," but the thematic title developed from it could be "Our Lives Depend on Trees" or "The Forest is our Pharmacy." An exhibit explaining research on ungulates in Alaska catches visitors with the title: "Poop: A Valuable Research Tool." Studies of exhibit titles at the Birmingham Zoo Predator House revealed theme titles, such as

"Animals that Eat Animals," had high attracting and holding powers. Researchers have found, however, that few people take the time to stop and read an entire exhibit.

A thematic title and strong graphic design can attract audience members to your exhibit, but it is the subtitles and body text that exert the holding power on the audience once they are near. Subtitles should be smaller than the main title, but they must clearly relate several separate ingredients of the theme. There should be fewer than five subtitles on any given exhibit panel. A coherent graphic design that physically separates the components of the theme while maintaining a consistent look, will aid the viewer in finding the subtitles and understanding the theme. The body text is placed near each subtitle. More detailed text and object labels also may be added, but should be placed away from the main themes and text of the exhibit panels.

Exhibit text for general, mixed audiences should be written at the level of 12–15-year-olds (sixth–eighth-grades), and lower for school age audiences. Many computer software programs can test readability. Alternatively, you can calculate readability using a method like the Fry test to see if your text meets the guidelines.

For the Fry Test, use a sample of several passages of your exhibit totaling about 300 words. Count the number of words, sentences, and syllables in the text. For text to target the 12–15-year-old level, there should be between 5 and 10 sentences per 100 words (10–20 words per sentence) and between 120 and 150 syllables per 100 words. It is better for text to fall at the lower end of these guidelines (Ambrose and Paine 1993).

Other factors also affect readability. The lighting and layout of the exhibit must be good, and the typeface must be large enough to read at an appropriate distance. For indoor exhibits, main titles should be a minimum of 2–3 cm tall and subtitles should be at least 1–2 cm tall. For outdoor exhibits to be read at a distance, titles should be 10–13 cm tall and subtitles 7–12 cm. Remember that these are *minimum* sizes—larger sizes will always be easier and more attractive for the audience to read (Figure 12.6).

Many exhibits convey a conservation message that encourages people to seek further information or take action. Audience members may be asked to think about a question, pick up trash during their trail hike, avoid trampling an endangered species, write to a legislator, volunteer for a conservation project, take home a brochure from an exhibit, or get more information at the front desk of a visitor center. The additional information also may take the form of a marketing device, for example, as a request for support or membership in an organization.

By following the guidelines for text development, readability, style, content, and exhibit development during the planning stage, you should be able to implement a display that has the power to attract and hold your audience and to deliver your conservation message. Once the exhibit is opened to the public, maintenance requires an ongoing plan. This ensures interactive displays continue to work, signs are clean and legible, and vegetation is growing as anticipated. Any problems with

Fig. 12.6 Large lettering and a bold image attract attention to this outdoor exhibit placed in moose habitat at Yellowstone National Park (photo by S. Jacobson).

litter, vandalism, or other negative visitor behaviors must be addressed immediately to ensure long-term success.

12.3.3 Evaluation

Evaluation should be infused into every stage of exhibit development. Methods of getting information to improve the exhibit design at the planning stage include portfolio reviews and focus groups with members of the audience and with experts. Ongoing monitoring during implementation helps ensure smooth functioning of the program. The completed exhibit should be assessed to determine its effectiveness and to make needed modifications. Methods used for the evaluation of exhibits include observational measures of behavior, before-and-after-surveys, and visitor interviews. While problems can develop at any stage of development, many can be corrected, taking into account visitor characteristics and exhibit objectives. Fixing problems depends on good evaluation data. For example, text can not be worded for the interests of visitors if those interests are unknown, or major eye contact points and visitor pathways must be observed by testing a mock-up exhibit or through observational techniques in order to design an effective exhibit.

Use Box 12.2 to check the style, content and structure of your exhibit to assess how to better attract and hold your target audience. A wayside exhibit at Glacier National Park's Bird Woman Falls explains the geology of the specific hanging valley that can be seen above the waterfall. An evaluation of the effectiveness of the exhibit revealed that visitors learned to recognize this glacial feature and looked for other hanging valleys during a tour of the park (Hoffman 1999). In a more quantitative assessment of an exhibit about the Great Bay Estuary in New Hampshire visitors were surveyed after exposure to the material. They were asked to describe characteristics of an estuary, and list two reasons why the Great Bay is valuable (Heffernan 1998).

12.4 Demonstration

Demonstration areas are used to show visitors the results of a variety of conservation activities. Urban dwellers can see for themselves how to plant a small butterfly garden. Farmers can see the benefits of leaving habitat for birds that eat pests. Forest owners can view improved silvicultural practices. Demonstration areas provide an opportunity to show the efforts required and the results of management actions in a more convincing manner than vicarious methods, such as reading or listening. A demonstration butterfly garden can provide signs that identify suitable plants, and explain the butterfly life cycle and the steps for planting native plants to attract butterflies. Visitors then can watch a myriad of butterflies sipping nectar or laying eggs, and make decisions about the aesthetic and economic impacts of developing their own butterfly garden (Chapter 9).

12.4.1 Planning

Pennsylvania State University established seven forest stewardship demonstration areas as an educational technique to encourage responsible forest resource management (Harmon *et al.* 1997). The main audience for the demonstrations is the half a million private forest landowners in the state. The forestry demonstration areas were developed in partnership with the state and federal forest management agencies. The demonstrations areas are small, about 5 ha, but show 6 silvicultural treatments: tree thinning from above, thinning from below, shelterwood, improvement thinning, clearcutting, and a control area that is not logged. Although not all these harvest alternatives represent good forestry practices, they are all used in Pennsylvania. The purpose of the demonstration forest program is to allow visitors to compare harvesting options and encourage dialogue about their positive and negative consequences (Broussard *et al.* 2001). The university extension service hosted workshops at the two of the demonstration areas to test the effectiveness of the program with forest landowners, timber harvesters, timber buyers, and foresters.

Planning demonstration areas requires careful attention to the goals, the audience, and the site itself. The initial expenditures for labor and materials are a key concern of any organization planning a demonstration area. Ongoing and long-term staff and funding requirements to properly care for the area are equally important constraints. Sometimes partnering with organizations with similar interests can help provide long-term labor or funding needs. For example, local garden clubs may be willing to volunteer to maintain a demonstration butterfly or native plant garden. Water management agencies may be willing to design demonstrations of xeroscaping home landscapes in order to promote reduction of water use in their district. Parents and local plant nurseries may pitch in labor and materials for the development of a schoolyard ecosystem demonstration.

Planning a demonstration site generally entails a team of individuals with expertise in the practical aspects of developing the site, educators with knowledge of how to use the site, members of the target audience to ensure the usefulness of the site, and individuals who can bring to bear other resources, such as funding, marketing, and other organizational support. The following steps were taken to create a schoolyard wildlife habitat demonstration (Flint 2002):

1. Meet with the principal, teachers, students, parents, school officers, and maintenance staff to discuss using the school grounds for a demonstration area.
2. Form a project team and involve as many groups as possible in the labor, as well as in the fun and learning.
3. Include the maintenance staff in every step of the planning and implementation to ensure the area is designed for easy maintenance.
4. Obtain or draw a map of the school and demarcate all the existing wildlife features, such as sources for food, shelter, and water.
5. Choose a site that is easily accessible to the target audience: students, teachers, and staff.
6. Have the team decide on wildlife habitat improvements or additions that should be made.
7. Plan the implementation phase, including a detailed diagram of the project, a comprehensive list of materials, a description of how to build the project, and a schedule of maintenance requirements.
8. Incorporate walkways, tables, shaded areas, and other elements into the demonstration area that will maximize the comfort and safety of visitors to the site to promote an environment that is conducive to learning.
9. Develop safety and use procedures with input from appropriate groups.
10. Schedule the necessary steps to build the project, from gathering materials to assigning workdays.
11. Plan and draft the education and outreach materials—guides, signs, brochures, or exhibits—that will support the demonstration area, as described previously.

12.4.2 Implementation

Implementing a demonstration area involves establishing a daily and seasonal maintenance schedule. Implementing the pedagogical activities for your demonstration area include pre-visit and post-visit activities as well as on-site activities.

Student activities associated with a demonstration pond may vary from observing aquatic life to learning about their regional watershed. The North Carolina State Museum of Natural Sciences suggests the following learner activities focused on a demonstration pond in a schoolyard (Flint 2002).

- Observe aquatic animals on the surface of the water and answer questions such as: what do you see, what is it doing, how does it move, eat, or breathe?
- Record and discuss observations about underwater organisms using underwater viewers.
- Catch aquatic animals with dip nets and identify them using guides and keys, then report interesting facts about them to the class.
- Make your own field guide to aquatic insects or plants.
- Compare water, air and ground temperatures, or water pH over time and develop hypotheses to explain differences.
- Study aquatic plants by measuring their growth over time and relate it to changes in water chemistry and temperature.
- Discuss natural wetlands and learn about the watershed in which students live.
- Tie the science-based exercises to other subjects, such as language arts and social studies, with environment-themed exercises in other classes (Chapter 5).

Demonstration areas for all audience—students, landowners, consumers—provide opportunities for first-hand observation and experiences. Depending on the goals, these sites provide information about the species present, equipment needed, and other logistical support to enable visitors to emulate the demonstration at home or work or to support similar activities in their communities.

12.4.3 Evaluation

"Mrs Williams and I were very surprised that this lesson had spilled over to the students' own yards. They went home and returned the next day so excited about the bugs they had found in their own backyard," reported teachers implementing a schoolyard demonstration program. (Flint 2002).

Evaluation of demonstration areas can use a number of techniques for collecting data. Ultimately, evidence should indicate if the visitors use the information or apply the skills learned at the demonstration site to different settings, such as these students in their own backyards. Data about the number of visitors attracted to a site, length of time spent at a site, visitor understanding of the information and objectives of the demonstration, and their acting upon the information, provide feedback for assessing the program and making improvements.

Informal interviews with visitors can be used to reveal if they understand:

- The main point of the demonstration.
- Aspects that are unclear or confusing.
- Whether the demonstration is interesting, attractive, and relevant to them.

- The usefulness of any supporting material, or additional materials needed.
- What would make their family, friends, or colleagues more likely to visit the demonstration area?

A more formal evaluation was conducted with visitors to the forest demonstration areas in Pennsylvania. They were tested with a knowledge and attitude questionnaire during their visit to a demonstration forest site. Participants were randomly assigned to one of three groups: exposure to a forest management slide-show, exposure to a slide-show and a demonstration forest tour, and a control group that was not exposed to the slide-show or tour. Visitors experiencing the demonstration forest scored significantly higher in both knowledge and attitudes about forestry practices than either other group. The forest demonstration areas were a valuable educational tool that contributed to landowners' knowledge and understanding of forest management beyond that achieved by an indoor learning experience (Harmon *et al.* 1997).

Other forest demonstration sites may have a more practical feedback mechanism. High school students in Yellow Springs, Ohio, developed an Evergreen Forest Management Demonstration area with the help of a professional forester. They held a Christmas tree festival to attract visitors to their demonstration and sold over 1200 trees, earning thousands of dollars and increasing public interest in local tree-farming (Ashbaugh 1963).

12.5 Nature awareness and study

Remember the curiosity of young children drawn to watching a bee hover around a flower? Nature awareness activities build on this appreciation of the natural world. Nature awareness ranges from energetic activities where young children simulate a "Bat and Moth" to reflective experiences for adults, such as a silent hike or sharing circle in the outdoors. Whether you work in a primary school in Kenya or for the USDA Forest Service, nature awareness activities will allow you to explore the environment with your participants and create a sense of joy in the outdoors. The theory behind nature awareness activities developed from the nature study movement in the early 1900. But the practice of nature awareness builds on the innate curiosity and joy that children experience playing outdoors (Cornell 1979).

On-site programs associated with parks and natural areas often offer a range of activities from nature awareness games to in-depth nature study opportunities for people seeking a scientific understanding of the natural environment. Nature study programs may focus on increasing knowledge or skills in a particular subject area targeted to a specific audience, such as botany for gardeners or bird identification for photographers. On-site awareness and study programs enhance visitors' experiences and simultaneously help build support for the specific site or park system. Programs can last from an afternoon to an entire summer and offer activities for which visitors and community members are often willing to pay, especially if experienced staff plan the activities or recognized experts teach the courses.

12.5.1 Planning

Planning on-site nature awareness and study activities involves identifying goals and objectives, identifying audiences, and developing or marketing the programs. Scheduling, staffing, and budgeting must all be determined and planned well in advance. Some on-site classes operate with the assistance of a cooperating association or "friends" group. Others are self-supporting, nonprofit organizations. The Teton Science School offers a range of residential programs for school groups in Grand Teton National Park in Wyoming. Their typical 3–5 day session includes activities, such as a natural history hike or ski, canoeing on a nearby lake, a lesson on animal tracks and signs, wildlife observation, assisting in ongoing field research in the park, and field guide use and map-reading skills. Planning these activities for school groups involves coordination with teachers and adherence to state academic standards.

Whether you are an educator planning a field trip to a natural area or park staff hosting schools and other groups at your site, planning usually requires consideration of the following steps and are discussed in Chapter 5 in more detail:

1. Get approval from principal, teachers, parents, students, school boards, or other participants. For school visits, this often entails showing how the trip is an integral part of the broader curriculum and addresses education standards.
2. Arrange transportation and housing.
3. Estimate total costs.
4. Obtain parental permission slips and medical information.
5. Provide a recommended clothing and gear list.
6. Develop fundraising activities if needed.
7. Design a detailed trip itinerary.
8. Clarify expectations of behavior and activities for participants.
9. Plan pre-trip activities to introduce new concepts to be covered, equipment to be used, and landscapes to be experienced (Giacalone 2003).

The novelty of a field setting affects learning. Settings that are too novel cause fear or nervousness; settings that are too familiar cause boredom. Learning is maximized when the field trip setting is of moderate novelty (Falk and Balling 1980). This can be accomplished by briefing participants in advance. Provide participants with pictures or a video of the site, locate the field trip route on a map, and provide details of the learning activities, and information about possible weather conditions, safety hazards, restrooms, meals, and other logistical support (Athman and Monroe 2002).

Pre-trip activities are key to accomplishing educational objectives. Teachers planning a field trip to Yellowstone National Park in Wyoming tied almost every element of their year-long science curriculum to the destination (Giacalone 2003). They studied geology, volcanoes, earthquakes, glaciers, erosion, food webs, ecosystems, succession, microbiology, chemical reactions, minerals, energy transfer, animal life, and the scientific method in the context of

Yellowstone. Students created their own guide to focus their study while visiting the park. The field guides contained plant keys, animal tracks, scavenger hunt questions, Yellowstone facts, an animal observation checklist, thermal feature comparisons, trail maps, questions for each stop, travel times between stops, and blank journal pages to draw and describe features. The students worked on the field guides together before the trip and received grades based on thoroughness and creativity.

Planning on-site activities at natural areas requires time regardless of whether the program involves in-depth study or a basic awareness activity. Planning on-site nature awareness activities requires careful attention to the sequencing of specific activities. Most people have experienced a facilitator who ignored the learning process, beginning a lesson with complex concepts and losing the audience shortly after. You can avoid that mistake by using a system for sequencing nature awareness activities, such as the four stages of Flow Learning (Cornell 1989b). Each stage provides general guidelines for choosing activities to ensure the activities build upon each other:

- STAGE 1: Awaken **enthusiasm**—Building enthusiasm creates the alertness and interest critical to learning.
- STAGE 2: Focus **attention**—Channeling that enthusiasm into a focused attention primes the participants for learning.
- STAGE 3: Direct **experience**—Calm attention allows a direct experience with nature.
- STAGE 4: Share **inspiration**—This direct experience opens the door to a deeper awareness and inspiration (Cornell 1989b).

The concepts of flow learning were initially used for planning an entire sequence of complementary activities and lessons. You can use these four stages, however, as flexible principles for planning almost any nature awareness activity.

12.5.2 Implementation

Many environmental awareness activities emphasize sensory experiences—seeing, touching, hearing, smelling, and tasting. This encourages learners to develop a personal awareness of nature. Sensory experiences help participants "experience" nature, in addition to studying it (Van Matre 1972). An outline of activities to explore a forest habitat might use the following sequence (adapted from Van Matre 1972):

Awake enthusiasm:

1. Provide participants with blindfolds and explain that they are going to "lose" one of their senses in order to increase their perceptions of their other senses. Explain how blind people often have more acute senses of hearing and touch than sighted people. Lead blindfolded participants to the site for exploration using a rope or railing to guide the way. This should stimulate participants to concentrate on their other senses and create a feeling of alertness and enthusiasm (Figure 12.7).

Fig. 12.7 A sensory awareness game allows participants to explore the natural environment using a fun blindfold experience (photo by S. Jacobson).

Focus attention and direct experience:

2. Give participants something to smell, such as an aromatic leaf or twig, as they begin their walk in the forest. This gets them started using nonvisual senses.

3. Ask participants to stand still and just listen. Ask them to think about how the noises change as they do this in several spots. They can notice changes in air temperature, hardness of the ground, and other things that stimulate their senses.

4. Have participants dig a hole and sift the soil through their fingers. They should notice the grainy feeling of the soil. They can check for moisture content, temperature, and smell—a sensory recognition of the soil. This can be compared with drier, sandier soil in other areas.

5. Have participants remove their blindfolds and look at the soil they have just felt. Have them look around the forest. They will notice that their vision, too, is more acute.

6. Have participants lightly touch and rub their fingers around a flower or a mushroom without crushing it. They *feel* what the flower or fungus is like while handling it with care.

Share inspiration:

7. Have the participants discuss their feelings about the results of these experiences. They also can continue their walk in the forest. When an animal, plant, or rock draws their attention, ask them to stop and write their

impressions in a journal, for their own reflection and inspiration, or to share with the group later.

8. Another activity to share inspiration, involves providing cards with inspiring sayings related to nature written on them. Include a brief activity with each meditation. Turn the cards over and have each participant choose a card and find a quiet place to be alone. For example, a card might say: *"Holy Earth Mother, the trees and all nature are witnesses of your thoughts and deeds."*— *Winnebago Indians.* Go on a walk and repeat these words of thanks for the Earth (Cornell 1979).

Nature study programs take advantage of natural areas as outdoor classrooms. Many on-site programs offer classes about a variety of subjects and activities. Mammoth Cave National Park cooperates with Western Kentucky University's Center for Cave and Karst Studies to offer a series of 1-week summer courses focusing on caves and karst landscapes. These intensive field courses combine daily lectures with field observations and excursions. Visiting professors who are authorities in caving and karst science teach the classes. Some courses require previous subject knowledge, while others are designed for people merely curious about caves and caving.

The North Cascades Institute works in cooperation with North Cascades National Park in Washington to provide a variety of nature awareness and environmental activities. With 60 programs for children and adults, they cover topics ranging from geology, history and wildlife to nature writing and paddling skills. Their hikes and paddles offer a blend of science, natural and cultural history, literature, and the arts. Box 12.3 describes their weekend class on edible wild

---

Box 12.3 Example of nature courses offered at North Cascades Institute in Washington, (North Cascades Institute 2005).

"Skagit Valley Foodshed: Wild Edibles"

Rediscover the ways Northwest cultures have sustained themselves physically and spiritually using resources near at hand. Join two local naturalists for a weekend of learning about wild plants. There will be lessons on plant identification, ethical gathering practices and traditional use of plants for food, fiber medicine and shelter. We'll explore meadows and woodlands, forage in nature's well-stocked pantry and learn how to collect in a safe and legal manner while preserving plant habitats. In the evening we'll share the tasty task of creating nettle lasagna for our potluck dinner.

Native Plants: Wild and Restored

With more than 1600 native species, North Cascades National Park harbors some of the most diverse plant life in North America. Spend the weekend studying native flora with the Institute's restoration coordinator and a plant specialist for the National park Service. Focusing on the area, we'll learn how to assess the terrain, treat soils, sow plants and monitor for biodiversity. Examining projects to replace exotic species with natives, we'll also tour the Park's greenhouse and investigate wetland sites threatened by reed canary grass. We'll camp in North Cascades National Park.

plants. Participants pay US$145 for this cultural learning experience. Another class, "Native Plants: Wild and Restored," provides an ecological look at plant life in the region.

Other sites, such as zoos and aquaria provide on-site study activities for a variety of audiences. The Tennessee Aquarium in Chattanooga TN, developed a program to enhance high school science curriculum taught to homeschool students (Matherly 2000). Through this program, students meet for a total of eight 2-hr sessions during the year. Laboratory sessions give students a chance to learn basic science skills using equipment and live specimens not available at home. Students observe the characteristics and behavior of live animals, participate in a dissection, learn how to use microscopes, and key out various plants and animals.

The aquarium program's outdoor sessions give students a chance to discover the science in their everyday lives and to interact with nature. Students collect and identify macroinvertebrates that are later used to determine the health of a local water source. Students practice observation skills on a guided hike and test soil samples and vegetation at different locations.

12.5.3 Evaluation

Each nature awareness activity should include some form of debriefing to evaluate the immediate impact of the activity. Even simple activities, such as the sensory exploration of a tree, can evoke a rich response from participants (Chapter 5). Debriefing is a method of asking questions to allow reflection and processing after the activity. When you first ask participants about an activity, start by asking questions that allow them to talk about what happened during the experience. "What did you just experience? What did you observe?" Then move to questions that allow the group to reflect on "so what does that mean?" Finally, prompt discussion from the group about applications of the activity to their lives. You can use debriefing as a quick check-in for evaluation or a more lengthy discussion. This method of debriefing follows the experiential learning cycle (Chapter 2).

Another form of evaluation suited to nature awareness activities is the use of pre- and post-drawings (Chapter 7). Drawings are simple to administer and easy to use with young participants. Drawings also provide direct feedback to students on their learning (Hein and Price 1994). Before your activities, ask participants to draw the setting in which you will be conducting the lessons, such as the schoolyard or national park, or the concept you are exploring, such as ecosystems. Then have them draw the same setting or concept at the conclusion of your curriculum or workshop. Changes in the details and elements included in the drawings can reveal specific changes in the level of awareness of students before and after the nature awareness activities.

Evaluation of informal courses usually includes an assessment form for participants to comment on whether the objectives were met, course content, and instructors (Box 12.4). For long courses, an evaluation form can be passed out halfway through the course to collect immediate feedback about the program. Results help guide any needed modifications during the remainder of the program. Peer review

---

Box 12.4 Examples of open- and close-ended questions included in a course evaluation form for an on-site nature class. Responses help assess if the class objectives are achieved.

---

1. On a scale from 1 to 5 (1 = very unsatisfied to 5 = very satisfied), please circle the number that corresponds to your satisfaction with the class:

A. Coverage of subject matter	1	2	3	4	5
B. Learning environment	1	2	3	4	5
C. Facilities for class	1	2	3	4	5
D. Class organization	1	2	3	4	5
E. Personal learning experience	1	2	3	4	5
F. Knowledge of instructor	1	2	3	4	5
G. Presentation by instructor	1	2	3	4	5
H. Group leadership	1	2	3	4	5
I. Take-home materials	1	2	3	4	5

2. What were the main strengths of the class?

3. What were the main weaknesses of the class?

4. How can we improve the class?

of a course by colleagues in the field, records of return attendees, and longer-term course popularity also provide information about the success of a class.

The Tennessee Aquarium collected feedback from students and their parents to assess their program. Parents felt their children gained technical skills in using laboratory and field equipment. They also felt the students had developed leadership skills as they worked together with other students on the activities, a positive, though unplanned, benefit of the program.

12.6 Facility design

Development of site plans for visitor activity centers is a key element to enhance the quality of experience and learning opportunities for the visitor while preserving the ecological and cultural integrity of the site. Although the design of this infrastructure requires the expertise of architects, engineers, recreational planners, educators, biologists, and audience members, the basic principles related to the land, buildings, and people are useful when considering almost any kind of structures on your site. Many parks and outdoor education facilities have visitor centers. From small community nature centers to large museums, visitor centers can play an essential role in connecting people with special places and/or experiences. Visitor centers make visitors feel welcome, provide for their basic comforts and needs, and orient, inform, and inspire them. At some national parks, visitors spend as much time in the visitor center as they do in the rest of the park!

12.6.1 Planning

The design of a visitor center includes attention to parking and walkways, basic visitor information needs, and educational media and programs. First you must answer universal planning questions, such as: Who is the audience and what are their needs and interests? What are the organizational mission, educational objectives, and outreach goals for the visitor center? What is significant about the site and what stories will convey its attributes and uniqueness? What are the staffing, budget, and other resource needs and constraints? The budget includes not only capital outlays for land, buildings, and equipment; but also annual operating costs, which include personnel services, utilities, supplies, and items such as insurance. A design team including staff with specific expertise as well as professionals and community members can provide a full range of ideas for consideration. Workshops bringing these various groups together during the planning stage help ensure implementation is as effective as possible.

The style of the entrance and building convey meaning to visitors (Bitgood 2002). Basic design elements to consider include the following (Gross and Zimmerman 2002):

Parking and entrance

- Design the road to the parking lot to follow the natural contours of the site and maintain the native vegetation.
- Unify style of entrance signs with other signs on site that reflect the center's themes.
- Place the parking lot to avoid detracting from the building.
- Provide a drop-off loop for buses and people with physical disabilities; a separate entrance for deliveries and an emergency drive may be needed for large centers.
- Plant shade trees and native landscaping in the parking lot.
- Ensure the visitor center is visible or its direction clearly demarcated.
- Construct buildings and associated visitor services so that they do not degrade the natural site.

Building design

- Design the entrance and architecture to fit into the environment and culture of the site (Figure 12.8).
- Research and select sustainable building materials to save money and resources over time, or to serve as a demonstration of locally appropriate materials.
- Use low maintenance materials and structures.
- Plan for additions and easy updates or repairs.
- Focus on function to ensure maximum use of the building.
- Invest in water saving and energy saving technologies.
- Site the building to optimize sunlight for heating and light, and shade for cooling.
- Consider restoring or renovating existing structures.

Fig. 12.8 The entrance to the Rome Biopark in Italy uses classical sculptures of animals to reflect the history of the area (photo by S. Jacobson).

Visitors' basic needs

- Provide after-hours information, maps to orient visitors, and an emergency phone or numbers on an information kiosk or introductory signs.
- Clearly mark trailheads.
- Provide basic comforts visitors expect, such as convenient toilets, and bench seating for tours.
- Design the lobby so that it is open and inviting with an information desk and other important destinations easily visible.
- Make signs concise and large, using international symbols where appropriate.
- Provide personal services at the information desk to complement brochures, maps, and interactive computers to meet the requirements of most visitors.

Education and outreach activities and programs

- Design the exhibit area to attract visitors to enter and exit at trailheads or interesting features.
- Clearly advertise activities, such as guided walks, family games, special lectures, and other interpretive activities.
- Design areas between panels to provide adequate spacing for busy times and to allow both rapid pacing and in-depth viewing by visitors with different interests.
- Design the auditorium or multi-purpose room to serve the educational objectives and audience needs.

- Place sales areas for souvenirs and educational materials to avoid impeding the traffic flow of information-seeking visitors.
- Include transitions between the visitor building and outdoors using observation windows, bird-watching stations, overlooks, and porches.
- Lead visitors from the building with exciting, inviting trailheads or outdoor exhibits.
- Ensure traffic flow of visitors within exhibits follows cultural norms for the region, such as the inclination of visitors to turn in a specific direction, move along a straight-line path, and exit out the first open door.
- Follow the guidelines for effective exhibits discussed earlier in the chapter.

12.6.2 Implementation

Once the planning process is complete, constructing a visitor facility can take from months to years, depending on the complexity. To create the Zion National Park Visitor Center in Utah, the National Park Service worked with the Department of Energy's Renewable Energy Laboratory to design a sustainable building that incorporated the area's natural features and energy-efficient building concepts into an attractive design (US National Park Service 2005). This has saved energy and operating expenses while protecting the environment. The 2.5 million annual visitors to Zion National Park are awed by the park's natural sandstone canyons, mesas, and rock sculptures. Since the time of the ancestral Pueblo Indians, the narrow canyon has provided shade and cool breezes in the summer and warm rock surfaces in the winter. The visitor center capitalizes on these features and serves as a model of how to protect precious resources through energy-saving activities and technologies.

The implementation of the new visitor center involved redesigning the transportation and parking experience as well as creating a new building and environs. Because Zion exists in a narrow canyon, automobile traffic causes air and noise pollution as well as congestion that are detrimental to the park's resources and visitor experience. With the new center, the park invested in clean-running propane buses to shuttle visitors to nine stops within the park and six stops in the nearby town. Visitors are asked to leave their vehicles at parking facilities outside the park.

The new visitor center incorporated low-energy design and renewable energy technologies, based on plans made by a multidisciplinary design team of engineers, architects, energy experts, and park staff. Some of the features of the building (US National Park Service 2005) include:

- Lighting. The primary source of light in the center is daylight. The building's energy management computer adjusts electric light as needed. Compact and T-8 fluorescent lamps are used because of their high efficiency.
- Windows. Windows placed high in the building are part of the lighting system as well as a part of the heating and cooling systems. Computer simulations helped size the windows to collect the right amount of light. The sun enters in the winter, helping to keep the space heated (passive solar heating), and roof overhangs shade the glass from the high summer sun. A coating on

the glass reduces heat loss in the winter. The high windows help cool the building by allowing hot air to escape. Low windows near the doors allow cool air in. The building's energy management computer controls the opening of windows and provides continuous natural ventilation.

- Location. The building was located to block the west windows from the summer sun. A tree canopy minimizes heat gain on summer afternoons. Windows on the west side of the building have glass that diverts the sun's heat.

- Insulation. The building is well insulated, designed to use 70% less energy than a typical building without costing more to build. The roof was made from structural-insulated panels of foam and strand board. Foam insulation in the wall cavities and insulated windows help keep the heat out in summer and in the building in winter.

- Cooling and Heating. When natural ventilation is inadequate, cooltowers help lower the temperature. Water sprayed on pads at the top of the tower evaporates, cooling the air, which is directed down and into the building or patio. A Trombe wall provides most of the heat for the building by trapping solar heat between a pane of glass and a black coating on a masonry wall. The heat is stored and released into the building to provide radiant comfort to visitors.

- Energy. Photovoltaic panels on the south roof provide the majority of the electricity needed by the building. The efficient building design minimized lighting needs and eliminated air-conditioning, two major electric loads for the area. Excess power produced by the solar panels is stored in batteries as well as sold back to the power company for use elsewhere. A computer ensures that all the energy-efficient features work together, collects weather data, and makes energy decisions about the building. Additionally, it controls the cooltowers, radiant ceiling panels, lighting, and windows.

- Landscaping. Landscaping helped create an extension of the visitor center with shade structures and existing tress. These outdoor rooms display permanent exhibits and allowed for a smaller building design. Irrigation ditches provide most of the water needed for landscaping, saving pumping energy, and water treatment (US National Park Service 2005).

Concomitant with planning a model energy-efficient building, the building also followed the criteria listed in the exhibit planning section to create an appealing, inspirational, and informative experience for the visitor. The inside space of visitor centers must be designed to ensure that the placement of doors, windows, columns, and other permanent fixtures will complement exhibits, sales, information, basic needs, and educational activities for the visitor numbers projected.

12.6.3 Evaluation

Evaluation of a facility design includes an assessment of the structural design of the facility as well as the visitor experience. In the case of the Zion National Park Visitor Center, the changes in transportation and building design resulted in significant improvements. The buses dramatically reduced automobile traffic in the park, protecting the park and providing a pleasant experience for visitors. The building was designed to use 70% less energy than a typical building without costing more to build (US National Park Service 2005). Performance data collected after the completion of the building helped document the savings. Low energy demands have reduced operating costs and provided a model for other agencies to emulate. Initial reactions from visitors were overwhelmingly positive. The structural elements, such as lighting for clear visibility and a floor plan that provided easy circulation by visitors were observed by staff. More formal types of feedback discussed in the section on exhibits will help with evaluation of the themes, content, and graphics filling the building.

12.7 Summary

On-site education and outreach activities can enhance first-hand experiences with natural areas by orienting, informing, and stimulating visitors. The development of on-site activities considers the visitor experience, resources of the site, and education and outreach objectives of the organization. An initial planning process at a site paves the way for developing trails, exhibits, demonstrations, nature awareness and study activities, and visitor centers to achieve the goals of an organization.

The expertise of educators, artists, architects, engineers, recreational planners, maintenance staff, and target audience members are often needed to plan and implement on-site activities. The activities and techniques discussed in this chapter help visitors relate the site to their personal experiences and prior knowledge. This is an important feat to accomplish if learning is to be long-term and meaningful. From private forest owners at a demonstration forest to family groups at a state-of-the-art visitor center, on-site activities help audiences to understand their surroundings, explore their personal feelings, and acquire new skills.

Further Reading

Sharing the Joy of Nature, by Joseph Cornell (1989). Dawn Publications, Nevada City, CA.
Communication Skills for Conservation Professionals, by Susan Jacobson (1999). Island Press, Washington, DC.
Interpretation of Cultural and Natural Resources, by Douglas Knudson, Ted Cable and Larry Beck (1995). Venture Press, State College, PA.
Interpretive Master Planning, by John Veverka (1994). Falcon Press Publishing Co., Helena, MT.
Interpretive Centers: The History, Design, and Development of Nature and Visitor Centers, by Michael Gross and Richard Zimmerman (2002). UW-SP Foundation Press, Inc. Stevens Point, WI.
Environmental Interpretation: A Practical Guide for People with Big Ideas and Small Budgets, by Sam Ham (1992). North American Press, Golden, CO.

References

3-D Life Adventures. 2003. *Preliminary Results of Analyses of Pretest-posttest Data for Appalachia to Atlanta and Carolina Immersion*. Unpublished document. Analysis by Martha Knight-Oakley, Warren Wilson College, NC.

Abbey, E. (1968). *Desert Solitaire: Season in the Wilderness*. Ballantine, New York, NY.

Abiator. (2005). *Abiator's Online Learning Styles Inventory*. Last updated January 31, 2005. Retrieved June 19, 2005, from http://www.berghuis.co.nz/abiator/lsi/lsitest1.html.

Ady, J.C. (1994). Teach about geese. In Bardwell, L.V., Monroe, M.C., and Tudor, M.T., eds. *Environmental Problem Solving: Theory, Practice, and Possibilities in Environmental Education*, pp. 123–124. NAAEE, Troy, OH.

AEEPE. (2005). Applied Environmental Education Program Evaluation Online Course Evaluation Responses: UWSP Students in Spring 2005. Unpublished evaluation report from University of Wisconsin, Stevens Point, WI.

afrolNews (2003). Community-based research nets results in Benin. Retrieved January 9, 2005, from http://www.afrol.com/articles/10547

Aipanjiguly, S., Jacobson, S.K., and Flamm, R. (2003). Conserving manatees: Knowledge, attitudes, and intentions of boaters in Tampa Bay, Florida. *Conservation Biology*, 17(4), 1098–1105.

Ajzen, I. (1985). From intentions to actions: A theory of planned behavior. In Kuhl, J. and Beckman, J., eds. *Action-control: From Cognition to Behavior*, pp. 11–39. Springer, Heidelberg, Germany.

Ajzen, I. (2002). Theory of Planned Behavior Web site. Retrieved August 4, 2004, from http://www-unix.oit.umass.edu/~aizen/tpb.diag.html.

Akers, C.L., Segrest, D.H., Kistler, M.J., Smith, J.H., Davis, C.S., and Baker, M. (2005). *Evaluating the Effectiveness of Texas Parks and Wildlife Hueco Tanks State Historic Site Orientation/Conservation Video: A Media Systems Dependency Theory Perspective*. Presentation to the Southern Association of Agricultural Scientists, Little Rock, AR, February 5–9, 2005. Retrieved May 3, 2005, from http://agnews.tamu.edu/saas/2005/akers.pdf

Alaska Native Knowledge Network. (1998). *Alaska Standards for Culturally Responsive Schools*. University of Alaska Fairbanks, Fairbanks, AK.

Allen, I. (2000). The Gambia Environmental Awards Scheme—Creating environmental awareness through participation. In Day, B.A. and Monroe, M.C., eds. *Environmental Education and Communication for a Sustainable World: Handbook for International Practitioners*, pp. 105–119. Academy for Educational Development, Washington, DC.

Ambach, J., Perrone, C., and Repenning, A. (1995). Remote exploratoriums: Combining network media and design environments. *Computers and Ethics*, 24(3), 163–176.

Ambrose, T. and Paine, C. (1993). *Museum Basics*. International Council of Museums (ICOM) supported by the Cultural Heritage Division of UNESCO and Paine, New York and London.

American Philosophical Association. (1990). *Critical Thinking: A Statement of Expert Consensus for Purposes of Educational Assessment and Instruction*. (ERIC Document No. ED 315 423)

Andrews, E., Stevens, M., and Wise, G. (2002). A model of community-based environmental education. In Dietz, T. and Stern, P., eds. *New Tools for Environmental Protection: Education, Information, and Voluntary Measures*, pp. 161–182. National Academy Press, Washington, DC.

Anthony, C. (2005). Just, green, and beautiful cities. *Yes! A Journal of Positive Futures*, 34, 12–15.

APPT. (2004). Advanced Practical Thinking Training, Inc. Edward de Bono and Lateral Thinking. Retrieved September 7, 2004, from http://www.aptt.com/ltmethods.htm.

Archie, M.L. (2003). *Advancing Education through Environmental Literacy*. Association for Supervision and Curriculum Development, Alexandria, VA.

Arcury, T., Quandt, S.A., and McCauley, L. (2000). Farmworkers and pesticides: Community-based research. *Environmental Health Perspectives*, 100(8), 787–792.

Armstrong, D.M. (1992). *Managing by Storying Around: A New Method of Leadership*. Bantam Doubleday Dell Publishing Group. Inc., New York, NY.

Armstrong, T. (1994). *Multiple Intelligences in the Classroom*. Association for Supervision and Curriculum Development, Alexandria, VA.

Aronson, E. and O'Leary, M. (1982–83). The relative effectiveness of models and prompts on energy conservation: A field experiment in a shower room. *Journal of Environmental Systems*, 12(3), 219–224.

ASCD (Association for Supervision and Curriculum Development) (Aug, 2001). Moving into the educational mainstream. *ASCD Infobrief*, 26, 1–10.

Ashbaugh, B. (1963). *Planning a Nature Center*. Information-Education Bulletin No. 2, National Audubon Society, NY.

Aschwanden, C. (2000). Six Steps that can Change Your Life. WebMDHealth. WebMD Feature Archive. Retrieved May 22, 2000, from http://my.webmd.com/content/article/12/1676_50967.htm.

Astleitner, H. (2002). Teaching critical thinking online. *Journal of Instructional Psychology*, 29 (2), 53–76.

Athman, J. and Monroe, M. (2002). Enhancing Natural Resource Programs with Field Trips. UF/IFAS EDIS Publication FR135, University of Florida. Retrieved April 6, 2005, from http://edis.ifas.ufl.edu/FR135.

Athman, J. and Monroe, M. (2004). The effects of environment-based education on students' achievement motivation. *Journal of Interpretation Research*, 9(1), 9–25.

Australian Department of Environment and Heritage (2005). *Coastal Creatures in Crisis*. Retrieved July 3, 2005, from http://www.deh.gov.au/education/publications/coastcare.html

Bachman, W. and Katzev, R. (1982). The effects of non-contingent free bus tickets and personal commitment on urban bus ridership. *Transportation Research*, 16A(2), 103–108.

Bandura, A. (1997). *Self-Efficacy: The Exercise of Control*. W. H. Freeman and Company, New York, NY.

Bang, M. (1997). *Common Ground: The Water, Earth and Air We Share*. Blue Sky Press, New York, NY.

Baral, H., Petheram, R.J., and Liffmann, P. (2004). A community-university-GIS partnership for weed mapping. *Agricultural Research and Extension Network Newsletter*, 50, 9. Retrieved January 15, 2005, from http://www.odi.org.uk/agren/papers/newsletter_50.pdf

Bardon, R. and Moore, S. (2005). Video-teleconferencing as an effective and economical tool for technology transfer. Presentation at the International conference on Transfer of Forest Science Knowledge and Technology, May 10–13, 2005, Troutdale, OR.

Barlow, Z. (2000). Ecoliteracy: Mapping the Terrain. *Ecoliteracy: Mapping the Terrain*, pp. 13–18. Learning in the Real World, Berkeley, CA.

Barnet, S. and Stubbs, M. (1977). *Practical Guide to Writing*. Little Brown & Co.

Bartels, W., Breeze, M., and Peterson, N. (2003) Evaluating Web sites through the use of focus group interviews. Fact Sheet CNR 14, Center for Natural Resources, University of Florida, Gainesville, FL. Retrieved June 1, 2005, from http://edis.ifas.ufl.edu/WC049

Batchelor, S.J., McKemey, K., and Sakyi-Dawson, O. (1999). *Barriers to the Adoptions of Efficient Energy Strategies in Northern Ghana*. Project Technical Report. Updated March 27, 2001. Retrieved January 14, 2004, from http://www.gamos.demon.co.uk/ just%20gamos%20homepage/ghana.htm.

Bauer, R. (2005). *How we Learn from the Viewpoint of Cognitive Neuroscience*. Seminar presented at University of Florida, January 31, 2005 through the University Center for Excellence in Teaching.

Beamish, R. (1995). *Getting the Word Out in the Fight to Save the Earth*. The Johns Hopkins University Press, Baltimore, MD.

Beaudin, B.P. and Quick, D. (1996). Instructional video evaluation instrument. *Journal of extension*, 34(3). Retrieved May 3, 2005, from http://www.joe.org/joe/1996june/a1.html.

Bélanger, P. (2003). Learning environments and environmental education. In Hill, L.H. and Clover, D.E., eds. *Environmental Adult Education: Ecological Learning, Theory, and Practice for Socioenvironmental Change*. New Directions for Adult and Continuing Education, 99, 79–88.

Beldon and Russonello. (1995). *Communicating Biodiversity: Focus Group Research Findings Conducted for the Consultative Group on Biological Diversity*. Beldon and Russonello Research and Communications Report, Washington, DC.

Bennett, D.B. (1988–89). Four steps to evaluating environmental education learning experiences. *Journal of Environmental Education*, 20(2), 14–21.

Berensohn, P. (2002). Environment and the arts: Interview on Earthbeat, June 1, 2002, ABC Radio National, Australia. Retrieved March 4, 2005, from www.abc.net.au/m/science/earth/stories/s570032.htm.

Berry, W. (2002). *The Art of the Commonplace: The Agrarian Essays of Wendell Berry*. Shoemaker & Hoard, Inc. US.

Berry, W. (1985). *Collected Poems*. North Point Press, San Francisco, CA.

Biodiversity Project. (2002). *Americans and Biodiversity: New Perspectives in 2002*. Belden, Russonello, and Stewart. Washington, DC. Retrieved June 2005, from www.biodiversityproject.org/o2toplines.pdf.

Bitgood, S. (2002). Environmental psychology in museums, zoos, and other exhibition centers. In Bechtel, R. and Churchman, A. eds. *Handbook of Environmental Psychology*, pp. 461–480. John Wiley and Sons, New York, NY.

Bitgood, S. and Patterson, D. (1987). Principles of exhibit design. *Visitor Behavior* 2(1), 4.

Blanchard, D. (2004). *Academic Service-Learning: The Reflection Concept*. Learning To Give of the Council of Michigan Foundations and Center on Philanthropy at Indiana University. Retrieved 9 January 2005, from http://www.learningtogive.org/papers/index.asp?bpid=1

Blanchard, K.A. and Monroe, M.C. (1990). Effective educational strategies for reducing population declines in seabirds. *Transactions of the North American Wildlife and Natural Resources Conference* 55, 108–117.

Blasch, K. and Polley, K.P. (2003). Advancing interpretation with real-time data. *Legacy*, 14(3), 18–21.

Blatner, A. (2002). *Role-playing in Education*. Retrieved February 25, 2005, from http://www.blatner.com/adam/pdntbk/rlplayedu.htm.

Block, L. (2003). *Helpful Hints for Clean Water for North Carolina Festival Volunteers*. Unpublished document. Clean Water for North Carolina, Asheville, NC.

Boaler, J. (1999). Commentary: Mathematics for the moment, or the millennium? *Education Week*, March 31, 1999. Retrieved July 3, 2005, from http://www.Edweek.org

Bostdorff, D.M. and Woods, J.L. (2003). Lessons from a failed PDR (purchase of development rights) campaign in Wayne County, Ohio. *Applied Environmental Education and Communication*, 2(3), 169–175.

Braus, J.A. and Monroe, M.C. (1994). *Designing Effective Workshops*. EE Toolbox, Workshop Resource Manual. National Consortium for Environmental Education, Ann Arbor, MI.

Braus, J.A. and Monroe, M. (1994). *EE Toolbox: Workshop Resource Manual: Designing Effective Workshops*. Kendall Hunt Publishing, Dubuque, IA.

Braus, J.A. and Wood, D. (1993). *Environmental Education in the Schools: Creating a Program that Works*. Peace Corps Information Collection and Exchange, Washington, DC.

Brehm, S. and Brehm, J.W. (1981). *Psychological Reactance: A Theory of Freedom and Control*. Academic Press, New York, NY.

Brew, A. (2003). *Writing Activities: A Primer for Outdoor Educators*. ERIC Digest, ED475390 2003-06-00 ERIC Clearinghouse on Rural Education and Small Schools, Charleston, WV.

Brigham, N. (1991). *How to Do Newsletters and Newspapers*. PEP Publishers, Detroit, MI.

Brody, M. (2002). Park visitors' understandings, values and beliefs related to their experience at Midway geyser Basin, Yellowstone National Park, USA. *International Journal of Science Education*, 24(11), 1119–1141.

Brooks, J.G. and Brooks, M.G. (1993). *In Search of Understanding: The Case for Constructivist Classrooms*. Association for Supervision and Curriculum Development, Alexandria, VA.

Broussard, S., Jones, S., Nielsen, L., and Flanagan, C. (2001). Forest stewardship education: Fostering positive attitudes in urban youth. *Journal of Forestry*, January, 37–42.

Brown, T. (1983a). *Tom Brown's Field Guide to Wilderness Survival*. The Berkley Publishing Group, New York, NY.

Brown, T. (1983b). *Tom Brown's Field Guide to Nature Observation and Tracking*. The Berkley Publishing Group, New York, NY.

Brown, T. (1984). *Tom Brown's Field Guide to City and Suburban Survival*. The Berkley Publishing Group, New York, NY.

Brown, T. (1985). *Tom Brown's Field Guide to Wild Edible and Medicinal Plants*. The Berkley Publishing Group, New York, NY.

Cable, T. and Ernst, T. (2003). Interpreting rightly in a left-brain world. *Legacy*, 14(5), 27–29.

Caduto, M.J. and Bruchac, J. (1989). *Keepers of the Earth: Native American Stories and Environmental Activities for Children*. Fulcrum, Inc., Golden, CO.

Caduto, M.J. and Bruchac, J. (1994). *Keepers of the Night: Native American Stories and Nocturnal Activities for Children*. Fulcrum, Inc., Golden, CO.

Caduto, M.J. and Bruchac, J. (1997). *Keepers of the Animals: Native American Stories and Wildlife Activities for Children*. Fulcrum, Inc., Golden, CO.

Cairn, R. (2003). *Partner Power and Service Learning: Manual for Community-based Organizations to Work with Schools*. ServeMinnesota! and Minnesota Department of Education, Minneapolis, MN.

Cairn, R. and Coble, T.L. (1993). *Learning by Giving: K-8 Service-Learning Curriculum Guide*. National Youth Leadership Council, St. Paul, MN, www.nylc.org.

Campbell, M.C. and Floyd, D.W. (1996). Thinking critically about environmental Mediation. *Journal of Planning Literature*, 10(3), 235–247.

CampSilos. (2002). *Why take Field Trips? Silos and Smokestacks National Heritage Area*. Retrieved February 25, 2005, from http://www.campsilos.org/excursions/hc/fieldtrip.htm

Campus Compact. (2005). *Syllabi by Discipline: Neighborhoods and Watersheds*. Retrieved June 8 2005, from http://www.compact.org

Capra, F. (2000). Ecoliteracy: A systems approach to education. *Ecoliteracy: Mapping the Terrain*, pp. 27–35. Learning in the Real World, Berkeley, CA.

Carey, H.A. (1996). Posters, *The Communicator's Handbook: Tools, Techniques, and Technology*, Third Edition, pp. 81–85. Maupin House, Gainesville, FL.

Carmichael, C. (2002). *Responsible Environmental Purchasing for Faith Communities*. Center for the New American Dream, Takoma Park, MD.

Case Studies in Science (2003). *Case Studies in Ecology/environment*. National Center for Case Study Teaching in Science. Retrieved February 25, 2005, from http://ublib.buffalo.edu/libraries/projects/cases/ubcase.htm#ecology

Cembalest, R. (1991). The Ecological Arts Explosion. *Art News*, 90, 96–105.

Center for Talented Youth (2002). *Lab Experiments: New Twists on Doing Experiments/labs*. Johns Hopkins University. Retrieved February 25, 2005, from http://www.jhu.edu/gifted/teaching/strategies/handson/labexperiments.htm

Chawla, L. (1998). Significant life experiences revisited: A review of research on sources of environmental sensitivity. *Environmental Education Research*, 4(4), 369–382.

Chawla, L. (1999). Life paths into effective environmental action. *Journal of Environmental Education*, 31(1), 15–26.

Citizen Science Toolbox (2004a). *Tool: Workshops*. Retrieved January 30, 2005, from http://www.coastal.crc.org.au/toolbox/details.asp?id=65.

Citizen Science Toolbox (2004b). *Tool: Displays and Exhibits*. Retrieved January 30, 2005, from http://www.coastal.crc.org.au/toolbox/alpha-list.asp.

Citizen Science Toolbox. (2004c). *Tool: Field Trips*. Retrieved February 25, 2005, from http://www.coastal.crc.org.au/toolbox/details.asp?id=14

Clean Marina Program. (2004). Florida Department of Environmental Protection, State of Florida. Last updated December 15, 2004. Retrieved December 23, 2004, from http://www.dep.state.fl.us/law/Grants/CMP/default.htm.

Clover, D.E. (2003). Environmental adult education: Critique and creativity in a globalizing world. In Hill, L.H. and Clover, D.E., eds. *Environmental Adult Education: Ecological Learning, Theory, and Practice for Socioenvironmental change*. New Directions for Adult and Continuing Education, 99, 5–15.

Coalition on the Environment and Jewish Life (n.d.). *To Till and to Tend: A Guide to Jewish Environmental Study and Action*. CEJL, New York, NY.

Coastal Zone Australia. (2004). *What is Citizen Science?* Retrieved January 12, 2005, from http://www.coastal.crc.org.au/citizen_science/

Cornell, J. (1979). *Sharing Nature with Children*. Dawn Publications, Nevada City, CA.

Cornell, J. (1989a). *Sharing Nature with Children II*. Dawn Publications, Nevada City, CA.

Cornell, J. (1989b). *Sharing the Joy of Nature*. Dawn Publications, Nevada City, CA.

Corredor, J. (2004). General and domain-specific influence of prior knowledge on setting of goals and content use of museum Web sites. *Computers and Education*. Retrieved June 30, 2005, from www.sciencedirect.com.

Coyle, J.R. and Thorson, E. (2001). The effects of progressive levels of interactivity and vividness in Web marketing sites. *Journal of Advertising*, 30(3), 65–76.

Creighton, J., Simon-Brown, V., and Sulzmann, J. (2004). *Saving Eden Creek: A Play about People and Forests*. Retrieved January 4, 2005, from http://www.eeesc.orst.agcomwebfile/edmat/html.

Crimmel, H. (2003). *Teaching in the Field: Working with Students in the Outdoor Classroom*. University of Utah Press, Salt Lake City, UT.

Crowfoot, J.E. and Wondolleck, J.M. (1990). *Environmental Disputes: Community Involvement in Conflict Resolution*. Island Press, Washington, DC.

Culen, G.R., Hungerford, H., and Volk, T. (2000). *Coastal Marine Environmental Issues: An Extended Case study for the Investigation of Issues of the Gulf Coast and Florida Peninsula*. Stipes Publishing, Champaign, IL.

Curtis, D. (2003a). Initial impressions on the role of the performing and visual arts in influencing environmental behavior. TASA Conference, University of New England. 4–6 December 2003, Australia.

Curtis, D. (2003b). The arts and restoration: a fertile partnership? *Ecological Management and Restoration*, 4(3), 163–169.

Czerniak, C.M. (2004). Wetlands: An interdisciplinary exploration. *Science Activities*, 41(2), 3–11.

Dallman-Jones, A. (1994). *The Expert Educator: A Reference Manual of Teaching Strategies for Quality Education*. Three Blue Herons Publishing, Inc., Fond du Lac, WI.

Davis, R. (2005). 16: Is it too young to drive a car? *USA Today*, March 2, 2005, page 1B.

Davis-Case, D. (1990). *The Community's Toolbox: The Idea, Methods, and Tools for Participatory Assessment, Monitoring, and Evaluation in Community Forestry*. Food and Agriculture Organization of the United Nations, Rome, Italy.

Day, B. and Monroe, M., eds. (2000). *Environmental Education and Communication for a Sustainable World: Handbook for International Practitioners.* Academy for Educational Development, Washington, DC.

Day, B.A. (2001). Environmental communication: Toward a model for policy adoption. In Lasker, G. and Day, B., eds. *Advances in Education*, vol. IV. Conference Proceedings of the International Institute for Advanced Studies in Systems Research and Cybernetics in Baden-Baden, Germany 2001. International Institute for Advanced Studies in Systems Research and Cybernetics, Windsor, Ontario, Canada.

Day, B.A. and Monroe, M.C., eds. (2000). *Environmental Education and Communication for a Sustainable World: Handbook for International Practitioners.* Academy for Educational Development, Washington, DC.

De Young, R. (1988–89). Exploring the difference between recyclers and non-recyclers: The role of intrinsic motivation. *Journal of Environmental Systems*, 15, 281–292.

De Young, R. (1993). Changing behavior and making it stick: The conceptualization and management of conservation behavior. *Environment and Behavior*, 25, 485–505.

De Young, R. (2000). Expanding and evaluating motives for environmentally responsible behavior. *Journal of Social Issues*, 56(3), 509–526.

Dearing, J. (2005). Accelerating the diffusion of effective techniques. Presentation at the International Conference on Transfer of Forest Science Knowledge and Technology, May 10–13, 2005, Troutdale, OR.

Decker, D.J., Brown, T.L., and Siemer, W.F. (2001). *Human Dimensions of Wildlife Management in North America.* The Wildlife Society, Bethesda, MD.

Department of Education, Queensland. (1993). *P-12 Environmental Education Curriculum Guide.* Publishing Services for Studies Directorate, Brisbane, Queensland.

DesMarais, J., Yang, T., and Farzanehkia, F. (2000). Service-learning leadership development for youths. *Phi Delta Kappan*, 81(9), 678–680.

Devney, D.C. (2001). *Organizing Special Events and Conferences: A Practical Guide for Busy Volunteers and Staff.* Second Edition. Pineapple Press, Inc., Sarasota, FL.

DeVries, H., Knapp, D.S., and Taricani, E. (1996). Navigating the Internet. In *The Communicator's Handbook: Tools, Techniques, and Technology*, Third Edition, pp. 220–229. Maupin House and Agriculture Communicators in Education, Gainesville, FL.

Dierking, L.D., Burtnyk, K., Büchner, K.S., and Falk, J.H. (2001–2002). *Visitor Learning in Zoos and Aquariums: A Literature Review.* Institute for Learning Innovation, Annapolis, MD.

Dillard, A. (1988). *Pilgrim at Tinker Creek.* Harper, New York, NY.

Dillman, D.A. (2000). *Mail and Internet Surveys.* John Wiley & Sons, New York, NY.

Disinger, J. and Monroe, M.C. (1994). *Defining Environmental Education.* EE Toolbox, Workshop Resource Manual. National Consortium for Environmental Education, Ann Arbor, MI.

Dobbins, R. and Pitman, B. (2001). *GreenWorks! Connecting Community Action and Service Learning Guide.* American Forest Foundation and Project Learning Tree, Washington, D.C. Retrieved January 9, 2005, from http://www.plt.org/greenworks/greenworksguide.pdf

Dodge, B. (2005). *The WebQuest Page.* Last updated March 25, 2005. Retrieved May 3, 2005, from www.webquest.org.

Driscoll, A., Holland, B., Gelmon, S., and Kerrigan, S. (2003). An assessment model for service-learning: comprehensive case studies of impact on faculty, students, community, and institution. In Campus Compact, eds. pp. 225–230. *Introduction to Service-Learning Toolkit: Readings and Resources for Faculty.* Second Edition. Campus Compact, Providence, RI.

Driver, B.L., Tinsley, H.E.A., and Manfredo, M.J. (1991). The paragraphs about leisure and recreation experience preference scales: Results from two inventories designed to assess the breadth of the perceived psychological benefits of leisure. In Driver, B.L., Brown, P.J., and Peterson, G.L., eds. *Benefits of Leisure*, pp. 261–286. Venture Publishing, State College, PA.

Dunwoody, S. (2003). Explaining Popular Science, unpublished presentation at Florida Museum of Natural History, Gainesville, FL. November 10, 2003.

Eagan, D.J. and Keniry, J. (1998). *Green Investment, Green Return: How Practical Conservation Projects Save Millions on America's Campuses*. National Wildlife Federation, Washington, DC.

Easton, J. and Monroe, M.C. (2001). *Enhancing Natural Resource Programs with Service Learning*. University of Florida Cooperative Extension Fact Sheet. School of Forest Resources and Conservation, Florida Cooperative Extension Service. Gainesville, FL.

Easton, J., Monroe, M.C., and Bowers, A. (1996). *Give Forests a Hand: a Youth Program for Environmental Action and Community Service*. School of Forest Resources and Conservation, University of Florida, Gainesville, FL.

EcoRecycle Victoria (2003–2005). EcoRecycle Victoria's Waste Wise Business Program. Last updated June 21, 2005. Retrieved June 22, 2005, from http://www.ecorecycle.vic.gov.au/www/html/503-waste-wise-business-program.asp?intSiteID=1

Ehlert, L. (1987). *Growing Vegetable Soup*. Harcourt, Inc., Orlando, FL.

Ehrlich, G. (1985). *The Solace of Open Spaces*. Penguin Books, New York, NY.

Eichler, M.L., Xavier, P.R., Costa Araújo, R., Castro Forte, R., and Del Pino, J.C. (2005). Carbopolis: A Java technology-based free software for environmental education. *Journal of Computers in Mathematics and Science Technology*, 24(1), 43–72. The simulation, *Carbopolis*, can be found at www.iq.ufrgs.br/aeq/carbop.htm.

Elder, J.L. (2003). *A Field Guide to Environmental Literacy: Making Strategic Investments in Environmental Education*. Environmental Education Coalition, Rock Springs, GA.

Environmental Working Group. (2005). *Body Burden*. Retrieved June 25, 2005, from http://www.ewg.org/reports/bodyburden/es.php.

Enviroschools Foundation. (2004). *Learning for a Sustainable Future: Enviroschools*. Retrieved June 3, 2005, from http://www.enviroschools.org.nz

Ernst, J. and Monroe, M. (2004). The effects of environment-based education on students' critical thinking skills and disposition toward critical thinking. *Environmental Education Research* 10(4), 507–522.

Ernst, J.M. and Heesacker, M. (1993) . Application of the Elaboration Likelihood Model of attitude change to assertion training. *Journal of Counseling Psychology*, 40(1), 37–45.

European Commission (1999). What do Europeans Think about the Environment? Office for the Official Publications of the European Communities. Retrieved May 2005, from europa.eu.int/comm./environment/barometer.

Evangelical Environmental Network (EEN). (2005). *Evangelical Environmental Network and Creation Care Magazine*. Retrieved January 30, 2005, from http://www.creationcare.org

Eyler, J. and Giles, J. (1999). *Where's the Learning in Service-learning?* Jossey-Bass, San Francisco, CA.

Eysenbach, G. and Kohler, C. (2002). How do consumers search for and appraise health information on the World Wide Web? Qualitative study using focus groups, usability tests, and in-depth interviews. *BMJ*, 324, 573–577.

Facione, P. (1998). *What it is and Why it Counts*. The California Academic Press, Millbrae, CA.

Falk, J. and Balling, J. (1980). The school field trip: where you go makes a difference. *Science and Children*, 6–8, 17(6), 6–8.

Fazio, J.R. and Gilbert, D.L. (2000). *Public Relations and Communications for Natural Resource Managers*. Kendall Hunt Publishing, Dubuque, IA.

Feather River Water and Air Quality Project. (2004). McKenzie-Mohr Associates, Community-Based Social Marketing, Retrieved June 16, 2004, from http://www.cbsm.com/discussion/community/members/go.lasso?-database=Cases.fp3&-response=%2fCasesDatabase%

2fdetail.lasso&-layout=Cases&-op=cn&category=Watershed%20Protection&-maxRecords=1&-skipRecords=1&-search

FEE (Foundations for Environmental Education) International Secretariat. (2001). *Eco-schools International Newsletter*. Retrieved 3 June 2005, from http://www.eco-schools.org/new/news/news2001.htm

Feuerstein, M.T. (1986). *Partners in Evaluation: Evaluating Development and Community Programs with Participants*. MacMillan Education Ltd, Hong Kong.

Fien, J., Scott, W., and Tilbury, D. (2002). Exploring principles of good practice: Learning from a meta-analysis of case studies on education within conservation across the WWF network. *Applied Environmental Education and Communication*, 1, 153–162.

Fishbein, M. and Ajzen, I. (1975). *Belief, Attitude, Intention and Behavior: An Introduction to Theory and Research*. Addison-Wesley, Reading, MA.

Fisher, R. and Ury, W. (1991). *Getting to Yes: Negotiating Agreement without Giving in*. Second Edition. Penguin Books, New York, NY.

Fiske, E.B. (2002). Learning in deed: The power of service-learning for American schools. The National Commission on Service Learning, Newton, MA. Retrieved January 9, 2005, from http://learningindeed.org/slcommission/report.html

Flint, C. (ed.) (2002). *Nature Neighborhood: Creating a Place for Wildlife and Learning*. North Carolina State Museum of Natural Sciences, Environmental Media Corporation, Port Royal, SC.

Fogg, B.J. (2002). *Stanford Guidelines for Web Credibility*. A Research Summary from the Stanford Persuasive Technology Lab. Stanford University. Retrieved May 1, 2005, from www.webcredibility.org/guidelines.

Ford Foundation. (2005). *Ford Foundation Report Winter*. Retrieved April 1, 2005, from http://www.ford.found.org.

Frazier, R. (2002). Singapore sensory trail. *Legacy* 13(3), 12–17.

Freire, P. (1982). *Pedagogy of the Oppressed*. Continuum, New York, NY.

Friedrich, M. J. (1999). The arts of healing. *Journal American Medical Association*, 281(19), 1779–1781.

Frost, S. (2003). Winging northward: Bringing Alaskan wildlife into classrooms around the world. *Legacy*, 14(2), 32–35.

Gallup International (2002). Voice of the People. Retrieved May 2005, from www.gallup-international.com.

Gardner, H. (1999). *The Disciplined Mind: What All Students Should Understand*. Simon and Schuster, New York, NY.

Gautier, C. (2003). Mock environmental summit. Retrieved February 25, 2005, from http://serc.carleton.edu/introgeo/roleplaying/examples/envsumit.html

Gelmon, S. (2003). How do we know that our work makes a difference? Assessment strategies for service-learning and civic engagement. In Campus Compact, eds. *Introduction to Service-Learning Toolkit: Readings and Resources for Faculty*. Second Edition, pp. 231–240. Campus Compact, Providence, RI.

Giacalone, V. (2003). How to plan, survive, and even enjoy an overnight field trip with 200 students. *Sciencescope*, 26(4), 22–26

Gill, J.D., Crosby, L.A., and Taylor, J.R. (2001). Ecological concern, attitudes, and social norms in voting behavior. *Public Opinion Quarterly*, 50, 537–554.

Giving Trash the Boot! (2004). German Fest & The Boot Mug, McKenzie-Mohr Associates, Community-Based Social Marketing, Retrieved June 16, 2004, from http://www.cbsm.com/go.lasso?-database=Cases.fp3&-response=%2fCasesDatabase%2fdetail.lasso&-layout=Cases&-op=cn&category=Source%20Reduction&-maxRecords=1&-skipRecords=6&-search

Gleason, J. and Holian, P. (1996). Video productions. In *The communicator's Handbook: Tools, Techniques, and Technology*, Third Edition, pp. 153–193. Maupin House and Agriculture Communicators in Education, Gainesville, FL.

GLEF (2001). Project-based learning research. *Edutopia On-line*. Retrieved January 15, 2005, from http://www.glef.org/php/print.php?id=Art_887&template=printarticle.php

Glendinning, A., Mahapatra, A., and Mitchell, C.P. (2001). Modes of communication and effectiveness of agroforestry extension in Eastern India. *Human Ecology*, 29(3), 283–305.

Gonzales, M.H., Aronson, E., and Costanzo, M.A. (1988). Using social cognition and persuasion to promote energy conservation: A quasi-experiment. *Journal of Applied Social Psychology*, 18(12), 1049–1066.

Grace, C.O. (2004). Medicine and metaphor: poetry as prescriptive. *Duke Magazine*, Nov.–Dec, 42–47.

Grant, P. (2001). Wild art at the world's end. *Artlink*, 21(1), 14–18.

Grant, T. and Littlejohn, G., eds. (2001). *Greening School Grounds: Creating Habitats for Learning*. New Society Publishers, Gabriola Island, BC, Canada.

Greenwald, A.G., Carnot, C.G., Beach, R., and Young, B. (1987). Increasing voting behavior by asking people if they expect to vote. *Journal of Applied Psychology*, 72(2), 315–318.

Grieser, M. (2000). Participation. In Day, B.A. and Monroe, M.C. eds. *Environmental Education and Communication for a Sustainable World*, pp. 17–22. Academy for Educational Development, Washington, DC.

Grieser, M. and Rawlins, R. (2000). Gender matters. In Day, B.A. and Monroe, M.C., eds. *Environmental Education and Communication for a Sustainable World*, pp. 23–31. Academy for Educational Development, Washington, DC.

Griggs, B. (2004). From Arctic refuge to Utah deserts, a focus on wild lands. *Salt Lake Tribune*. D4.

Gross, M. and Zimmerman, R. (2002). *Interpretive Centers: The History, Design, and Development of Nature and Visitor Centers*. UW-SP Foundation Press, Inc. Stevens Point, WI.

Gurevitz, R. (2000). Affective approaches to environmental education: going beyond the imagined worlds of childhood? *Ethics, Place and Environment*, 3, 253–268.

Guynup, S. (2004). Toxins accumulate in Arctic peoples, animals, study says. *National Geographic News*, August 27, 2004. Retrieved February 26, 2005, from http://news.nationalgeographic.com/news/2004/08/0827_040827_tvarctic_toxins.html

Haines, S. (2003). Informal life science: Incorporating service learning components into Biology education. *Journal of College Science Teaching*, 32(7), 440–442.

Ham, S. and Sewing, D. (1987/88). Barriers to environmental education. *Journal of Environmental Education*, 19(2), 17–24.

Ham, S.H. (1992). *Environmental Interpretation: A Practical Guide for People with Big Ideas and Small Budgets*. North American Press, Golden, CO.

Hammond, B. (1995). Engaging students in wildlife-focused action projects in Florida: a 35-year perspective. In Jacobson, S.K., ed. *Conserving Wildlife: International Education and Communication Approaches*, pp. 198–218. Columbia University Press, New York, NY.

Hammond, B. (1996/97). Educating for action: A framework for thinking about the place of action in environmental education. *Green Teacher*, 50, 6–14.

Harmon, A., Jones, B. and Finley, J. (1997). Encouraging private forest stewardship through demonstration. *Journal of Forestry* (Jan.), 21–25.

Harris, R. (1998). *Introduction to Creative Thinking*. VirtualSalt. Retrieved September 4, 2004, from http://www.virtualsalt.com/crebook1.htm.

Hassrick, P. (2002). *Drawn to Yellowstone: Artists in America's First National Park*. Autry Museum of Western Heritage, Los Angeles, CA.

Haury, D.L. and Rillero, P. (1994). What is hands-on learning, and is it just a fad? In *Perspectives of Hands-On Science Teaching*. The ERIC Clearinghouse for Science, Mathematics, and

Environmental Education, Columbus, OH. Retrieved February 25, 2005, from http://www. ncrel.org/sdrs/areas/issues/content/cntareas/science/eric/eric-1.htm

Hawthorne, W. (2004). Comparison and development of tropical forest plant field guide formats with a handbook to assist production of field guides: Phase 2. Dept. Plant Sciences, Oxford University, UK Retrieved February 5, 2005, from (www.herbaria.plants.ox.ac.uk).

Heffernan, B.M. (1998). Evaluation techniques for the Sandy Bay Point Discovery Center, Great Bay National Estuarine Research Reserve. *Journal of Environmental Education* 30(1), 25–33.

Heffernan, K. (2001). *Fundamentals of Service-Learning Course Construction*. Campus Compact, Providence, RI.

Heimlich, J. (2003). Environmental educators on the Web: Results of a national study of users and nonusers. *Journal of Environmental Education*, 34(3), 4–11.

Hein, G.E. and Price, S. (1994) Active assessment for active science: A guide for elementary school teachers. Heinemann, Portsmouth, NH.

Heinrich, B. (1994). *A Year in the Maine Woods*. Addison Wesley, Boston, MA.

Henderson, K. and Tilbury, D. (2004). *Whole-school Approaches to Sustainability: an International Review of Sustainable School Programs*. Report prepared by the Australian Research Institute in Education for Sustainability (ARIES) for the Department of the Environment and Heritage, Australian Government.

Henderson, R. (2005). Hot tips for networking at trade expos. *Positive Path Network*. Retrieved January 30, 2005, from http://positivepath.net/ideasRH2.asp

Hernández, O. (2000). Formative research. In Day, B.A. and Monroe, M.C., eds. *Environmental Education and Communication for a Sustainable World: Handbook for International Practitioners*, pp. 47–56. Academy for Educational Development, Washington, DC.

Herreid, C.F. (1994). Case studies in science: A novel method of science education. *Journal of College Science Teaching*, 23, 221–229.

Hill, R.J. (2003). Environmental justice: Environmental adult education at the confluence of oppressions. In Hill, L.H. and Clover, D.E., eds. *Environmental Adult Education: Ecological Learning, Theory, and Practice for Socioenvironmental change*. New Directions for Adult and Continuing Education, 99, 27–38.

Hoban, T.J. (2000). *Managing Conflict: A Guide for Watershed Partnerships*. Retrieved January 1, 2005, from http://www.ctic.purdue.edu/KYW/Brochures/ManageConflict.html

Hoffman, R. (1999). Wayside exhibit planning: Focusing on significant terrain. *Legacy*, 10(2), 17–19 and 32.

Hollweg, K.S. (1997). *Are we Making a Difference? Lessons Learned from VINE program Evaluations*. NAAEE, Washington, DC.

Hough, R.R. and Day, B.A. (2000). Addressing the social dimension: An application of systems thinking. In Day, B.A. and Monroe, M.C., eds. *Environmental Education and Communication for a Sustainable World: Handbook for International Practitioners*, pp. 33–37. Academy for Educational Development, Washington, DC.

Hubbell, S. (1987). *A Country Year: Living the Questions*. Perennial Library, NY.

Huitt, W. and Hummel, J. (2003). Piaget's theory of cognitive development. *Educational Psychology Interactive*. Valdosta, GA: Valdosta State University. Retrieved November 15, 2004, from http://chiron.valdosta.edu/whuitt/col/cogsys/piaget.html.

Hunecke, M., Blöbaum, A., Matthies, E., and Höger, R. (2001). Responsibility and environment: Ecological norm orientation and external factors in the domain of travel mode choice behavior. *Environment and Behavior*, 33(6), 830–852.

Hungerford, H.R. and Volk, T.L. (1990). Changing learner behavior through environmental education. *Journal of Environmental Education*, 21(3), 8–22.

Hungerford, H.R., Litherland, R.A., Peyton, R.B., Ramsey, J.M., and Volk, T.L. (1985). *Investigating and Evaluating Environmental Issues and Action: Skill Development Modules.* Stipes Publishing, Champaign, IL.

Hungerford, H.R., Volk, T., Ramsey, J.M., Litherland, R.A., and Peyton, R.B. (2003). *Investigating and Evaluating Environmental Issues and Actions: Skill Development Program.* Stipes Publishing, Champaign, IL.

Jackson, E. (2005). Literacy Beds. Unpublished document by the Appalachian Sustainable Agriculture Project, Asheville, NC.

Jacobson, S. (1992). *The Bay Islands: Nature and People/Las Islas de la Bahia: La Naturaleza y la Gente.* Bay Island Conservation Association Publication, Roatan, Honduras.

Jacobson, S. (1999). *Communications Skills for Conservation Professionals.* Island Press, Washington, DC.

Jacobson, S.K. (2005). Communications for Wildlife Professionals. In Braun, C.E., ed. *Techniques for Wildlife Investigations and Management,* Sixth Edition, pp. 24–42. The Wildlife Society, Bethesda, MD.

Jacobson, S.K., Monroe, M.C., and Marynowski, S. (2001). Fire at the wildland interface: The influence of experience and mass media on public knowledge, attitudes, and behavioral intentions. *Wildlife Society Bulletin,* 29(3), 929–937.

James, K. (1993). *A Qualitative Study of Factors Influencing Racial Diversity in Environmental Education.* Unpublished doctoral dissertation. University of Minnesota.

Jensen, E. (1998). *Teaching With The Brain In Mind.* Association for Supervision and Curriculum Development, Alexandria, VA.

Job, D. (1996). Geography and environmental education—an exploration of perspectives and strategies. In Kent, A., Lambert, D., Naish, M., and Slater, F., eds. Geography in education: viewpoints on teaching and learning, pp. 22–49. Cambridge University Press, Cambridge, UK.

Johnson, D.W., Johnson, R.T., and Holubec, E.J. (1994). *The New Circles of Learning: Cooperation in the Classroom and School.* Association for Supervision and Curriculum Development, Alexandria, VA.

Jones, J. (1999). *Development of a Self-assessment Method for Patients to Evaluate Health Information on the Internet.* Proc AMIA Symp, 1999, 540–544.

Kaplan, S. (2000). Human nature and environmentally responsible behavior. *Journal of Social Issues,* 56(3), 491–508.

Kaplan, S. and Kaplan, R. (1982). *Cognition and Environment: Functioning in an Uncertain World.* Praeger, New York, NY.

Kasowitz, A. (2000). Tools for automating instructional design. *Educational Media and Technology Yearbook,* pp. 49–52. Volume 25. ERIC Clearinghouse on Information and Technology and the Association for Educational Communications and Technology. Libraries Unlimited, Englewood, CO.

Kassirer, J. (1999). *Norway Public School Litterless Lunch, Tools of Change Web site.* Cullbridge Marketing and Communications, 61 Forest Hill Avenue, Ottawa ON, Canada Updated August 2004. Retrieved December 13, 2004 from, http://www.toolsofchange.com/English/CaseStudies/default.asp?ID=16.

Katz, A. (1995). Impact of environmental education classes at Missouri Botanical Garden on attitude and knowledge change of elementary school children. *HortTechnology,* 5(4), 338–340.

Katz, L.G. (1987). *What should Young Children be Learning? ERIC Digest: ERIC Clearinghouse on Elementary and Early Childhood Education.* Retrieved February 25, 2005, from http://readyweb.crc.uiuc.edu/library/pre1990/katz87.html

Kellert, S.R. (1996). *The Value of Life: Biological Diversity and Human Society.* Island Press, Washington, DC.

Kelsey, K.D. (2000). Impact of communication apprehension and communication skills training on interaction in a distance education course. *Journal of Applied Communication*, 84(4), 7–21.

Kenney, J.L., Militana, H.P., and Donohue, M.H. (2003). Helping teachers to use their school's backyard as an outdoor classroom: A report on the Watershed Learning Center program. *The Journal of Environmental Education*, 35, 18–26.

Kent, T.W. and McNergney, R.F. (1999). *Will Technology Really Change Education?* Corwin Press, Thousand Oaks, CA.

Kirkpatrick, D. (1998). *Evaluating Training Programs: the Four Levels*. Berrett-Koehler Publishing, San Francisco, CA.

Klingemann, H.D. and Rommele, A. (2002). *Public Information Campaigns and Opinion Research*, pp. 130. Sage Publications Ltd. Thousand Oaks, CA.

Knudson, D.M., Cable, T.T., and Beck, L. (1995). *Interpretation of Cultural and Natural Resources*. Venture Publishing, Inc., State College, PA.

Kolb, A.Y. and Kolb, D.A. (2004). *Experience Based Learning Systems Inc.* Last modified April 9, 2004. Retrieved September 1, 2004, from http://www.learningfromexperience.com/ FAQ.htm.

Kotler, P., Roberto, N., and Lee, N. (2002). *Social Marketing: Improving the Quality of Life*, Second Edition. Sage Publishing, Thousand Oaks, CA.

Krajnc, A. (2002). Conservation biologists, civic science and the preservation of BC Forests. *Journal of Canadian Studies*, 37(3), 219–238.

Lawrence, A. and Hawthorne, W. (2005). Plant identification, conservation and management: methods for producing user-friendly field guides. Earthscan: London, UK. Retrieved May 5, 2005, from www.eci.ox.ac.uk/humaneco/fieldguides.html.

LdPride. (2005). *What's Your Learning Style?* Retrieved June 19, 2005, from http://www.ldpride. net/learning_style.html

Learning Skills Program. (2003). *University of Victoria, Counselling Services*. Retrieved September 1, 2004, from http://www.coun.uvic.ca/learn/program/hndouts/bloom.html.

Leave No Trace. (2005). *Leave No Trace Principles*. Leave No Trace Center for Outdoor Ethics. Retrieved 15 March 2005, from http://www.lnt.org.

Leinhardt, G. (1992). What research on learning tells us about teaching. *Educational Leadership*, 49(7), 20–25.

Lennox, C. (2004). Discovering on-line learning. *Legacy*, 15(3), 38–44.

Lenton, P. (2002). Tapping music to encourage environmental literacy (or why we do what we do!). *Connections: The Newsletter of the Global, Environmental and Outdoor Education Council of Alberta*, 26(1), 1–5.

Leopold, A. (1966). *A Sand County Almanac*. Ballantine, New York, NY.

Levin, G. (1993). Too green for their own good. *Advertising Age*, 64(15), 29.

Levinthal, C. (1988). *Messengers of Paradise: Opiates and the Brain*. Doubleday, New York, NY.

Lewis, P. (1993). *Creative Transformation*. Chiron Publications, Wilmette, IL.

Lieberman, G.A. and Hoody, L.L. (1997). *Putting the Pieces Together: Improving Student Learning with the Environment as an Integrating Context*. State Education and Environment Roundtable, Pew Charitable Trust.

Lieberman, G.A. and Hoody, L.L. (1998). *Closing the Achievement Gap: Using the Environment as an Integrating Context for Learning*. State Education and Environment Roundtable, Science Wizards, Poway, CA.

Lindbergh, A.M. (1978). *Gift from the Sea*. Vintage, New York, NY.

Lindemann, E. and Anderson, D. (2001). *A Rhetoric for Writing Teachers*. Fourth Edition. Oxford University Press, NY.

Lindén, A. and Carlsson-Kanyama, A. (2003). Environmentally friendly disposal behavior and local support systems: Lessons from a metropolitan area. *Local Environment*, 8(3), 291–301.

Lockhard, J., Abrams, P.D., and Many, W.A. (1994). *Microcomputers for Twenty-first Century Educators*. HarperCollins College Publishers, New York, NY.

Londo, A.J. and Gaddis, D.A. (2003). Evaluating Mississippi non-industrial private forest landowners acceptance of an interactive video short course. *Journal of Extension*, 41(5). Retrieved April 15, 2005, from http://www.joe.org/joe/2003october/rb4.shtml.

Loney, B. (Fall/Winter 2000–2001). Middle school and high school students map a community: Using service-learning to meet and measure standards. *Continuance*, Fall/Winter, 26.

Lowery, L. (1998). How new science curriculums reflect brain research. *Educational Leadership*, 56(3), 26–30.

Lueck, T.J. and Lee, J. (2004). No fighting the co-op board, even with talons. *New York Times*, December 11, 2004, Page A-1.

Luvmour, S. (1990). *Everyone Wins!: Cooperative Games and Activities*. New Society Publishers, Gabriola Island, British Columbia.

MacGregor, S.K., and Lou, Y. (2004–2005). Web-based learning: How task scaffolding and Web site design support knowledge acquisition. *Journal of Research on Technology in Education*, 37(2), 161–175.

Machlis, G. and Field, D., eds. (1984). *On Interpretation: Sociology for Interpreters of Natural and Cultural History*. Oregon State University Press, Corvallis, OR.

Maclean, N. (1983). *A River Runs Through It*. University of Chicago Press, Chicago, IL.

Malone, J. (1999). *Wild Adventures: a Guidebook of Activities for Building Connections with Others and the Earth*. Pearson Custom Publishing, Needham Heights, MA.

Marcinkowski, T.J. (2004). *Using a Logic Model to Review and Analyze an Environmental Education Program*. North American Association for Environmental Education, Washington, DC.

Markham, T., Larmer, J., and Ravitz, J. (2003). *Project-based Learning Handbook*. Second Edition. Buck Institute for Education, Novato, CA.

Marks, L., Stein, T., Jacobson, S., Gape, L., and Sweeting, M. (2004). *Recreational Management Planning for Abaco National Park: Initial Results from Stakeholder Meetings*. Unpublished report to the Bahamas National Trust.

Martin, C. (2003). Geocaching: Technology meets the great outdoors. *Legacy*, 14(2), 15–19.

Martin, L.C. (1984). *Wildflower Folklore*. The Globe Pequot Press, Chester, CT.

Martin, L.C. (1992). *The Folklore of Trees and Shrubs*. The Globe Pequot Press, Chester, CT.

Martinez, M.D. and Scicchatano, M.J. (1998). Who listens to trash talk?: Education and public media effects on recycling behavior. *Social Science Quarterly*, 79(2), 287–300.

Matherly, C. (2000). Exploring nature from the inside out: Homeschooling opportunities at informal-learning facilities. *Legacy*, 11(4), 14–20.

Matsuoka, B.M. (Executive Producer). (2004). Concept to Classroom, Inquiry-based Learning. NY: Educational Broadcast Corporation. Retrieved June 10, 2004, from http://www.thirteen.org/edonline/concept2class/month6/index_sub1.html

Matthews, C. and Bennett, K. (2002). Naturalist writers and environmental sentiments. *Sciencescope* 26(3), 22–27.

McDuff, M.D. (2000). Thirty years of environmental education in Africa: The role of the Wildlife Clubs of Kenya. *Environmental Education Research*, 6(4), 383–396.

McDuff, M.D. (2001). Building the capacity of grassroots conservation organizations to conduct participatory evaluation. *Environmental Management*, 27(5), 715–727.

McKenzie-Mohr, D. (2000). Fostering sustainable behavior through community-based social marketing. *American Psychologist*, 55(5), 531–537.

McKenzie-Mohr Associates and LURA Consulting. (2001). *Turn it Off: Reducing Vehicle Engine Idling, Final Report*. Retrieved November 21, 2004, from http://www.cbsm.com/Reports/IdlingSummary.pdf

McKenzie-Mohr, D. and Smith, W. (1999). *Fostering Sustainable Behavior: An Introduction to Community-based Social Marketing*. New Society Publishers, Gabriola Island, BC, Canada.

McKibben, B. (2003). My mileage is better than your mileage. *Orion*, 22(1), 80–81.

McNeil, L. (2001). Goodbye honebuckets. Retrieved October 12, 2005 from http://serc.carlton.edu/introgeo/icbl/strategy1.html.

Meadows, D.H. (1991). *The Global Citizen*. Island Press, Washington, DC.

Meadows, R. (2000). Tips from the media committee: how to write and place an op-ed. *Society for Conservation Biology Newsletter* 7(3), 13.

Meany, J. and Shuster, K. (2002). *An Introductory Guide to Effective Public Presentations*. Retrieved February 26, 2005, from http://etsportal.mckenna.edu/FITness/Features/docs/meany_presentation_guide.pdf

Mehers, G., ed. (2000). *Environmental Communcation Planning Handbook for the Mediterranean Region*. International Academy of the Environment, Geneva, Switzerland.

Melchoir, A. and Ballis, L.N. (2002). In Turco, A. and Billig, S.H., eds. *Service-Learning: the Essence of the Pedagogy*, pp. 201–222. Information Age Publishing, Greenwich, CT.

Midden, C.J., Meter, J.E., Weenig, M.H., and Zieverink, H.J. (1983). Using feedback, reinforcement and information to reduce energy consumption in households: A field-experiment. *Journal of Economic Psychology*, 3(1), 65–86.

Millennium Ecosystem Assessment (2005). *Ecosystems and Human Well-being Synthesis*. Island Press, Washington, DC. Retrieved June 2005, from www.millenniumassessment.org.

Milone, M. (1996). *How to Use Technology to Improve Student Learning: Problems and Solutions*. American Association of School Administrators, Washington, DC.

Ministry of Education, Wellington, New Zealand (1999). *Guidelines for Environmental Education in New Zealand Schools*. Learning Media Limited, Wellington, New Zealand. Retrieved January 2, 2005, from http://www.tki.org.nz/r/environ_ed/guidelines/ plan_enviro_e.php

Minnesota Office of Environmental Assistance. (2000). *GreenPrint, Second Edition: State Plan for Environmental Education*. Minnesota Office of Environmental Education, St. Paul, MN.

Minnesota Office of Environmental Assistance. (2002). *Environmental Literacy Scope and Sequence: Providing a Systems Approach to Environmental Education in Minnesota*. Minnesota Office of Environmental Assistance, St. Paul, MN.

Monmoto, J. (1991). *Kenju's Forest*. Harper Collins Publishers, New York, NY.

Monroe, M.C. (2003). Two avenues for encouraging conservation behaviors. *Human Ecology Review*, 10(2), 113–125.

Monroe, M.C. (2005). Addressing Misconceptions about Wildland-Urban Interface Issues. Fact sheet FOR 108, Gainesville, FL: University of Florida. Retrieved May 16, 2005, from http://edis.ifas.ufl.edu.

Monroe, M.C. and Cappaert, D. (1994). *EE Toolbox—Workshop Resource Manual. Integrating Environmental Education into the School Curriculum*. Kendall Hunt Publishing Company, Dubuque, IA.

Monroe, M.C. and Kaplan, S. (1988). When words speak louder than actions: Environmental problem-solving in the classroom. *Journal of Environmental Education*, 19(3), 38–41.

Monroe, M. and Weaver, L. (2005). Enhancing Natural Resource Programs: Designing Effective Brochures. University of Florida, EDIS publication 1436.

Moore, R.C. and Wong, H.H. (1997). *Natural Learning: Creating Environments for Rediscovering Nature's Way of Teaching*. MIG Communications, Berkley, CA.

Morgan, J. (2001). Popular culture and geographic education. *International Research in Geographic and Environmental Education*, 10(3), 284–297.

Muir, J. (1911). *My First Summer in the Sierra*. Penguin, New York, NY.

Murali, N.S., Secher, B.J.M., Rydahl, P., and Andreasen, F.M. (1999). Application of information technology in plant protection in Denmark: From vision to reality. *Computers and Electronics in Agriculture*, 22, 109–115.

Murdoch, K. (1994). *Ideas for Environmental Education: In the Elementary Classroom*. Heinemann, Portsmouth, NH.

Murphy, T. (2004). *Second Minnesota Report Card on Environmental Literacy*. Minnesota Office of Environmental Assistance, St. Paul, MN.

Murray, J.A. (1995). *The Sierra Club Nature Writing Handbook: A Creative Guide*. Sierra Club Books. San Francisco, CA.

NAAEE (2004). *Environmental Education Materials: Guidelines for Excellence*. National Project for Excellence in Environmental Education. Retrieved July 2, 2005, from http://www. naaee.org/pages/npeee/index.html.

NAAEE. (2005). National Project for Excellence in Environmental Education. North American Association for Environmental Education, Washington, DC. Retrieved July 3, 2005, from http://naaee.org/pages/npeee/index.html

Nabhan, G.P. and Trimble, S. (1994). *The Geography of Childhood: Why Children Need Wild Places*. Beacon Press, Boston, MA.

National Conservation Training Center. (2004). *Public Outreach and Education: Overview and Planning*. Division of Education Outreach, National Conservation Training Center, US Fish and Wildlife Service, Shepherdstown, WV.

National Environmental Education and Training Foundation. (2000). *Environment-based Education: Creating High Performance Schools and Students*. NEETF, Washington, DC.

National Environmental Education and Training Foundation. (2002). *Environmental Education and Educational Achievement: Promising Programs and Resources*. NEETF, Washington, DC.

National Opinion Research Center. (1995). *Environmental and Scientific Knowledge Around the World*. National Opinion Research Center, University of Chicago, IL.

National Park Service (n.d.). Retrieved April 6, 2005, from http://www.cr.nps.gov/museum exhibits/moran/yellow5.htm.

National Park Service (2005). *The National Park Service's Interpretive Development Plan*. Retrieved June 28, 2005, from http://www.nps.gov/idp/interp/101/whatisit.htm

National Service-Learning Clearinghouse (NSLC). (2004). *Service Learning is...* Retrieved January 9, 2005, from http://www.servicelearning.org/article/archive/35/

Nelkin, D. (1995). *Selling Science: How the press covers science and technology*. W.H. Freeman, New York, NY.

Neumann, S. (2000). Ecoliteracy: Practicing a systemic approach to education. In *Ecoliteracy: Mapping the Terrain*, pp. 65–71. Learning in the Real World, Berkeley, CA.

New South Wales Department of Education and Training (2001). *Environmental Education Policy for Schools*. Retrieved June 1, 2005, from http://www.curriculumsupport.nsw.edu.au/enviroed/files/Env_ee_policy.pdf

New Zealand Department of Conservation (2005). Build a Tree Activity. Retrieved July 3, 2005, from http://www.doc.govt.nz/community/001%7efor-schools/004-Activities/index.asp

New Zealand Ministry of Education (2004). *Environmental Education in New Zealand Schools: Research into Current Practice and Future Possibilities*. Retrieved January 2, 2005, from http://www.minedu.govt.nz/index.cfm?layout=document&documentid=9102&data=1

Newell, R.J. (2003). *Passion for Learning: How Project-based Learning Meets the Needs of 21st Century Students*. The Scarecrow Press, Inc., Lanham, MD.

Newstrom, J.H. and Lengnick-Hall, M.L. (1991). One size does not fit all. *Training and Development*, 45(6), 43–48.

Norlund, A.M. and Garvill, J. (2003). Effects of values, problem awareness, and personal norm on willingness to reduce personal car use. *Journal of Environmental Psychology*, 23, 339–347.

North Cascades Institute (2005).*Catalog*. North Cascades Institutesort, Bellingham, WA (www.ncascades.org).

Norton, P. and Sprague, D. (2001). *Technology for Teaching*. Allyn and Bacon, Boston, MA.

Null, E.H. (2002). East Feliciana parish schools embrace place-based education as a way to lift scores on Louisiana's high-stakes tests. The Rural School and Community Trust, Washington, DC. Retrieved February 25, 2005, from http://www.ruraledu.org/projects/efeltxt2a.html

Orleans, D. (2004) *Earthsinging: the Use of Music in Environmental Education*, pp. 28–29. Retrieved January 20, 2005, from www.geocities.com/RainForest/Vines/2400/earthsinging.html.

Palmer, J.A. (1993). Development of concern for the environment and formative experiences of educators. *Journal of Environmental Education*, 24, 26–30.

Pardee, M. (2005). River of words. *Volunteer Monitor* (Winter), 17–19.

Pennsylvania State Board of Education (2005). *Pennsylvania State Board of Education Academic Standards*. Retrieved June 2, 2005, from http://www.pde.state.pa.us/stateboard_ed

Petty, R.E. and Cacioppo, J.T. (1981). *Attitudes and Persuasion: Classic and Contemporary Approaches*. Wm C. Brown, Dubuque, IA.

Petty, R.E. and Priester, J.R. (1994). Mass media attitude change: Implications of the elaboration likelihood model of persuasion. In Bryant, J. and Zillmann, D., eds. *Media Effects: Advances in Theory and Research*, pp. 91–122. Lawrence Erlbaum, Hillsdale, NJ.

Pigozzi, M.J. (2003). UNESCO and The International Decade of Education for Sustainable Development (2005–2015). Connect. 28, 1–7.

Pomeroy, R., Parks, J., and Watson, L., (2004) *How is your MPA doing? A Guidebook of Natural and Social Indicators for Evaluating Marine Protected Area Management Effectiveness*. IUCN—The World Conservation Union, Gland, Switzerland.

PON (Program on Negotiation at Harvard Law School) (2004). Program on negotiation at Harvard Law School clearinghouse: A resource center for negotiation education. Retrieved June 25, 2005, from http://www.pon.org/catalog.

Powell, J. and Bails, J.D. (2000). Measuring the soft stuff: Evaluating public involvement in an urban watershed restoration. Proceedings from Watershed 2000 Management Conference, July 9–12, 2000 in Vancouver, BC, Canada. Water Environment Federation, Alexandria, VA.

Powell, K. and Wells, M. (2002). The effectiveness of three experiential teaching approaches on student science learning in fifth-grade school classrooms. *Journal of Evironmental Eucation*, 33(2), 33–38.

Pretty, J.N., Guijt, I., Scoones, I., and Thompson, J. (1995). *A Trainer's Guide for Participatory Learning and Action*. IIED, London, United Kingdom.

Price, S.L. and Hersh, W.R. (1999). Filtering Web pages for quality indicators: An empirical approach to finding high quality consumer health information on the World Wide Web. Presentation at the American Medical Informatics Association Conference. Retrieved July 1, 2005, from http://www.amia.org/pubs/symposia/D005524.PDF

Pringle, R., Hakverdi, M., Cronin-Jones, L., and Johnson, C. (2003). Zoo school for preschoolers: Laying the foundation for environmental education. Presented at the Annual Meeting of The American Educational Research Association (AERA), Chicago, IL, April 21–25, ERIC Document Reproduction Service No. ED 475 663.

Prochaska, J.O., DiClemente, C.C., and Norcross, J.C. (1992). In search of how people change: Applications to addictive behaviors. *American Psychologist*, 47(9), 1102–1114.

Project WILD (1992). *Aquatic Activity Guide*. Council for Environmental Education, Houston, TX.

Project WILD (1995). *Taking Action: An Educator's Guide to Involving Students in Environmental Action Projects*. Council for Environmental Education/Project WILD, Gaithersburg, MD.

Project WILD (2000). *K-12 Curriculum and Activity Guide*. Council for Environmental Education, Houston, TX.

Projects International (2002). *Mekong River: Education for Sustainability Project*. Retrieved January 22, 2005, from http://www.e-o-n.org/Projects_International/mekongefsproj.htm

Prysby, M.D. (2001). *Temporal and Geographical Variation in Monarch Egg and Larval Densities (Danaus plexippus): An Ecological Application of Citizen Science*. Master's thesis. University of Minnesota.

Prysby, M.D. and Super, P. (2006). *Best Practices in Citizen Science for Environmental Learning Centers*. A collaborative document produced from the Citizen Science Forum, November 13–16, 2003, Great Smoky Mountains Institute at Tremont, US.

Ramsey, D. (2002). The role of music in education: Lessons from the cod fishery crisis and the dust bowl days. *Canadian Journal of Environmental Education*, 7(1), 183–198.

Ramsey, J. (1993). The effects of issue investigation and action training on environmental behavior. *Journal of Environmental Education*, 24(3), 31–36.

Ramsey, J. and Hungerford, H. (1989). The effect of issue investigation and action training on environmental behavior of seventh grade students. *Journal of Environmental Education*, 20(4), 29–34.

Ramsey, J., Hungerford, H.R., and Tomera, A.N. (1981). The effects of environmental action and environmental case study instruction on the overt environmental behavior of eighth-grade students. *Journal of Environmental Education*, 13(1), 24–29.

RARE (2003). *Website for Rare Pride Campaign*. Retrieved February 13, 2005, from http://www.rareconservation.org/programs_radio_casestudy.htm and http://rareconservation.org/rare_pride.htm

Reich, J.W. and Robertson, J.L. (1979). Reactance and norm appeal in anti-littering messages. *Journal of Applied Social Psychology*, 9(1), 91–101.

Reynolds, J.I. (1980). Case types and purposes. In Reynolds, R.I. ed., *Case Method in Management Development: Guide for Effective Use*. Management Development Series No. 17, International Labour Office, Geneva, Switzerland.

Rhoades, E. (2005). *Distance education in the Agricultural communication realm: A Synthesis of Research*. Presentation to the Southern Association of Agricultural Scientists, Little Rock AR, February 5–9, 2005. Retrieved 3 May 2005, from http://agnews.tamu.edu/saas/2005/rhoades.pdf

Ribeiro, J. *et al.* (1999). *Flora da Resera Ducke: Guia de Identificacao das Palntas Vasculares de Uma Floresta de Terra-Firme na Amazonia Central*. INPA, Brazil.

Richmond, B. (2003). *Ten C's for Evaluating Internet Sources*. University of Wisconsin-Eau Claire. Retrieved May 15, 2005, from http://www.uwec.edu/library/guides/tencs.html

River of Words (2005). Retrieved June 19, 2005, from www.riverofwords.org.

Robottom, I. and Hart, P. (1995). Viewpoint: Behaviourist EE research: environmentalism as individualism. *The Journal of Environmental Education*, 26(2), 5–9.

Rocha, L.M. and Jacobson, S.K. (1998). Partnerships for conservation: Protected areas and nongovernmental organizations in Brazil. *Wildlife Society Bulletin*, 26(4); 937–946.

Rodekohr, J. (1996). Campaign communications: Public information campaigns. In *The Communicator's Handbook: Tools, Techniques, and Technology*, Third Edition, (Agricultural Communicators in Education) pp. 120–127. Maupin House, Gainesville, FL.

Rogers, A. and Andres, Y.M. (2004). Harnessing the power of the Web: A tutorial for collaborative project-based learning. *Section A: Introduction to NetPBL: On-line Project-Based Learning*. Global SchoolNet Foundation, Retrieved January 9, 2005, from http://www.gsn.org/web/pbl/whatis.htm.

Rogers, E.M. (1995). *Diffusion of Innovations*. Free Press, New York, NY.

Rogers, N. (1993). *The Creative Connection: Expressive Arts as Healing*. Science & Behavior Books, Inc., Palo Alto, CA.

Rohwedder, W.J. and Alm, A. (1994). Using computers in environmental education: Interactive multimedia and on-line learning. In Monroe, M.C. and Cappaert, D., eds., *EE Toolbox, Workshop Resource Manual*. National Consortium for Environmental Education and Training, University of Michigan, Ann Arbor, MI.

Roper-Starch Worldwide (2000). *The National Report Card on Environmental Knowledge, Attitudes, and Behaviors: The Ninth Annual Survey of Adult Americans*. National Environmental Education and Training Foundation, Washington, DC.

Roth, K.J. (1989). Science education: it's not enough to 'do' or 'relate.' *American Educator*, 13(4), 16–22.

Rothstein, R.N. (1980). Television Feedback Used to Modify Gasoline Consumption. *Behavior Therapy*, 11, 683–688.

Rous, E.W. (2000). *Literature and the Land: Reading and Writing for Environmental Literacy*, pp. 7–12. Boynton/Cook Publishers, Portsmouth, NH.

Russell, H.R. (2001). *Ten-Minute Field Trips: A Teacher's Guide To Using The Schoolgrounds For Environmental Studies*. NSTA Press, Washington, DC.

Russo, T. and Benson, S. (2005). Learning with invisible others: Perceptions of online presence and their relationship to cognitive and affective learning. *Educational Technology and Society*, 8(1), 54–62.

Saguaro—A Science Education Solutions Product. (2004). *High School Curriculum, 5E Learning*. This page last updated March 11, 2005 by Science Education Solutions. Retrieved May 4, 2005, from http://www.scieds.com/saguaro/5e_learning.html.

Salzman, J. (1998). *Making the News: A Guide For Nonprofits and Activists*. Westview Press, Boulder, CO.

Schultz, P.W. (1999). Changing behavior with normative feedback interventions: A field experiment on curbside recycling. *Basic and Applied Social Psychology*, 21(1), 25–37.

Schultz, P.W. (2002). Knowledge, information, and household recycling: Examining the knowledge-deficit model of behavior change. In Dietz, T. and Stern P.C., eds. *New tools for Environmental Protection: Education, Information, and Voluntary Measures*, pp. 67–82, National Academy Press, Washington, DC.

Schulz, C. (2000). *School Environmental Clubs in Wisconsin: 2000 and Beyond*. Wisconsin Center for Environmental Education, Stevens Point, WI.

Schwartz, S. (1977). Normative influences on altruism. In Berkowitz, L., ed. *Advances in Experimental Social Psychology*, 10, pp. 221–279. Academic Press, New York, NY.

Schwarzwald, J., Raz, M., and Zvibel, M. (1979). The efficacy of the door-in-the-face technique when established behavioral customs exist. *Journal of Applied Social Psychology*, 9(6), 576–586.

Sclove, R.E., Scammell, M.L., and Holland, B. (1998). *Community-based Research in the United States: An Introductory Reconnaissance, including 12 Organizational Case Studies and Comparison with the Dutch Science Shops and the Mainstream American Research System*. The Loka Institute, Amherst, MA.

Scott, W. and Reid, A. (1998). The revisioning of environmental education: A critical analysis of recent policy shifts in England and Wales. *Education Review*, 50(3), 213–223.

Screven, C. (1990). Uses of evaluation before, during and after exhibit design. *ILVS Review: A Journal of Visitor Behavior*, 1(2), 36–66.

SCWUIRI. (2005). Firewise home. Southern Center for Wildland-Urban Interface Research and Information, USDA Forest Service, Southern Research Station, Gainesville, FL. Retrieved June 25, 2005, from http://www.interfacesouth.org/fire/firewisehome/

SEADSS (2005). Southeastern Agroforestry Decision Support System. Center for Subtropical Agroforestry. University of Florida, Gainesville, FL. Retrieved July 1, 2005, from http://cstaf.ifas.ufl.edu/SEADSS.htm.

Seed, J. (2001). *The Council of All Beings*. Retrieved June 26, 2005, from http://www.rainforestinfo.org.au/deep-eco/council.htm.

Seng, P. and Rushton, S., eds. (2003). *Best Practices Workbook for Boating, Fishing, and Aquatic Resources Stewardship Education*. Recreational Boating and Fishing Foundation. Alexandria, VA.

Shumer, R., Duttweiler, P., Furco, A., Hengel, M., and Willems, G. (2000). *Shumer's Self-Assessment for Service-learning*. Center for Experiential and Service-Learning, University of Minnesota, St. Paul, MN. Retrieved January 9, 2005, from http://www.servicelearning.org/filemanager/download/3/

Silverstein, S. (1964). *The Giving Tree*. Harper Collins Publishers, New York, NY.

Sischy, I. (2003). The Smithsonian's big chill. *Vanity Fair*, December 242–256.

Slocum, R., Wichhart, L., Rocheleau, D., and Thomas-Slaytor, B. (1995). *Power, Process, and Participation: Tools for Change*. Intermediate Technology Publications, London, UK.

Smith, J. (1995). *The New Publicity Kit*. John Wiley, New York, NY.

Smith, W.A. (1995). Behavior, social marketing and the environment. In Palmer, J., Goldstein, W., and Curnow, A., eds. *Planning Education to Care for the Earth. International Union for Conservation of Nature and Natural Resources*, pp. 9–20. Gland, Switzerland.

Snyder, G. (1974). *Turtle Island*. New Directions Publishing Co., New York, NY.

Sobel, D. (1996). *Beyond Ecophobia: Reclaiming the Heart in Nature Education*. The Orion Society and The Myrin Institute, Great Barrington, MA.

Sobel, D. (1998). *Mapmaking with Children: Sense of Place Education for the Elementary Years*. Heinemann, Portsmouth, NH.

Speisman, S. (2005). *Ten Tips for Successful Business Networking*. Business kno-how. Retrieved January 30, 2005, from http://www.businessknowhow.com/tips/networking.htm

Spigner-Littles, D. and Anderson, C.E. (1999). Constructivism: A paradigm for older learners. *Educational Gerontology*, 25, 203–209.

St. Clair, R. (2003). Word for the world: Creating critical environmental literacy for adults. In Hill, L.H. and Clover, D.E., eds. *Environmental Adult Education: Ecological Learning, Theory, and Practice for Socioenvironmental Change. New Directions for Adult and Continuing Education*, 99, 69–78.

Stahl, R.J. (1994). *The Essential Elements of Cooperative Learning in the Classroom*. ERIC Digest. ED370881. ERIC Clearinghouse for Social Studies/Social Science Education, Bloomington, IN,.

Stanley, E. and Waterman, M. (2002). Student Survey: *Using Investigative Cases*. Retrieved February 25, 2005 http://serc.carleton.edu/files/introgeo/icbl/StudentSurvey.doc.

Stanley, E. and Waterman, M. (2002). *Using Investigative Cases in Geoscience*. Starting Point-Teaching Entry Level Geoscience. Retrieved February 25, 2005, from http://serc.carleton.edu/introgeo/icbl/index.html

Stapp, W.B. and Wals, A.E.J. (1994). An action research approach to environmental problem solving. In Bardwell, L., Monroe, M., and Tudor, M., eds. *Environmental Problem Solving: Theory, Practice and Possibilities in Environmental Education*. North American Association for Environmental Education, Troy, OH.

Stapp, W.B., Cromwell, M.M., and Wals, A. (1995). The global rivers environmental education network. In Jacobson, S.K., ed. *Conserving Wildlife: International Education and Communication Approaches*, pp. 177–197. Columbia University Press, New York, NY (Also, see http://www.igc.apc.org/green)

Stapp, W., Wals, A., and Stankorb, S., eds. (1996a). *Environmental Education for Empowerment*. Kendall/Hunt, Dubuque, IA.

Stapp, W.B., Cromwell, M.M., Schmidt, D.C., and Alm, A.W. (1996b). *Investigating Streams and Rivers*. Kendall-Hunt, Dubuque, IA.

State Education and Environment Roundtable. (2000). *California Student Assessment Project: the Effects of Environment-based Education on Student Achievement*. SEER, San Diego, CA. Retrieved June 1, 2005, from http://www.seer.org/pages/csap.pdf.

Stern, P.C. (2000). Toward a coherent theory of environmentally significant behavior. *Journal of Social Issues*, 56(3), 407–424.

Stern, P.C., Dietz, T., and Kalof, L. (1993). Value orientations, gender, and environmental concern. *Environment and Behavior*, 25(3), 322–348.

Stevenson, R.B. (2004). Electronic field guides and citizen science: Steering society in a more earth-friendly direction. Conservation Perspectives: *The On-line Journal of NESCB*. Retrieved January 12, 2005, from http://www.nescb.org/epublications/winter2004/ stevenson.html

Stoecker, R. (2001). *Community-based Research: The Next New Thing*. A report to the Corella and Bertram F. Bonner Foundation and Campus Compact. Retrieved January 9, 2005, from http://coserver.sa.utoledo.edu/drafts/cbrreportb.htm

Stoecker, R. (2002). Practices and challenges of community-based research. *Journal of Public Affairs*, 6, 219–239.

Storey, M.A., Phillips, B., Maczewski, M., and Wang, M. (2002). Evaluating the usability of Web-based learning tools. *Educational Technology and Society*, 5(3), 91–100.

Strand, K., Marullo, S., Cutforth, N., Stoecker, R., and Donohue, P. (2003a). Principles of best practice for community-based research. *Michigan Journal of Community Service Learning* 9(3), 5–15.

Strand, K., Murullo, S., Cutforth, N., Stoecker, R. and Donohue, P. (2003b). *Community-based Research and Higher Education*. Jossey-Bass, San Francisco, CA.

Strunk Jr., W. and White, E. (1979). *The Elements of Style*, Third Edition. McMillan Publishing Co., New York, NY.

Survey Research Center. (2000). *Environmental Studies in the K-12 Classroom: A Teachers' View*. North American Association for Environmental Education, Washington, DC.

Sylwester, R. (1994). How emotions affect learning. *Educational Leadership*, 52(2), 60–66.

Sylwester, R. (2000). Unconscious emotions, conscious feelings, and curricular challenges. *Educational Leadership*, 58(4), 46–51.

Teaching and Learning with Technology, (2003). *Using Cases in Teaching: Writing the Case*. Penn State University, PA, Retrieved February 25, 2005, from http://tlt.its.psu.edu/ suggestions/cases/write.html

Teed, R. (2003). *Role-playing Exercises*. Starting Point-Teaching Entry Level Geoscience. Retrieved February 25, 2005, from http://serc.carleton.edu/introgeo/roleplaying/ howto.html

Teed, R. (2004). *Game-based Learning*. Starting Point-Teaching Entry Level Geoscience. Retrieved February 25, 2005, from http://serc.carleton.edu/introgeo/games/index.html

Telg, R. (1999a). Instructional methods for distance education. Fact sheet AEC 345, Gainesville FL: University of Florida. Retrieved May 16, 2005, from http://edis.ifas.ufl.edu.

Telg, R. (1999b). Introduction to distance education. Fact sheet AEC 344, University of Florida, Gainesville, FL. Retrieved May 16, 2005, from http://edis.ifas.ufl.edu.

Telg, R. (2004). Producing an educational video. Fact sheet AEC 343, University of Florida, Gainesville, FL. Retrieved May 16, 2005, from http://edis.ifas.ufl.edu.

Thayer, R. (1989). *The Biopsychology of Mood and Arousal*. Cambridge University Press, New York, NY.

The Nature Conservancy (2005). Web-site at www.nature.org.

The Nature Conservancy (2004). *Wild New York: creating a field guide for urban environments* (www.nature.org). Society for Conservation Biology Meeting, July 2004, Columbia University, New York, NY.

The Watercourse and the Council for Environmental Education. (1995). *Project WET K-12 Curriculum and Activity Guide.* The Watercourse, Bozeman MT, and CEE, Houston, TX.

Thoreau, H.D. (1854). *Walden.* Konemann Publishers, Boston, MA.

Thoreau, H.D. (1906). *The Writings of Henry David Thoreau.* Boston, Houghton Mifflin. Reprinted, (1984). *The Journal of Henry David Thoreau* v. 8. Gibbs M. Smith Inc., Layton, UT. pp. 4.

Tilden, F. (1956). *Interpreting our Heritage.* The University of North Carolina Press, Chapel Hill, NC.

Traffic International. (2003) WWF and TRAFFIC UK Wildlife Trade Campaign Wins Public Affairs Award. Last updated June 17, 2005. Retrieved June 23, 2005 from http://www.traffic.org/news/award.html

Trapp, S., Gross, M., and Zimmerman, R. (1994). *Signs, Trails, and Wayside Exhibits: Connecting People and Places,* Second Edition. Interpreter's Handbook Series, UW-SP Foundation Press, Inc., Stevens Point, WI.

Trotter, A. (2002). Internet access has no impact on test scores, study says. *Education Week,* 22(1), 10. Retrieved July 1, 2005, from http://www.edweek.org.

Trowbridge, L.W. and Bybee, R.W. (1986). *Becoming a Secondary School Science Teacher,* Fourth Edition. Merrill Publishing Company, Columbus, OH.

Tufte, E. (1990). *Envisioning Information.* Graphics Press, Cheshire, CT.

Turner, K. and Freedman, B. (2004). Music and Environmental Studies. *Journal of Environmental Education,* 36(1), 45–52.

US Department of Education, National Center for Education Statistics. (2002). *The Condition of Education 2002* (NCES 2002–451). US Government Printing Office, Washington, DC.

Ulstrup, S. (2001). Nature Guidance and Guidance in Friluftsliv. *Pathways,* 13(3), 28–30.

UNESCO (1978). *Final report. Intergovernmental Conference on Environmental Education,* United Nations Educational, Scientific, and Cultural Organization with United Nations Environment Program in Tbilisi, USSR, 14–16 October 1977. UNESCO, ED/MD/49, Paris, France.

Union of Concerned Scientists (2000). *The Great Green Web Game.* Retrieved June 26, 2005, from http://www.ucsusa.org/game/index.html.

United Nations Department of Economic and Social Affairs (2003). *Catalysing Community Participation: Water Quality Monitoring Programme in the Philippines.* Retrieved January 23, 2005, from http://www.un.org/esa/sustdev/mgroups/success/phil_pro.htm

University of British Columbia Campus Sustainability Office (2005). *A Vision of Sustainability.* Retrieved January 2, 2005, from http://www.sustain.ubc.ca/.

US National Park Service (2005). *Zion National Park Website.* Retrieved May 2005, from http://www.nps.gov/zion and http://www.nrel.gov/buildings/highperformance.

Van Matre, S. (1972). *Acclimatization: A Sensory and Conceptual Approach to ecological Involvement.* American Camping Association, Martinsville, IN.

Veverka, J.A. (1994). *Interpretive Master Planning.* Falcon Press Publishing Co., Helena, MT.

Vining, J. and Ebreo, A. (2002). Emerging theoretical and methodological perspectives on conservation behavior. In Bechtel, R. and Churchman, A., eds. *Handbook of Environmental Psychology,* pp. 541–558. John Wiley, New York, NY.

Voordouw, J.J. (1987). *Youth in Environmental Action: An International Survey.* International Youth Foundation. International Union for the Conservation of Nature. United Nations Environment Programme, Gland, Switzerland.

WCK (1997). *Trees, Myths, and Medicines: a Collection of Stories by Children of the Wildlife Clubs of Kenya.* Jacaranda a Designs Limited. Nairobi, Kenya.

Weick, K.E. (1984). Small wins: Redefining the scale of social problems. *American Psychologist,* 39(1), 40–49.

Weiss, R.P. (2000). The wave of the brain. *Training and Development,* 7, 20–23.

WGBH Educational Foundations (2000). *Doing Hands-on Science with Kids*. Retrieved January 7, 2005, from http://www.pbs.org/wgbh/buildingbig/educator/pla_hos.

Wholey, J.S., Hatry, H.P., and Newcomer, K.E., eds. (1994). *Handbook of Practical Program Evaluation*. Jossey-Bass Publisher, San Francisco, CA.

Williams, R. and Tollett, J. (1998). *The Non-designer's Web Book*. Peachpit Press, Berkeley CA.

Wilson, J.R. and Monroe, M.C. (2005). Biodiversity curriculum that supports education reform. *Applied Environmental Education and Communication*, 4(2), 125–138.

Winther, A.A. (2001). Investigating and evaluating environmental issues and actions. In Hungerford, H., Bluhm, W.J., Volk, T.L., and Ramsey, J.M., eds. *Essential Readings in Environmental Education*, pp. 155–171. Stipes Publishing, Champaign, IL.

Withrow-Robinson, B., Broussard, S., Simon-Brown, V., Engle, M., and Reed, A (2002). Seeing the forest: art about forests and forestry. *Journal of Forestry*, Dec, 8–14.

Woods, K. (2003). *Seven Steps to a Strong Opinion*. posted February 2003. Poynteronline. Retrieved July 7, 2004, from http://www.poynteronline.com.

World Resources Institute. (2003). *Ecosystems and Human Well-being: A Framework for Assessment*. Island Press, Washington, DC.

WWF-UK. (2004). Stop illegal wildlife trade. Retrieved December 13, 2004, from http://www.wwf.org.uk/wildlifetrade/.

Xerces Society for Invertebrate Conservation (2005). Native Bee Pollination of Watermelon. Retrieved May 5, 2005, from www.xerces.org

Yaffe, S.L., Phillips, A.F., Frentz, I.C., Hardy, P.W., Maleki, S.M., and Thorpe, B.E. (1996). *Ecosystem Management in the United States: An Assessment of Current Experience*. Island Press, Washington, DC.

Young, J. (2001). *Exploring Natural Mystery: Kamana One*. Owlink Media, Santa Cruz, CA.

Zinsser, W. (1985). *On Writing Well: An Informal Guide to Writing Nonfiction*. Harper & Row Publishers, New York, NY.

Zuckerman, M., Lazzaro, M.M., and Waldgeir, D. (1979). Undermining effects of the foot-in-the-door technique with extrinsic rewards. *Journal of Applied Social Psychology*, 9(3), 292–296.

Index